D0259954

PASW STATISTICS BY SPSS

A PRACTICAL GUIDE

VERSION 18.0

PETER ALLEN
KELLIE BENNETT

CENGAGE
Learning

Australia • Brazil • Japan • Korea • Mexico • Singapore • Spain • United Kingdom • United States

PASW Statistics by SPSS: A practical guide, version 18.0
1st Edition
Peter Allen
Kellie Bennett

Publishing manager: Alison Green
Publishing editor: Ann Crabb
Project editor: Tanya Simmons
Publishing assistant: Miriam Allen
Editor and indexer: Julie King
Cover design: Olga Lavecchia

Any URLs contained in this publication were checked for currency
during the production process. Note, however, that the publisher
cannot vouch for the ongoing currency of URLs.

First edition published in 2010

© 2010 Cengage Learning Australia Pty Limited

For product information and technology assistance,
in Australia call 1300 790 853;
in New Zealand call 0800 449 725

For permission to use material from this text or product, please email
aust.permissions@cengage.com

National Library of Australia Cataloguing-in-Publication Data
Author: Allen, Peter James, 1978-
Title: PASW statistics by SPSS: a practical guide: version 18.0/ Peter
Allen, Kellie Bennett.
Edition: 1st ed.
ISBN: 9780170188555 (pbk.)
Notes: Includes index.
Subjects: SPSS (Computer file)--Handbooks, manuals, etc PASW
(Computer file), Social sciences--Statistical methods.
Other Authors/Contributors: Bennett, Kellie
Dewey Number: 300.727

Cengage Learning Australia
Level 7, 80 Dorcas Street
South Melbourne, Victoria Australia 3205

Cengage Learning New Zealand
Unit 4B Rosedale Office Park
331 Rosedale Road, Albany, North Shore 0632, NZ

For learning solutions, visit cengage.com.au

Printed in China by RR Donnelley Asia Printing Solutions Limited.
2 3 4 5 6 7 8 14 13 12 11 10

Table of Contents

Chapter 7: One-Way Between Groups ANOVA.....................75

Chapter 8: Factorial Between Groups ANOVA 91

Chapter 9: One-Way Repeated Measures ANOVA...........109

Chapter 10: One-Way Analysis of Covariance (ANCOVA)............... 125

Chapter 11: Multivariate Analysis of Variance (MANOVA) 143

Chapter 12: Correlation **165**

Chapter 13: Multiple Regression **177**

Chapter 14: Factor Analysis **197**

Chapter 15: Reliability Analysis 209

Chapter 16: Non-Parametric Procedures 223

Chapter 17: Working with Syntax 279

References 285

Index 287

Preface

Predictive Analytics Software (PASW) by SPSS, an IBM Company[*], is a flexible set of data analytic tools used throughout many disciplines. Its roots can be traced back as far as 1968, when a small group of Stanford University doctoral students began developing the *Statistical Package for the Social Sciences* in response to their own need for a software system that would allow them to efficiently analyse the large amounts of data they were amassing at that time.

This program quickly grew beyond the cloistered confines of Stanford, with the publication of the first user manual in 1970. *PASW Statistics by SPSS: A Practical Guide* continues this long tradition of opening statistical analysis up to students and early career researchers from a wide range of applied and academic disciplines. We hope you find it useful!

About PASW (Predictive Analytics Software) Statistics

SPSS released PASW Statistics 18 in late 2009. It is packaged in a variety of flavours, ranging from the PASW Statistics Student Version (which is a limited version of PASW Statistics) through to the full PASW Statistics Family, which includes the complete version of PASW Statistics, along with a range of add-on modules. PASW Statistics is also available for a number of operating systems, including Microsoft Windows, Macintosh OSX, and several Linux distributions.

If you bought a "shrink-wrapped" or "bundled" copy of PASW Statistics with this text, it is probably the student version, developed for the Windows operating system. It will allow you to perform most of the procedures featured in this book (and many more beyond), and should be adequate for most undergraduate statistics courses and research projects. It is, however, limited in a few important respects.

First, it is probably licensed for you to use for a specific period of time (e.g., a semester, a year, or the length of your course). Second, it will limit the number of variables (to 50) and cases (to 1,500) you can include in a single data file. Third, it won't read the command syntax discussed in chapter 17. Finally, it doesn't include a couple of the more advanced procedures covered in this text (specifically, repeated measures ANOVA and MANOVA).

In most circumstances, you won't notice these limitations. If/when you do, your university probably licenses most (if not all) of the PASW Statistics Family, and makes them available to students through campus computer laboratories.

For more information about PASW Statistics, visit http://www.SPSS.com/

About This Book

With *PASW Statistics by SPSS: A Practical Guide* we aim to introduce readers to a range of commonly used statistical procedures that can be performed

---

[*] SPSS was acquired by IBM in October 2009, after PASW Statistics version 18 was released. The company has indicated that subsequent versions of the software will be called IBM SPSS Statistics.

with PASW Statistics, and are typically included in the curricula of undergraduate applied statistics and research methodology units.

Our approach is unashamedly practical, and highly visual. We take a hands-on approach to our subject matter, and work through each procedure in an illustrated, step-by-step fashion. Beyond the necessary focus on "doing", we emphasise interpretation and reporting throughout this text, which contains hundreds of tables of tightly annotated PASW Statistics output, and dozens of examples of how specific research findings can be communicated clearly and concisely.

Other prominent features of this text include:

- Illustrated examples of statistically "significant" and "non-significant" findings, recognising that real-life data does not always support our hypotheses.
- Guidelines for calculating and interpreting effect sizes for most of the inferential procedures discussed in this book.
- An ongoing emphasis on assumption testing, with recommendations for dealing with violated assumptions.
- An extensive section on non-parametric procedures.
- Online resources including datasets and syntax files, which are available at http://www.cengage.com.au/allenbennettv18/

The screen captures used throughout the book were taken with PASW Statistics version 18 for Windows. If you're using an older (or newer) version of SPSS/PASW Statistics, a Student Version, or PASW for Macintosh OSX or Linux, you will probably notice small differences between our captures and the windows and dialogue boxes on your own screen. In virtually all instances, these differences will be slight, and should not hinder your ability to follow our worked examples. However, if you do notice any substantial differences between your version of PASW Statistics and ours, please let us know, and we'll be sure to make a note of it in the next edition of this text.

We do not anticipate that you will read this text from cover to cover. Rather, we hope you'll be able to pick it up and identify quickly the sections you need to "get the job done". To make this task easier, we've divided our content into 17 conceptually distinct chapters.

Chapters 1 through 3 introduce new users to the PASW Statistics interface, and to some of the many ways PASW Statistics can be used to manipulate, summarise and display data. Chapters 4 through 16 are dedicated to specific inferential procedures, including:

- *t* tests (one sample; independent samples; and paired samples).
- Analysis of variance (one-way and factorial; between groups and repeated measures).
- Analysis of covariance.
- Multivariate analysis of variance.
- Bivariate and partial correlation.
- Multiple regression analyses (both standard and hierarchical).
- Factor analysis.
- Reliability analysis.
- Non-parametric alternatives (including chi-square tests for goodness of fit and contingencies; Mann-Whitney's *U*; McNemar's test of change; Wilcoxon's signed rank test; the Kruskal-Wallis ANOVA; the Friedman ANOVA; Cramer's *V*; Spearman's rho; and Kendall's tau-b).

Within each of these chapters, we outline the purpose of the test(s), and illustrate the types of research questions they can be used to address. We then step the reader through one or more illustrated examples, from initial assumption testing through to follow-up analyses and effect-size estimation. Each example then concludes with an annotated APA (American Psychological Association) style results section, demonstrating exactly how research findings can be clearly communicated in reports, assignments and poster presentations. All the examples and research findings discussed in this text are for illustrative purposes only. The datasets have been created by the authors and are not based on actual research studies.

Finally, chapter 17 looks at the use of PASW Statistics command syntax, and the flexibility and efficiencies it can offer more advanced users.

Some Final Comments

This is not a statistics textbook, and we are not mathematicians. We use statistics as a means to an end; as a tool for managing and making sense of the data we collect as part of our efforts – through our research – to better understand the world around us. We're assuming that most of our readers feel the same way, and are using PASW Statistics to take some of the time-intensive number-crunching out of data analysis (thus freeing us all up to do more important things, like interpreting our findings and communicating them to our colleagues and beyond).

Having said that, we do assume that you have a basic understanding of applied statistics, as well as issues of research design more broadly. If not (or if you're just feeling a bit rusty), we recommend using our text alongside books that do justice to these important issues. There are many such books available. For introductory level statistics, we recommend Gravetta and Wallnau's (2008) *Essentials of Statistics for the Behavioral Sciences* and Howell's (2010a) *Fundamental Statistics for the Behavioral Sciences*. At a slightly more advanced level, try Howell's (2010b) *Statistical Methods for Psychology* or Clark-Carter's (2009) *Quantitative Psychological Research*. For books that integrate both basic statistics and research methodology, consider Rosnow and Rosenthal's (2008) *Beginning Behavioral Research: A Conceptual Primer* or Langdridge and Hagger-Johnson's (2009) *Introduction to Research Methods and Data Analysis in Psychology*. Full citations for these texts can be found in our reference list.

Happy analysing!

Peter Allen
School of Psychology and Speech Pathology
Curtin University of Technology

Kellie Bennett
School of Psychiatry and Clinical Neurosciences
University of Western Australia

January 2010

Acknowledgements

The publisher and authors would like to thank the academics who reviewed chapters of this text and provided feedback, including Lynne Roberts and Nick Barrett, Curtin University of Technology; and all the students and colleagues who provided feedback on the previous version of this text including Harold Hill, University of Wollongong; Einar Thorsteinsson, University of New England and others.

Providing Feedback

Your feedback is important, and helps us continue developing and improving on subsequent editions of this text. Please do let us know if you notice any errors or omissions in this text, or if there are PASW Statistics procedures or pedagogic features you would like to see included in future editions. You can contact us at aust.highereducation@cengage.com

About the Authors

Peter Allen is a PhD candidate and Associate Lecturer in the School of Psychology and Speech Pathology at Curtin University of Technology in Perth, Western Australia. As a teacher, he works daily with undergraduate students to solve problems and answer practical questions using SPSS/PASW Statistics. As a researcher, Peter is interested in the psychology of Internet behaviour, and is currently investigating the psychological aspects of online copyright infringement.

Dr Kellie Bennett is an Associate Professor in the School of Psychiatry and Clinical Neurosciences at the University of Western Australia, where she teaches in the medical program. Kellie has lectured in research methodology for undergraduate and postgraduate students. Her main research interests involve communication in health, complementary medicine, statistical analysis and research methodology. In the past, Kellie has conducted research in the areas of Attention Deficit Hyperactivity Disorder, behavioural genetics and educational psychology.

Chapter 1: Getting Started With PASW Statistics

Chapter Overview

1.1. Introduction

This chapter has two purposes: (a) to introduce the **PASW Statistics Data Editor**; and (b) to step you through the process of setting up a basic PASW Statistics data file.

1.2. PASW Statistics Data Editor

When you first open PASW Statistics, you'll probably be greeted with the question, "**What would you like to do?**" and a list of options:

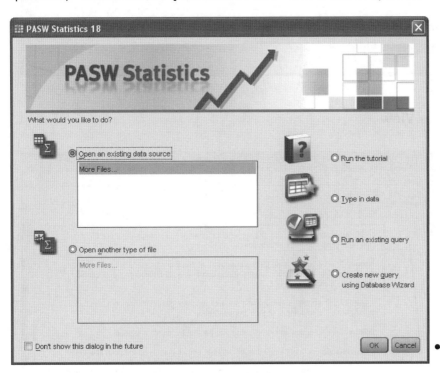

In most instances, you'll want to **Type in data**, or **Open an existing data source**.

However, these are not the only possibilities. You can also:

- **Run the tutorial.** PASW Statistics provides walk-through tutorials for a range of common processes.

- **Run an existing query.** To import data from a non-PASW Statistics data source.

- **Create a new query using Database Wizard.** To create a database query.

- **Open another type of file.** To open other PASW Statistics file types, like output files (see chapter 3) or syntax files (chapter 17).

Make your selection, then click **OK**.

Alternatively, click **Cancel** or the close (⊠) button to exit this dialogue and start working in the **PASW Statistics Data Editor**.

1.2.1. Data View

The drop-down menus provide access to the full range of options and analyses available in PASW Statistics.

These shortcut buttons provide quick access to a number of commonly performed operations (such as **Open**, **Save**, **Print** and **Find**). Hover your mouse cursor over each for further information.

▶▶ Link:
The **Data Editor** is just one component of the PASW Statistics working environment. Other components discussed in this book include:

- **Viewer** (ch 3)
- **Pivot Table Editor** (ch 3)
- **Chart Editor** (ch 3)
- **Syntax Editor** (ch 17)

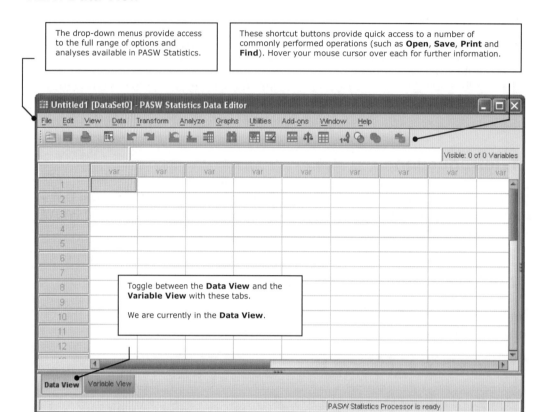

Toggle between the **Data View** and the **Variable View** with these tabs.

We are currently in the **Data View**.

1.2.2. Variable View

ⓘ Tip:
Work in rows in the **Variable View**, where each row defines one variable in your PASW Statistics data file.

ⓘ Tip:
Nominal data are categorical, and the values we assign to levels of a nominal variable are nothing more than shorthand labels. They have no true numeric properties.

Ordinal data can be ranked (i.e., an ordinal scale has magnitude), but it lacks any other numeric properties.

Interval and ratio data (which are both referred to as **Scale** data in PASW Statistics) have the properties we tend to associate with "real numbers".

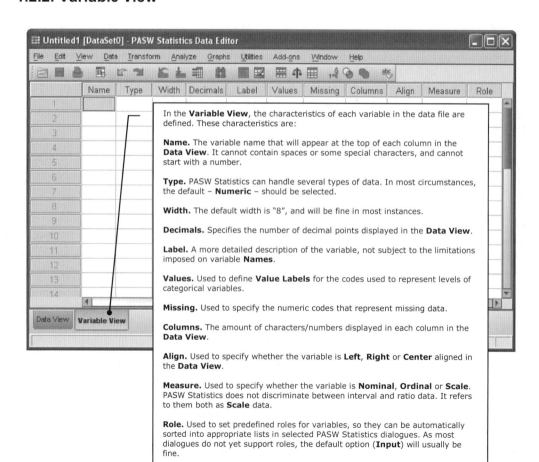

In the **Variable View**, the characteristics of each variable in the data file are defined. These characteristics are:

Name. The variable name that will appear at the top of each column in the **Data View**. It cannot contain spaces or some special characters, and cannot start with a number.

Type. PASW Statistics can handle several types of data. In most circumstances, the default – **Numeric** – should be selected.

Width. The default width is "8", and will be fine in most instances.

Decimals. Specifies the number of decimal points displayed in the **Data View**.

Label. A more detailed description of the variable, not subject to the limitations imposed on variable **Names**.

Values. Used to define **Value Labels** for the codes used to represent levels of categorical variables.

Missing. Used to specify the numeric codes that represent missing data.

Columns. The amount of characters/numbers displayed in each column in the **Data View**.

Align. Used to specify whether the variable is **Left**, **Right** or **Center** aligned in the **Data View**.

Measure. Used to specify whether the variable is **Nominal**, **Ordinal** or **Scale**. PASW Statistics does not discriminate between interval and ratio data. It refers to them both as **Scale** data.

Role. Used to set predefined roles for variables, so they can be automatically sorted into appropriate lists in selected PASW Statistics dialogues. As most dialogues do not yet support roles, the default option (**Input**) will usually be fine.

1.3. Creating a Data File

To illustrate the process of creating and setting up a data file, we developed a simple survey, and asked 15 students in the Health Sciences café to complete it. We've reproduced four of the completed surveys in full, and the data that was collected with the remaining 11 are summarised in Table 1.1.

🖥 **Data:**
This is data file **data_1_1.sav** on the companion website.

Participant 1

> **A Quick Student Satisfaction Survey!**
>
> 1. Gender: Male ☑   Female ☐   *(Please tick)*
>
> 2. Age: _17_
>
> 3. Course of Study: *Speech therapy*
>
> *On a scale from 1(strongly disagree) to 5 (strongly agree), please indicate the extent to which you agree with each of the following statements:*
>
> 4. I am enjoying my course.
>
>   *SD   1   2   3   4   ⑤   SA*
>
> 5. It is easy to get good grades in my course.
>
>   *SD   1   2   3   4   ⑤   SA*
>
> *Thanks for completing our survey!*

Participant 2

> **A Quick Student Satisfaction Survey!**
>
> 1. Gender: Male ☑   Female ☐   *(Please tick)*
>
> 2. Age: _____
>
> 3. Course of Study: Physiotherapy
>
> *On a scale from 1(strongly disagree) to 5 (strongly agree), please indicate the extent to which you agree with each of the following statements:*
>
> 4. I am enjoying my course.
>
>   *SD   1   2   3   ④   5   SA*
>
> 5. It is easy to get good grades in my course.
>
>   *SD   1   2   ③   4   5   SA*
>
> *Thanks for completing our survey!*

Participant 3

> **A Quick Student Satisfaction Survey!**
>
> 1. Gender: Male ☐   Female ☑   *(Please tick)*
>
> 2. Age: _18_
>
> 3. Course of Study: _OT_
>
> *On a scale from 1(strongly disagree) to 5 (strongly agree), please indicate the extent to which you agree with each of the following statements:*
>
> 4. I am enjoying my course.
>
>   *SD   1   2   3   4   ⑤   SA*
>
> 5. It is easy to get good grades in my course.
>
>   *SD   1   2   3   ④   5   SA*
>
> *Thanks for completing our survey!*

Participant 4

> **A Quick Student Satisfaction Survey!**
>
> 1. Gender: Male ☒   Female ☐   *(Please tick)*
>
> 2. Age: _17_
>
> 3. Course of Study: PHYSIO.
>
> *On a scale from 1(strongly disagree) to 5 (strongly agree), please indicate the extent to which you agree with each of the following statements:*
>
> 4. I am enjoying my course.
>
>   *SD   1   2   3   ④   5   SA*
>
> 5. It is easy to get good grades in my course.
>
>   *SD   1   2   3   ④   5   SA*
>
> *Thanks for completing our survey!*

Table 1.1

Data Collected From Participants 5-15 With "A Quick Student Satisfaction Survey"

ID	Q1	Q2	Q3	Q4	Q5
5	Female	20	Physiotherapy	4	4
6	Female	21	Psychology	3	3
7	Female	19	Speech Therapy	3	2
8	Male	18	Speech Therapy	2	1
9	Male	19	Nursing	3	3
10	Female	21	Public Health	5	4
11	Female	24	Occupational Therapy	4	3
12	Female	17	Occupational Therapy	1	1
13	Female	38	Nursing	3	3
14	Male	19	Occupational Therapy	4	5
15	Female	18	Occupational Therapy	5	4

(i) Tip:
Some dialogues in PASW Statistics 18+ support **Roles**.

When you use one of these dialogues, your variables will be automatically sorted into lists according to their defined roles.

There are six possible roles to select from:

Input. Should be selected for predictor or independent variables.

Target. Should be selected for output or dependent variables.

Both. Should be selected for variables that will be used as both predictor and output variables.

None. Should be selected for variables that you do not want to assign a specific role to.

Partition. Used to partition the data file into separate samples.

Split. Included for compatibility with PASW Modeler.

As most PASW Statistics dialogues currently ignore variable **Roles**, you can too!

Begin in the **Variable View** by defining each variable. We will need six rows: one for each of our five questions, plus a sixth – *ID* – to help us keep track of our participants.

Work in rows, defining one variable at a time.

Click on a **Measure** cell, and select the appropriate measurement scale from the drop-down menu.

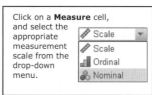

Click on a **Role** cell to select from the six available roles.

The default variable **Role** is **Input**. In most instances this does not need to be changed.

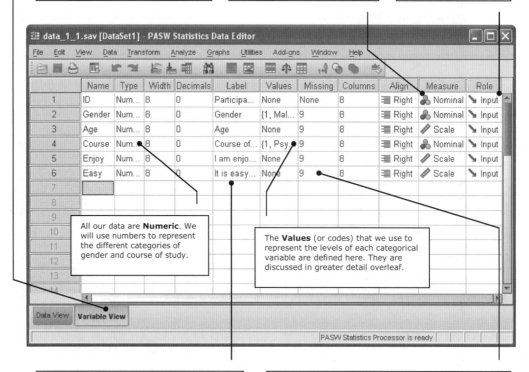

All our data are **Numeric**. We will use numbers to represent the different categories of gender and course of study.

The **Values** (or codes) that we use to represent the levels of each categorical variable are defined here. They are discussed in greater detail overleaf.

Labels can be used to provide more detailed descriptions of variables than are permitted in the **Name** field. If you provide a variable **Label**, it will be used throughout your PASW Statistics output.

Sometimes, research participants miss or elect not to answer certain questions (as was the case with our second participant). At other times, recording equipment may fail, or render some of your data unintelligible. In circumstances like these, we can use **Missing** values codes, which are described in more detail overleaf.

Value Labels

In PASW Statistics, we often use numeric **Values** (or codes) to represent levels of categorical (or **Nominal**) variables, like gender or course of study.

We define these codes in the **Value Labels** dialogue.

To open the **Value Labels** dialogue, select a **Values** cell in the **Variable View**, then click ▣.

Type the number used to represent the first category (or level) in the **Value** box, and the category name in the **Label** box.

Click the **Add** button to add them to the **Value Labels** list.

Repeat this process for the remaining categories, then click **OK** to close the **Value Labels** dialogue.

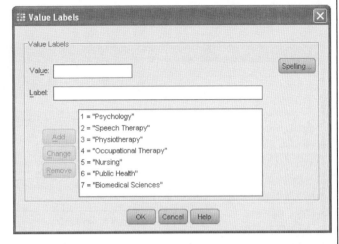

Here, we have used the values "1" through "7" to represent the seven courses that our student participants are enrolled in.

Missing Values

It is inevitable that, from time to time, you will be confronted with missing data. Participants sometimes miss questions, recording equipment fails, coffee spills smudge answers, and so on. We can record unfortunate situations like these with **Missing Values** codes.

At the simplest level, we can use one numeric code to represent all types of missing data. Or, we can discriminate between different sorts of missing data (e.g., questions that participants refused to answer, versus questions which were missed due to equipment failure) by specifying up to three unique **Missing Values** codes.

You can use any numeric code(s) to represent missing data, provided they are outside the range of your actual data. You can also use different codes on each variable, or the same code(s) throughout the entire data file.

We've selected "9" as our **Missing Values** code because it can't be confused with "real" data on any of our research variables. (For example, none of our participants will have a gender of "9", or will have responded with "9" to any of our 5-point rating scales.)

The **Missing Values** dialogue is accessed by selecting a cell under **Missing** in the **Variable View**, then clicking the ▣ button. **Missing Values** codes must be specified separately for each variable.

ⓘ Tip:
If you ever need to discriminate between more than three types of missing data, you can select **Range plus one optional discrete missing value** and enter a range of values you want PASW Statistics to treat as "missing".

After defining each variable, you can begin typing in data. In the **Data View**, each row represents a case (i.e., a participant). For example, row 3 contains the data provided by participant 3, an 18-year-old Occupational Therapy student.

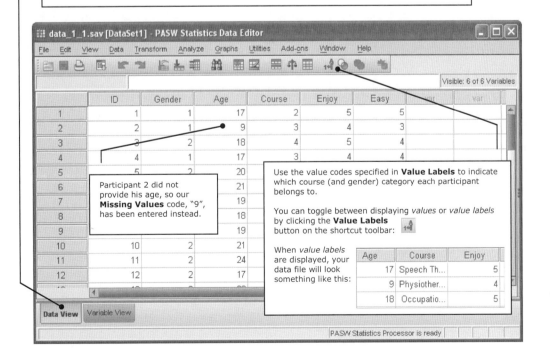

Participant 2 did not provide his age, so our **Missing Values** code, "9", has been entered instead.

Use the value codes specified in **Value Labels** to indicate which course (and gender) category each participant belongs to.

You can toggle between displaying *values* or *value labels* by clicking the **Value Labels** button on the shortcut toolbar:

When *value labels* are displayed, your data file will look something like this:

Age	Course	Enjoy
17	Speech Th...	5
9	Physiother...	4
18	Occupatio...	5

1.4. Conclusion

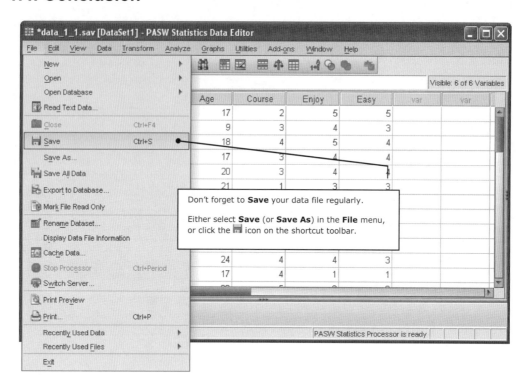

With the data saved, we can begin working with it, summarising it and displaying it. These are the topics covered in chapters 2 and 3.

Chapter 2: Working With Data

Chapter Overview

2.1. Introduction

In chapter 1 we described a short survey used to collect some basic demographic and course satisfaction data from 15 Health Sciences students. In the current chapter, we will continue using this data to illustrate how a data file can be manipulated in PASW Statistics.

The data we collected are reproduced in Table 2.1.

Table 2.1

Data Collected From 15 Participants With "A Quick Student Satisfaction Survey"

Data:
This is data file **data_2_1.sav** on the companion website.

ID	Q1[a]	Q2	Q3[b]	Q4	Q5
1	Male	17	Speech Therapy	5	5
2	Male		Physiotherapy	4	3
3	Female	18	Occupational Therapy	5	4
4	Male	17	Physiotherapy	4	4
5	Female	20	Physiotherapy	4	4
6	Female	21	Psychology	3	3
7	Female	19	Speech Therapy	3	2
8	Male	18	Speech Therapy	2	1
9	Male	19	Nursing	3	3
10	Female	21	Public Health	5	4
11	Female	24	Occupational Therapy	4	3
12	Female	17	Occupational Therapy	1	1
13	Female	38	Nursing	3	3
14	Male	19	Occupational Therapy	4	5
15	Female	18	Occupational Therapy	5	4

Note. Q1 = Gender; Q2 = Age; Q3 = Course of study; Q4 = I am enjoying my course (from 1 = strongly disagree to 5 = strongly agree); Q5 = It is easy to get good marks in my course (from 1 = strongly disagree to 5 = strongly agree).

[a] Value labels for gender are 1 = male; 2 = female.

[b] Value labels for course of study are 1 = Psychology; 2 = Speech Therapy; 3 = Physiotherapy; 4 = Occupational Therapy; 5 = Nursing; 6 = Public Health; 7 = Biomedical Sciences.

2.2. Compute

Compute Variable allows you to create a new variable from one or more existing variables. You can also use it to alter an existing variable.

2.2.1. Illustrated Example

Most commonly, we use **Compute Variable** to sum or average participants' responses to questionnaire items. Here, we'll create a new variable – *course satisfaction* – by averaging our participants' answers to Q4 and Q5.

▶▶| *Link:*
Compute Variable is used in chapter 6 (section 6.3.2.2) to create a difference scores variable.

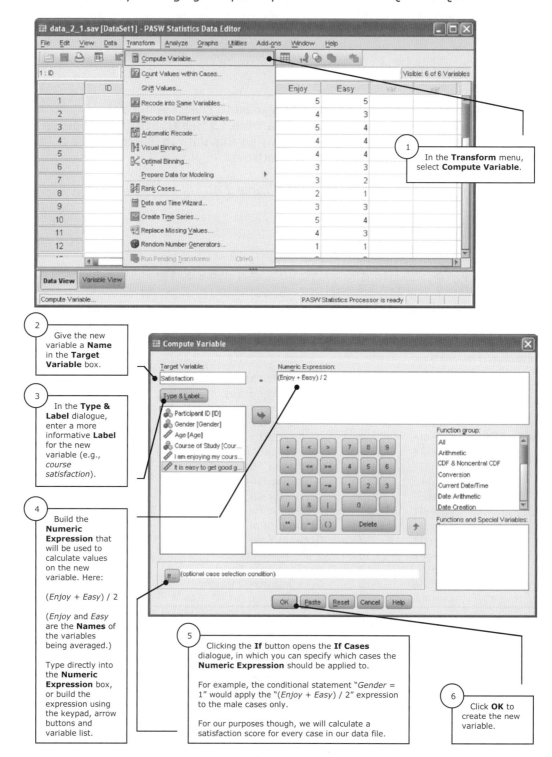

1 In the **Transform** menu, select **Compute Variable**.

2 Give the new variable a **Name** in the **Target Variable** box.

3 In the **Type & Label** dialogue, enter a more informative **Label** for the new variable (e.g., *course satisfaction*).

ⓘ *Tip:*
PASW Statistics includes many functions that can be used to compute new variables.

To access these functions, select a **Function group** and then choose from the options available in the **Functions and Special Variables** list.

When a selection is made, the function's description is provided in the space to the right of the variable list.

If the selected function meets your requirements, it can be moved into the **Numeric Expression** box with the arrow button:

4 Build the **Numeric Expression** that will be used to calculate values on the new variable. Here:

(*Enjoy* + *Easy*) / 2

(*Enjoy* and *Easy* are the **Names** of the variables being averaged.)

Type directly into the **Numeric Expression** box, or build the expression using the keypad, arrow buttons and variable list.

5 Clicking the **If** button opens the **If Cases** dialogue, in which you can specify which cases the **Numeric Expression** should be applied to.

For example, the conditional statement "*Gender* = 1" would apply the "(*Enjoy* + *Easy*) / 2" expression to the male cases only.

For our purposes though, we will calculate a satisfaction score for every case in our data file.

6 Click **OK** to create the new variable.

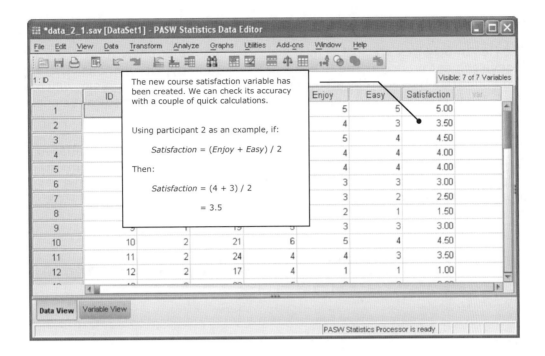

2.3. Recode

With **Recode** you can change specific values or ranges of values on one or more variables. This feature can be used to:

a. *Collapse continuous variables into categories.*
   We could collapse the *satisfaction* variable into two categories ("not satisfied" or "satisfied") by recoding scores of 3 or less as "1" (where 1 = not satisfied) and scores higher than 3 as "2" (where 2 = satisfied).

b. *Combine or merge values.*

   We could merge the courses into two categories – those that require new enrollees to have studied high-school human biology, and those that don't.

c. *Reverse negatively scaled questionnaire items.*

   Oftentimes, questionnaires contain both positively and negatively worded items. For example:

1.	I am enjoying my course.	*Strongly Disagree 1 2 3 4 5 Strongly Agree*
2.	I hate my course.	*Strongly Disagree 1 2 3 4 5 Strongly Agree*

   On this questionnaire, a participant who answers "5" to the first question would likely answer "1" to the second question, and summing or averaging these responses would make very little sense. However, by reversing his or her response to the second question (by recoding 1 as 5, 2 as 4, and so on) we are able to calculate a meaningful total or average, which can then be used in subsequent analyses.

d. *Replace missing values.*

e. *Bring outliers and extreme scores closer to the rest of the distribution.*

▶▶ *Link:*
Recode is used to reverse negatively scaled questionnaire items in chapter 15.

2.3.1. Illustrated Example

The Health Sciences precinct is midway through a major physical redevelopment program. The Schools of Physiotherapy, Occupational Therapy and Public Health were relocated to their new buildings late last year, whereas the Schools of Psychology, Speech Therapy, Nursing and Biomedical Sciences are yet to move.

To find out whether students in the new buildings are more or less satisfied with their courses than students in the old buildings, we could perform an independent samples *t* test (see chapter 5). Before performing this test though, we'll need to merge our seven courses into the two building categories ("old", which we'll code as "1"; and "new", which we'll code as "2").

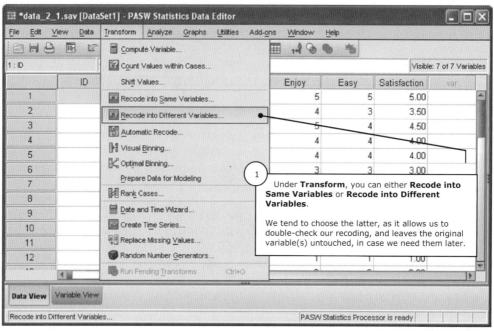

① **Tip:**
By moving multiple variables into the **Numeric Variable -> Output Variable** list, you can recode them simultaneously.

(This is particularly handy if you need to recode responses to several negatively worded questionnaire items.)

Don't forget that you will need to provide a unique **Name** for each **Output Variable**.

① **Tip:**
Clicking the **If** button opens the **If Cases** dialogue, where you can specify the conditions that cases need to meet before being recoded.

Cases that do not meet these conditions are ignored by PASW Statistics during the recoding process.

② Move the variable to be recoded into the **Numeric Variable -> Output Variable** list using the arrow button.

We are recoding the *course* variable.

③ Under **Output Variable**, provide a **Name** and **Label** for the new variable. As our **Output Variable** will indicate whether each participant is studying in the old or the new building, we've called it *building*.

Click the **Change** button, and the **Name** of the **Output Variable** is added to the **Numeric Variable -> Output Variable** list.

④ Click **the Old and New Values** button to open the **Old and New Values** dialogue.

In this dialogue, we will specify how each value on the course of study variable will be changed during the recoding process.

See overleaf for details.

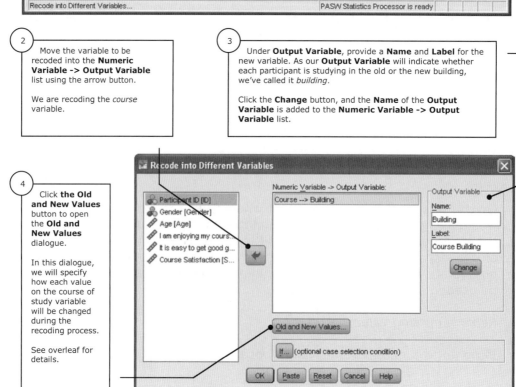

5

In the **Recode into Different Variables: Old and New Values** dialogue, specify which values on the **Numeric Variable** (the course of study variable) should be changed (the **Old Values**), and what they should be changed to (the **New Values**). **Add** each change to the **Old -> New** list before clicking **Continue**.

We will make the following changes:

Course	Old Value	New Value	Building Type
Psychology	1	1	Old
Speech Therapy	2	1	Old
Physiotherapy	3	2	New
Occupational Therapy	4	2	New
Nursing	5	1	Old
Public Health	6	2	New
Biomedical Sciences	7	1	Old

(i) Tip:

Reversing responses to negatively worded questionnaire items is commonly done with PASW Statistics **Recode.**

To reverse responses to a five-point Likert scale, the following **Old** and **New Values** would be used.

Old	->	New
1	->	5
2	->	4
3	->	3
4	->	2
5	->	1

Even though "3" will remain unchanged, you must still provide it with a **New Value**. If you don't, your new variable will have an empty cell (referred to as **System Missing**) where each "3" should be.

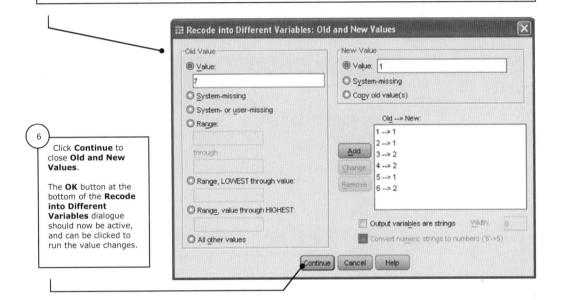

6

Click **Continue** to close **Old and New Values**.

The **OK** button at the bottom of the **Recode into Different Variables** dialogue should now be active, and can be clicked to run the value changes.

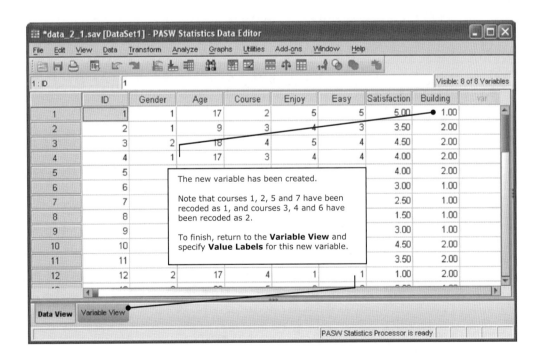

The new variable has been created.

Note that courses 1, 2, 5 and 7 have been recoded as 1, and courses 3, 4 and 6 have been recoded as 2.

To finish, return to the **Variable View** and specify **Value Labels** for this new variable.

2.4. Replace Missing Values

Participants sometimes miss questions (accidentally or deliberately), recording equipment sometimes fails, and coffee spills sometimes render answers unreadable. In situations like these, you are confronted with the problem of missing data/values, and decisions about how they are best handled.

▶▶| *Link:*
Although popular, mean substitution is regarded as one of the crudest methods for dealing with missing data.

For more sophisticated alternatives (including **PASW Missing Values**) see Tabachnick and Fidell (2007b).

One simple way of dealing with missing values is mean substitution, where missing values are replaced with a **Series** (i.e., variable) **mean**.

2.4.1. Illustrated Example

When completing *A Quick Student Satisfaction Survey* participant 2 did not provide his age. We can replace this missing value with the mean age of the participants who did answer this question (i.e., the **Series mean**).

▶▶| *Link:*
When setting up **data_2_1.sav**, we used "9" as a missing value code for each variable. For more information, refer to chapter 1 (section 1.3).

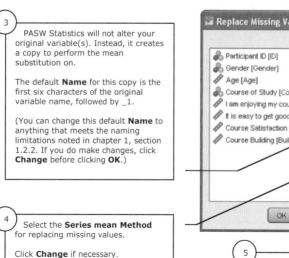

(i) *Tip:*
Using the arrow button, you can move multiple variables into the **New Variable(s)** list.

(i) *Tip:*
In addition to **Series mean**, there are four additional **Methods** for replacing missing values. For further information, click the **Help** button to open the **Replace Missing Values** help page and then follow the **Estimation Methods for Replacing Missing Values** link at the bottom of the page.

2 In the **Replace Missing Values** dialogue, move the age variable into the **New Variable(s)** list using the arrow button.

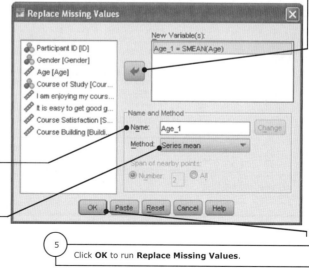

3 PASW Statistics will not alter your original variable(s). Instead, it creates a copy to perform the mean substitution on.

The default **Name** for this copy is the first six characters of the original variable name, followed by _1.

(You can change this default **Name** to anything that meets the naming limitations noted in chapter 1, section 1.2.2. If you do make changes, click **Change** before clicking **OK**.)

4 Select the **Series mean Method** for replacing missing values.

Click **Change** if necessary.

5 Click **OK** to run **Replace Missing Values**.

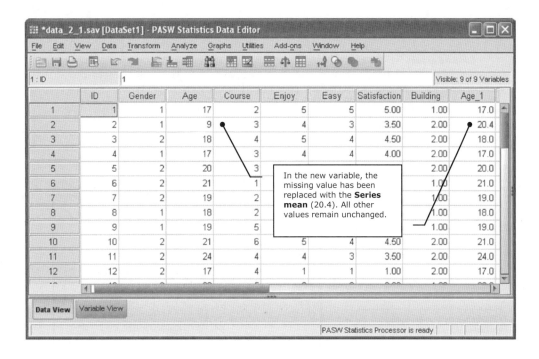

In the new variable, the missing value has been replaced with the **Series mean** (20.4). All other values remain unchanged.

2.5. Split File

Split File allows researchers to temporarily divide a data file into subgroups, which share one or more common characteristics. These subgroups can then be analysed separately (facilitating comparisons between subgroups etc.).

2.5.1. Illustrated Example

If we split **data_2_1.sav** by gender, we can calculate separate descriptive statistics for the male ($n = 6$) and female ($n = 9$) subgroups.

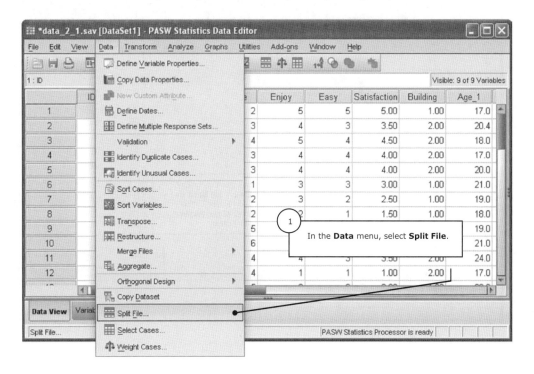

In the **Data** menu, select **Split File**.

2

In the **Split File** dialogue, there are three options:

1.   **Analyze all cases, do not create groups** (the default).
2.   **Compare groups**.
3.   **Organize output by groups**.

The only difference between option 2 and option 3 is the order in which the results of subsequent analyses are presented. As the names would suggest, **Compare groups** facilitates the direct comparison of results across subgroups, whereas **Organize output by groups** produces a completely separate set of results for each subgroup.

For this illustration, we'll **Organize output by groups**.

3

After selecting either **Compare groups** or **Organize output by groups**, you will be able to move the splitting variable into the **Groups Based on** list.

We are splitting the file by *gender*.

(i) *Tip:*
By moving additional variables into the **Groups Based on** list, you can split on multiple variables simultaneously.

For example, if we split on *gender* and *course of study*, we would get separate output/results for each gender-course combination in any subsequent analyses we perform.

That is, we'd get a set of results for: the male Psychology students; the female Psychology students; the male Speech Therapy students; the female Speech Therapy students; and so on.

4

Sort files by the grouping variables does exactly as the name suggests.

It should be used unless your data file is already sorted by the splitting variable(s).

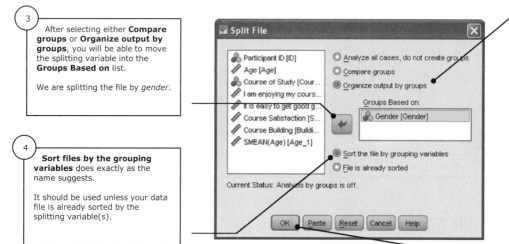

5

Click **OK** to activate **Split File**.

Note that **Split File** will remain active until you return to this dialogue and select **Analyze all cases, do not create groups**.

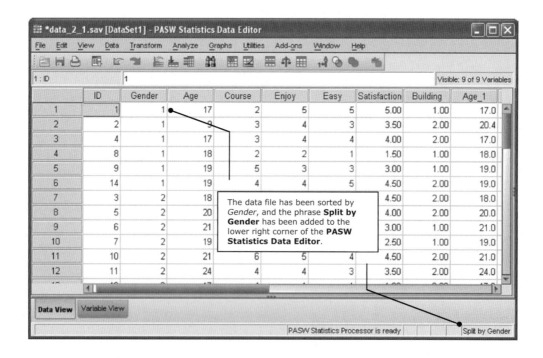

The data file has been sorted by *Gender*, and the phrase **Split by Gender** has been added to the lower right corner of the **PASW Statistics Data Editor**.

Any analyses we now perform will be split by gender. We can illustrate this with some simple descriptive statistics:

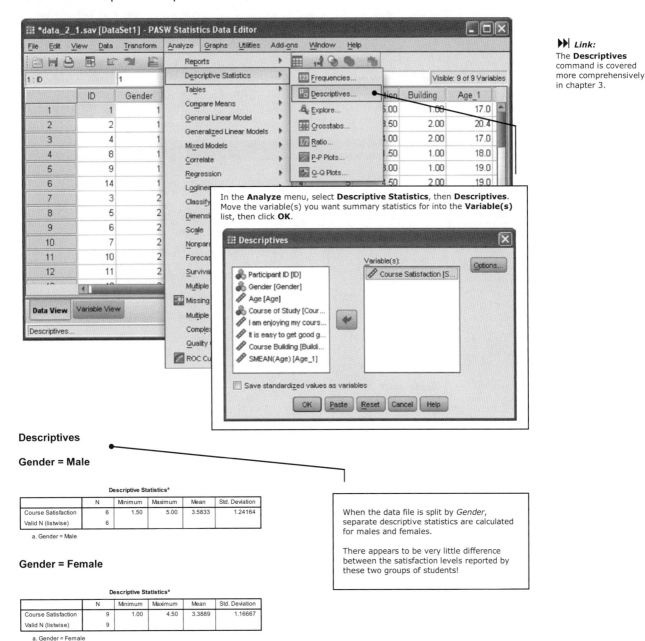

▶▶ *Link:*
The **Descriptives** command is covered more comprehensively in chapter 3.

In the **Analyze** menu, select **Descriptive Statistics**, then **Descriptives**. Move the variable(s) you want summary statistics for into the **Variable(s)** list, then click **OK**.

Descriptives

Gender = Male

Descriptive Statistics[a]

	N	Minimum	Maximum	Mean	Std. Deviation
Course Satisfaction	6	1.50	5.00	3.5833	1.24164
Valid N (listwise)	6				

a. Gender = Male

Gender = Female

Descriptive Statistics[a]

	N	Minimum	Maximum	Mean	Std. Deviation
Course Satisfaction	9	1.00	4.50	3.3889	1.16667
Valid N (listwise)	9				

a. Gender = Female

When the data file is split by *Gender*, separate descriptive statistics are calculated for males and females.

There appears to be very little difference between the satisfaction levels reported by these two groups of students!

Remember, **Split File** will remain active until you return to the **Split File** dialogue (in the **Data** menu), select **Analyze all cases, do not create groups**, and click **OK**.

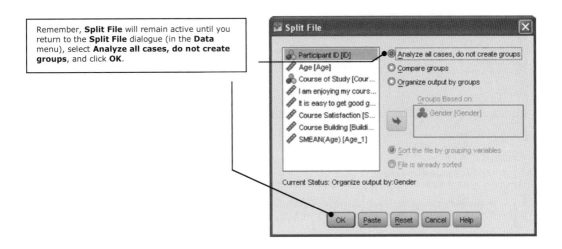

2.6. Select Cases

Oftentimes, researchers want to analyse part of a data file, rather than the whole lot. This can be achieved using **Select Cases**.

2.6.1. Illustrated Example

With **Select Cases**, we can restrict our analyses to those participants aged 18 and over.

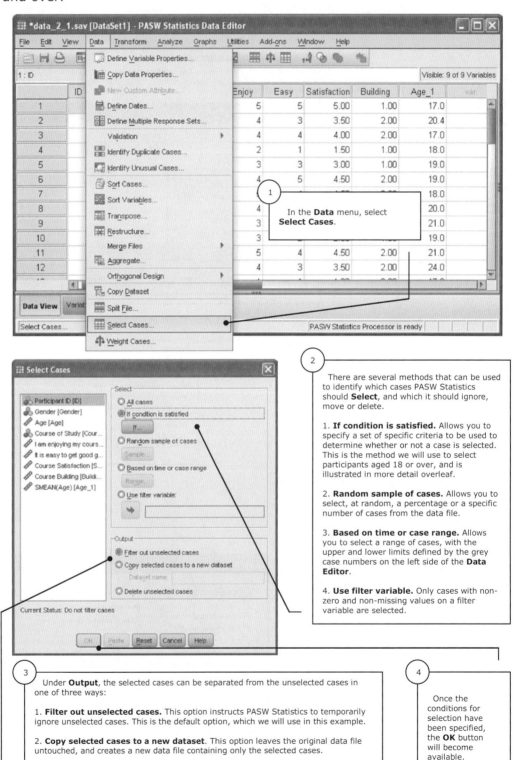

▶▶ Link:
The final item in the **Data** menu is **Weight Cases**. Its use is illustrated on several occasions in chapter 16.

1 In the **Data** menu, select **Select Cases**.

2 There are several methods that can be used to identify which cases PASW Statistics should **Select**, and which it should ignore, move or delete.

1. **If condition is satisfied.** Allows you to specify a set of specific criteria to be used to determine whether or not a case is selected. This is the method we will use to select participants aged 18 or over, and is illustrated in more detail overleaf.

2. **Random sample of cases.** Allows you to select, at random, a percentage or a specific number of cases from the data file.

3. **Based on time or case range.** Allows you to select a range of cases, with the upper and lower limits defined by the grey case numbers on the left side of the **Data Editor**.

4. **Use filter variable.** Only cases with non-zero and non-missing values on a filter variable are selected.

3 Under **Output**, the selected cases can be separated from the unselected cases in one of three ways:

1. **Filter out unselected cases.** This option instructs PASW Statistics to temporarily ignore unselected cases. This is the default option, which we will use in this example.

2. **Copy selected cases to a new dataset**. This option leaves the original data file untouched, and creates a new data file containing only the selected cases.

3. **Delete unselected cases.** Warning! This option is not reversible!

4 Once the conditions for selection have been specified, the **OK** button will become available.

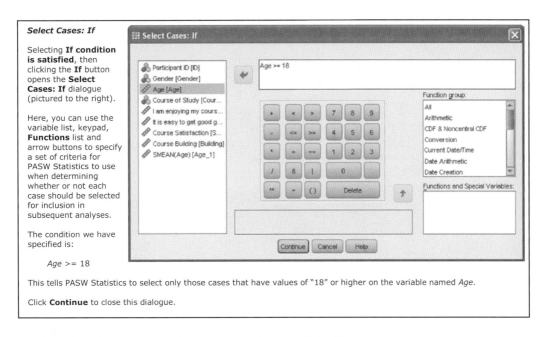

Select Cases: If

Selecting **If condition is satisfied**, then clicking the **If** button opens the **Select Cases: If** dialogue (pictured to the right).

Here, you can use the variable list, keypad, **Functions** list and arrow buttons to specify a set of criteria for PASW Statistics to use when determining whether or not each case should be selected for inclusion in subsequent analyses.

The condition we have specified is:

   Age >= 18

This tells PASW Statistics to select only those cases that have values of "18" or higher on the variable named *Age*.

Click **Continue** to close this dialogue.

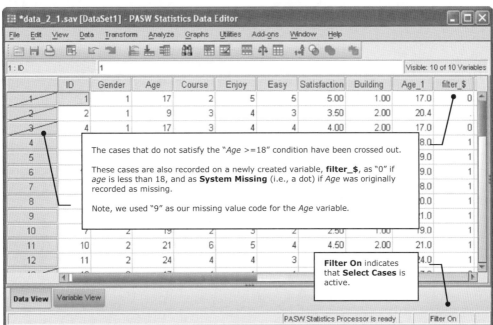

The cases that do not satisfy the "*Age* >=18" condition have been crossed out.

These cases are also recorded on a newly created variable, **filter_$**, as "0" if *age* is less than 18, and as **System Missing** (i.e., a dot) if *Age* was originally recorded as missing.

Note, we used "9" as our missing value code for the *Age* variable.

Filter On indicates that **Select Cases** is active.

If we now calculate some simple descriptive statistics (as we did in section 2.5.1), we end up with:

Descriptives

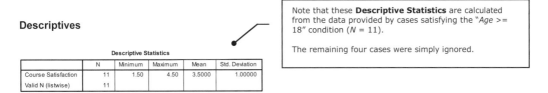

Note that these **Descriptive Statistics** are calculated from the data provided by cases satisfying the "*Age* >= 18" condition (*N* = 11).

The remaining four cases were simply ignored.

Descriptive Statistics

	N	Minimum	Maximum	Mean	Std. Deviation
Course Satisfaction	11	1.50	4.50	3.5000	1.00000
Valid N (listwise)	11				

2.7. Conclusion

This chapter has illustrated just a few of the many ways you can manipulate a data file in PASW Statistics. Most of these procedures will be put to use again in the analyses described throughout the remainder of this book.

Chapter 3: Summarising and Displaying Data

Chapter Overview

3.1. Introduction

In chapter 1, we used data collected with a short survey to illustrate the steps involved in setting up a PASW Statistics data file. In chapter 2, we used that data file to demonstrate some of the common data manipulation procedures available in PASW Statistics. In the current chapter, we will again draw on this questionnaire data (reproduced in Table 3.1) to illustrate how PASW Statistics can be used to efficiently summarise and display data.

Table 3.1

Data Collected With "A Quick Student Satisfaction Survey" (N = 15)

Data:
This is data file
data_3_1.sav
on the companion website.

ID	Q1[a]	Q2	Q3[b]	Q4	Q5
1	Male	17	Speech Therapy	5	5
2	Male		Physiotherapy	4	3
3	Female	18	Occupational Therapy	5	4
4	Male	17	Physiotherapy	4	4
5	Female	20	Physiotherapy	4	4
6	Female	21	Psychology	3	3
7	Female	19	Speech Therapy	3	2
8	Male	18	Speech Therapy	2	1
9	Male	19	Nursing	3	3
10	Female	21	Public Health	5	4
11	Female	24	Occupational Therapy	4	3
12	Female	17	Occupational Therapy	1	1
13	Female	38	Nursing	3	3
14	Male	19	Occupational Therapy	4	5
15	Female	18	Occupational Therapy	5	4

Note. Q1 = Gender; Q2 = Age; Q3 = Course of study; Q4 = I am enjoying my course (from 1 = strongly disagree to 5 = strongly agree); Q5 = It is easy to get good marks in my course (from 1 = strongly disagree to 5 = strongly agree).

[a] Value labels for gender are 1 = male; 2 = female.

[b] Value labels for course of study are 1 = Psychology; 2 = Speech Therapy; 3 = Physiotherapy; 4 = Occupational Therapy; 5 = Nursing; 6 = Public Health; 7 = Biomedical Sciences.

3.2. Frequencies

Frequencies can be used to produce summary statistics and basic graphs for a wide range of variables.

3.2.1. Illustrated Example

We will use **Frequencies** to generate frequency tables, and measures of central tendency and dispersion for a nominal variable (*course of study*) and an interval variable (*course satisfaction*).

3.2.1.1. PASW Statistics Procedure

Syntax:
Run these analyses with **syntax_3_1.sps** on the companion website.

Link:
We used **Compute** to create the course *satisfaction* variable (see chapter 2, section 2.2).

This variable is an average of participants' responses to Q4 and Q5 on our student satisfaction survey (see Table 3.1).

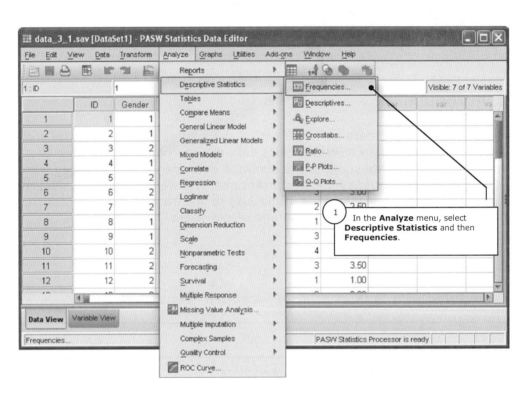

1 In the **Analyze** menu, select **Descriptive Statistics** and then **Frequencies**.

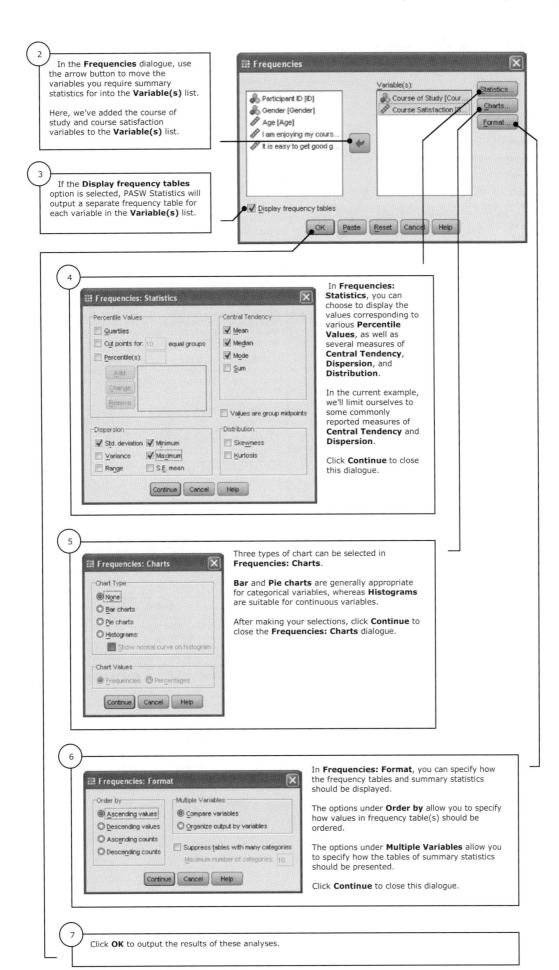

2 In the **Frequencies** dialogue, use the arrow button to move the variables you require summary statistics for into the **Variable(s)** list.

Here, we've added the course of study and course satisfaction variables to the **Variable(s)** list.

3 If the **Display frequency tables** option is selected, PASW Statistics will output a separate frequency table for each variable in the **Variable(s)** list.

4 In **Frequencies: Statistics**, you can choose to display the values corresponding to various **Percentile Values**, as well as several measures of **Central Tendency**, **Dispersion**, and **Distribution**.

In the current example, we'll limit ourselves to some commonly reported measures of **Central Tendency** and **Dispersion**.

Click **Continue** to close this dialogue.

5 Three types of chart can be selected in **Frequencies: Charts**.

Bar and **Pie charts** are generally appropriate for categorical variables, whereas **Histograms** are suitable for continuous variables.

After making your selections, click **Continue** to close the **Frequencies: Charts** dialogue.

6 In **Frequencies: Format**, you can specify how the frequency tables and summary statistics should be displayed.

The options under **Order by** allow you to specify how values in frequency table(s) should be ordered.

The options under **Multiple Variables** allow you to specify how the tables of summary statistics should be presented.

Click **Continue** to close this dialogue.

7 Click **OK** to output the results of these analyses.

3.2.1.2. PASW Statistics Output

Frequencies

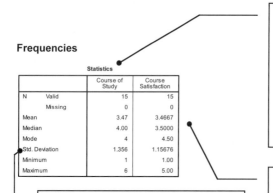

Statistics

		Course of Study	Course Satisfaction
N	Valid	15	15
	Missing	0	0
Mean		3.47	3.4667
Median		4.00	3.5000
Mode		4	4.50
Std. Deviation		1.356	1.15676
Minimum		1	1.00
Maximum		6	5.00

The **Statistics** table contains the statistics we requested in the **Frequencies: Statistics** dialogue.

In **Frequencies: Format** we selected **Compare variables**, and thus the statistics for the two variables are presented side-by-side for ease of comparison. Had we selected **Organize output by groups**, PASW Statistics would have produced two separate **Statistics** tables: one for each variable.

The first two rows of this table indicate how many cases the subsequent statistics are based on, and whether any cases had missing values on either variable.

Std. (Standard) **Deviation**, **Minimum** and **Maximum** are referred to as measures of dispersion.

Standard Deviation is the average amount by which scores in a distribution deviate from the mean of the distribution. In practical terms, it is a measure of how wide or spread out the distribution is. A large standard deviation indicates more variability in the group of scores than a smaller standard deviation. Standard deviation scores are only interpretable in the context of interval and ratio data.

Minimum and **Maximum** are the smallest and largest values in the distribution respectively. These values only make sense in the context of ranked data (that is, ordinal, interval and ratio data).

Mean, **Median** and **Mode** are all measures of central tendency.

The **Mean** of a distribution is simply the average. It is calculated by summing all the scores, then dividing by the total number of scores. Note that although a mean can be calculated for any array of numeric data, it only makes sense if those data are measured on an interval or ratio scale. In the current example, it makes sense to report that the mean course satisfaction was 3.4667; but not that the mean course of study was 3.47.

The **Median** is the middle score of a ranked distribution. It is the most appropriate measure of central tendency for ordinal data, but is also useful for skewed interval or ratio distributions.

The **Mode** is the most frequently occurring score in a distribution. It is seen most commonly in the context of nominal data. Here, the mode course of study was Occupational Therapy (coded as "4").

Frequency Table

Course of Study

		Frequency	Percent	Valid Percent	Cumulative Percent
Valid	Psychology	1	6.7	6.7	6.7
	Speech Therapy	3	20.0	20.0	26.7
	Physiotherapy	3	20.0	20.0	46.7
	Occupational Therapy	5	33.3	33.3	80.0
	Nursing	2	13.3	13.3	93.3
	Public Health	1	6.7	6.7	100.0
	Total	15	100.0	100.0	

Course Satisfaction

		Frequency	Percent	Valid Percent	Cumulative Percent
Valid	1.00	1	6.7	6.7	6.7
	1.50	1	6.7	6.7	13.3
	2.50	1	6.7	6.7	20.0
	3.00	3	20.0	20.0	40.0
	3.50	2	13.3	13.3	53.3
	4.00	2	13.3	13.3	66.7
	4.50	4	26.7	26.7	93.3
	5.00	1	6.7	6.7	100.0
	Total	15	100.0	100.0	

These **Frequency Tables** indicate how many times each value appeared in each distribution.

They are **Ordered by Ascending values**, as specified in the **Frequencies: Format** dialogue.

Looking at the **Course of Study** table, we can see that just one participant reported studying Psychology, three reported studying Speech Therapy, and so on.

The three Speech Therapy students comprise 20% of the total sample, as indicated in the **Percent** column of the **Course of Study** table.

In the **Course Satisfaction** frequency table, we can see scores on this variable ranged from 1 to 5, with the most frequently occurring value being 4.5. The four people reporting course satisfaction of 4.5 represent 26.7% of the entire sample.

3.2.1.3. The PASW Statistics Viewer

PASW Statistics output (i.e., tables, charts, statistics etc.) is presented in the **PASW Statistics Viewer**. As you work with PASW Statistics you will become very familiar with moving between this window and the **PASW Statistics Data Editor**, which was introduced in chapter 1.

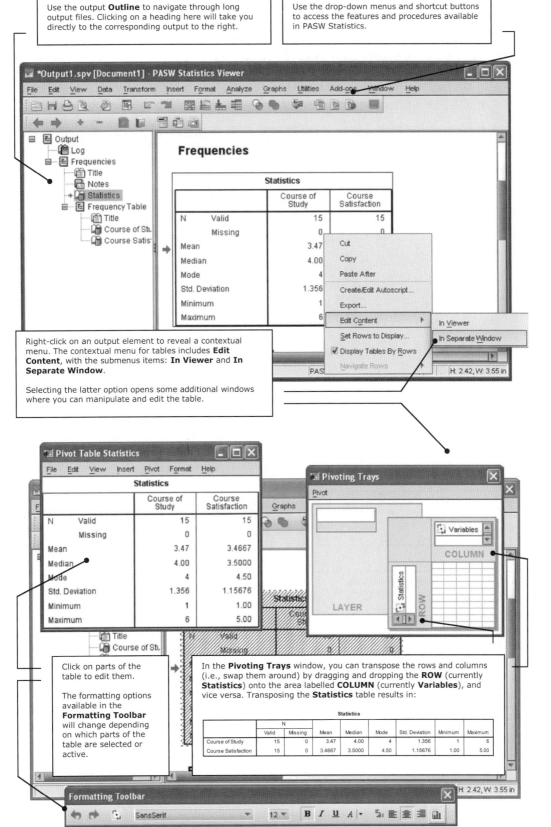

Use the output **Outline** to navigate through long output files. Clicking on a heading here will take you directly to the corresponding output to the right.

Use the drop-down menus and shortcut buttons to access the features and procedures available in PASW Statistics.

Right-click on an output element to reveal a contextual menu. The contextual menu for tables includes **Edit Content**, with the submenus items: **In Viewer** and **In Separate Window**.

Selecting the latter option opens some additional windows where you can manipulate and edit the table.

ⓘ **Tip:**
You can also begin editing an output element by double-clicking it.

Click on parts of the table to edit them.

The formatting options available in the **Formatting Toolbar** will change depending on which parts of the table are selected or active.

In the **Pivoting Trays** window, you can transpose the rows and columns (i.e., swap them around) by dragging and dropping the **ROW** (currently **Statistics**) onto the area labelled **COLUMN** (currently **Variables**), and vice versa. Transposing the **Statistics** table results in:

ⓘ **Tip:**
If the **Pivoting Trays** and/or **Formatting Toolbar** do not open automatically, you can open them manually via the drop-down menus of the **Pivot Table** window.

Open the **Pivoting Trays** by selecting **Pivoting Trays** in the **Pivot** menu.

Open the **Formatting Toolbar** by selecting **Toolbar** in the **View** menu.

3.3. Descriptives

Descriptives can be used to compute summary statistics generally suitable for interval and ratio variables. It can also be used to calculate a standardised score (or *z*-score) for each raw score in a distribution.

3.3.1. Illustrated Example

In this example, we will compute measures of central tendency, dispersion, skewness and kurtosis for two interval variables.

3.3.1.1. PASW Statistics Procedure

Syntax:
Run these analyses with **syntax_3_2.sps** on the companion website.

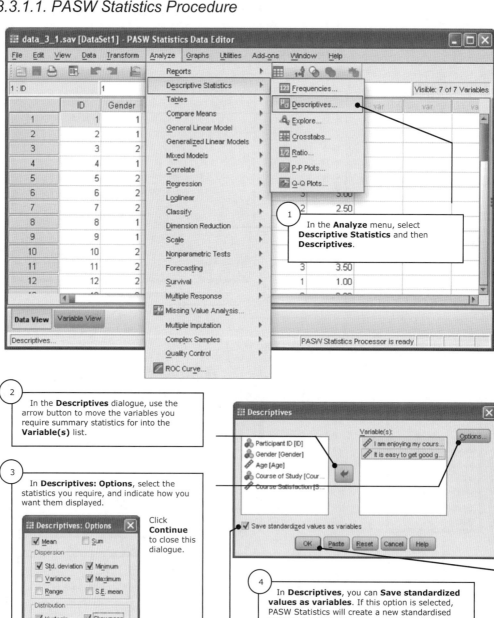

1   In the **Analyze** menu, select **Descriptive Statistics** and then **Descriptives**.

2   In the **Descriptives** dialogue, use the arrow button to move the variables you require summary statistics for into the **Variable(s)** list.

3   In **Descriptives: Options**, select the statistics you require, and indicate how you want them displayed.

Click **Continue** to close this dialogue.

4   In **Descriptives**, you can **Save standardized values as variables**. If this option is selected, PASW Statistics will create a new standardised variable for each raw score variable in the **Variable(s)** list. These will be appended to the data file.

5   Click **OK** to output the requested statistics and (if applicable) create the new standardised variables.

3.3.1.2. PASW Statistics Output

Descriptives

Descriptive Statistics

	N	Minimum	Maximum	Mean	Std. Deviation	Skewness		Kurtosis	
	Statistic	Statistic	Statistic	Statistic	Statistic	Statistic	Std. Error	Statistic	Std. Error
I am enjoying my course	15	1	5	3.67	1.175	-.767	.580	.367	1.121
It is easy to get good grades in my course	15	1	5	3.27	1.223	-.587	.580	-.088	1.121
Valid N (listwise)	15								

Although the statistics reported in the **Descriptive Statistics** table can be computed for any numeric data, they are generally only interpretable when calculated with data measured on interval and ratio scales.

The statistics reported in this table are:

N. The number of scores in each distribution. (Or, in the context of our research, the number of participants who answered each question.)

Minimum. The smallest value in each distribution.

Maximum. The largest value in each distribution.

Mean. The average of the scores in each distribution. The mean is calculated by summing all the scores, then dividing by the total number of scores.

Std. (Standard) **Deviation.** The average amounts by which the scores in each distribution deviate from their respective means.

Skewness Statistic. Skewness is a measure of the symmetry of a distribution of scores. When the skewness statistic is zero, the distribution is perfectly symmetrical. Skewed distributions can be either positive or negative. Both are illustrated below:

Positive Skew   Negative Skew

Skewness Std. (Standard) **Error.** A measure of how much the **Skewness Statistic** is expected to vary from sample-to-sample.

Kurtosis Statistic. Kurtosis is a measure of how peaked or flat a distribution of scores is. A kurtosis statistic of zero (along with a skewness statistic of zero) indicates that the distribution is normally distributed.

Kurtosis Std. (Standard) **Error.** A measure of how much the **Kurtosis Statistic** is expected to vary from sample-to-sample.

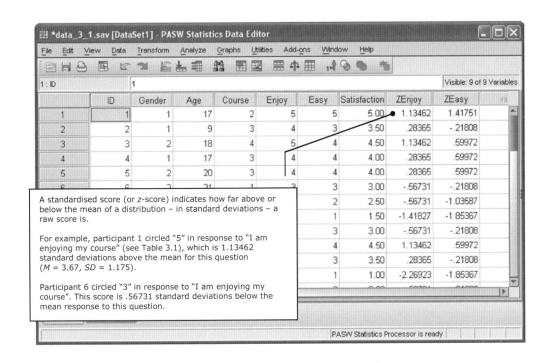

A standardised score (or z-score) indicates how far above or below the mean of a distribution – in standard deviations – a raw score is.

For example, participant 1 circled "5" in response to "I am enjoying my course" (see Table 3.1), which is 1.13462 standard deviations above the mean for this question (*M* = 3.67, *SD* = 1.175).

Participant 6 circled "3" in response to "I am enjoying my course". This score is .56731 standard deviations below the mean response to this question.

ⓘ *Tip:*

The formula for converting raw scores into z-scores is quite simple:

$$z = \frac{X - M}{SD}$$

In this formula, *X* is the score to be converted, and *M* and *SD* are the mean and standard deviation of the distribution the score belongs to.

3.4. Explore

Explore can be used to produce common summary statistics, graphs and tests of normality for several variables simultaneously. These can be produced for the entire sample, or for each subgroup within a sample.

3.4.1. Illustrated Example

We'll use **Explore** to examine the course satisfaction variable at each level of gender.

3.4.1.1. PASW Statistics Procedure

Syntax:
Run these analyses with **syntax_3_3.sps** on the companion website.

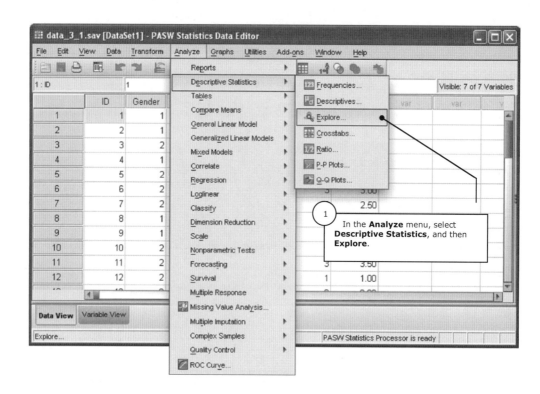

2 In **Explore**, move the variable(s) you require statistics, graphs and tests for into the **Dependent List**.

Generally speaking, these should be interval or ratio variables.

3 If you require separate sets of statistics etc. for subgroups within your sample, move the variable(s) that define these subgroups into the **Factor List**.

Generally speaking, these variable(s) will be categorical, like gender.

If you don't need to divide the sample into subgroups, simply leave the **Factor List** empty.

4 You can limit your output to just **Statistics** or **Plots**. In this example though, we'll generate **Both**.

5 In the **Explore: Statistics** dialogue, **Descriptives** are selected by default.

Click **Continue** to close this dialogue.

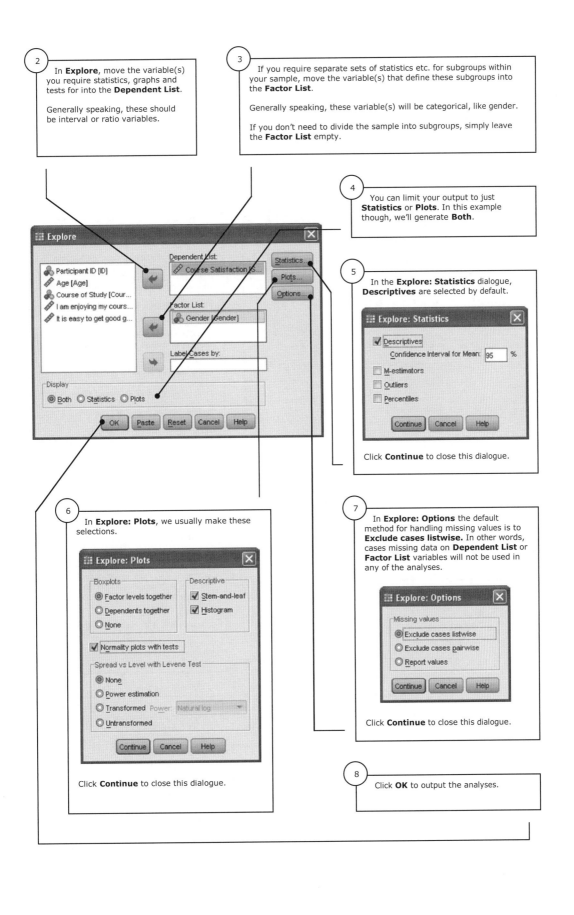

6 In **Explore: Plots**, we usually make these selections.

Click **Continue** to close this dialogue.

7 In **Explore: Options** the default method for handling missing values is to **Exclude cases listwise**. In other words, cases missing data on **Dependent List** or **Factor List** variables will not be used in any of the analyses.

Click **Continue** to close this dialogue.

8 Click **OK** to output the analyses.

3.4.1.2. PASW Statistics Output

Explore

Gender

> The **Case Processing Summary** shows how many cases were analysed, and whether any were excluded due to missing data.
>
> As gender was included in the **Factor List** in the **Explore** dialogue, we are provided with essentially two sets of output, one for each level of gender.

Case Processing Summary

	Gender	Cases					
		Valid		Missing		Total	
		N	Percent	N	Percent	N	Percent
Course Satisfaction	Male	6	100.0%	0	.0%	6	100.0%
	Female	9	100.0%	0	.0%	9	100.0%

> The **Descriptives** table provides summary statistics for each level of gender.
>
> For example, the course satisfaction mean for males ($n = 6$) was 3.5833, whereas for females ($n = 9$) it was 3.3889.
>
> Many of the statistics in this table should already be familiar to you (from, for example, section 3.3 of this chapter). A few you may not yet have encountered include:
>
> **95% Confidence Interval for Mean.** If estimating population parameters based on our sample data, we can be 95% confident that the interval between the **Lower Bound** and **Upper Bound** values includes the true population mean.
>
> **5% Trimmed Mean.** PASW Statistics ignores the highest and lowest 5% of scores to calculate the **5% Trimmed Mean**, which may be a better index of central tendency for distributions containing outliers.
>
> **Variance. Standard Deviation** squared.
>
> **Interquartile Range.** The distance between the first quartile (or 25th percentile) and third quartile (75th percentile).

Descriptives

	Gender			Statistic	Std. Error
Course Satisfaction	Male	Mean		3.5833	.50690
		95% Confidence Interval for Mean	Lower Bound	2.2803	
			Upper Bound	4.8864	
		5% Trimmed Mean		3.6204	
		Median		3.7500	
		Variance		1.542	
		Std. Deviation		1.24164	
		Minimum		1.50	
		Maximum		5.00	
		Range		3.50	
		Interquartile Range		2.00	
		Skewness		-.871	.845
		Kurtosis		.735	1.741
	Female	Mean		3.3889	.38889
		95% Confidence Interval for Mean	Lower Bound	2.4921	
			Upper Bound	4.2857	
		5% Trimmed Mean		3.4599	
		Median		3.5000	
		Variance		1.361	
		Std. Deviation		1.16667	
		Minimum		1.00	
		Maximum		4.50	
		Range		3.50	
		Interquartile Range		1.75	
		Skewness		-1.023	.717
		Kurtosis		.915	1.400

Tests of Normality

	Gender	Kolmogorov-Smirnov[a]			Shapiro-Wilk		
		Statistic	df	Sig.	Statistic	df	Sig.
Course Satisfaction	Male	.153	6	.200[*]	.957	6	.794
	Female	.170	9	.200[*]	.883	9	.170

a. Lilliefors Significance Correction

*. This is a lower bound of the true significance.

> Parametric statistics assume that data are normally distributed. The **Kolmogorov-Smirnov** and **Shapiro-Wilk** tests formally assess this assumption. Although useful, Tabachnick and Fidell (2007a) warn that "they are very sensitive and often signal departures from normality that do not really matter" (p. 46).
>
> Consequently, it is recommended that you interpret the results of these tests alongside more visual indicators of normality (such as histograms or stem-and-leaf plots).

> The vertical axis of each **Histogram** represents the frequency of each band (or range) of course satisfaction scores, which are ordered from lowest to highest along the horizontal axis.
>
> If your intention is to compare histograms, first ensure that the scales of their axes are identical (which PASW Statistics does not do by default, as illustrated here). See section 3.4.1.3 for instructions.

Course Satisfaction

Histograms

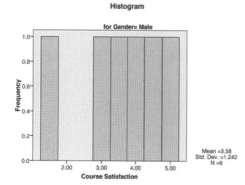

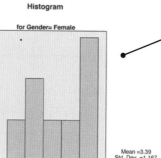

Stem-and-Leaf Plots

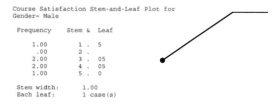

```
Course Satisfaction Stem-and-Leaf Plot for
Gender= Male

 Frequency    Stem &  Leaf

     1.00        1 .  5
      .00        2 .
     2.00        3 .  05
     2.00        4 .  05
     1.00        5 .  0

 Stem width:       1.00
 Each leaf:        1 case(s)
```

```
Course Satisfaction Stem-and-Leaf Plot for
Gender= Female

 Frequency    Stem &  Leaf

     1.00        1 .  0
     1.00        2 .  5
     3.00        3 .  005
     4.00        4 .  0555

 Stem width:       1.00
 Each leaf:        1 case(s)
```

Stem-and-Leaf Plots are very similar to histograms (rotated 90° clockwise), although they contain more information about the specific values in a distribution.

In a **Stem-and-Leaf Plot**, the "stem" contains the first digit(s) of each score, while the "leaves" are the trailing digits.

```
        2.00        4 .  05
```

Looking at the above extract from the male **Stem-and-Leaf Plot**, we can see that one participant had a course satisfaction score of 4.0 (where "4" is the stem, and ".0" is the leaf) and one participant had a score of 4.5.

Normal Q-Q Plots

A **Normal Q-Q** (Quantile-Quantile) **Plot** essentially graphs the observed sample data against the values we would expect if the data were normally distributed.

If a group of scores are normally distributed, the points should cluster tightly around the diagonal line.

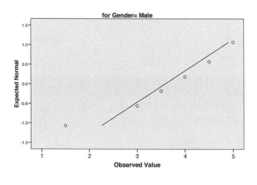

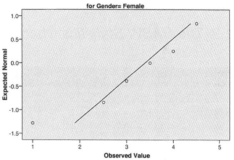

Detrended Normal Q-Q Plots

A **Detrended Normal Q-Q Plot** graphs the deviations from the diagonal line in a corresponding **Normal Q-Q Plot**.

If the data are normally distributed, we should see a roughly even spread of points above and below the horizontal line.

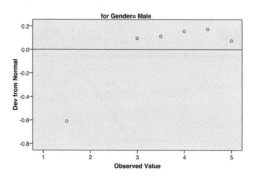

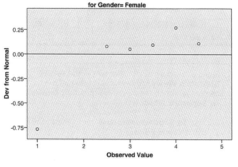

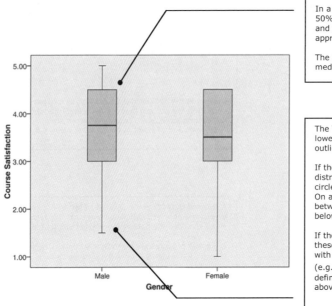

In a **Boxplot**, the "box" contains the middle 50% of scores (i.e., the interquartile range), and is bounded by Tukey's hinges, which approximate the 25th and 75th percentiles.

The line in the middle of the box is the median.

The "whiskers" extend to the highest and lowest scores in the distribution that are not outliers or extreme scores.

If there were *outliers* in either of these distributions, they would be denoted with a circle and data file row number. (e.g., o^{18}). On a boxplot, an outlier is defined as a score between 1.5 and 3 box lengths above or below the box boundaries.

If there were *extreme scores* in either of these distributions, they would be denoted with an asterisk and a data file row number (e.g., $*^{22}$). On a boxplot, an extreme score is defined as a score greater than 3 box lengths above or below the box boundaries.

3.4.1.3. The PASW Statistics Chart Editor

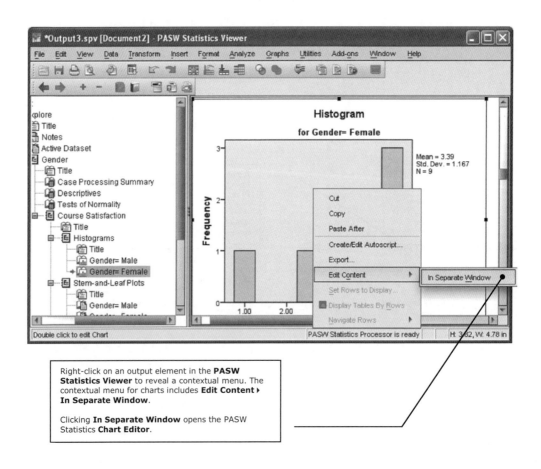

ⓘ Tip:
You can also open the PASW Statistics **Chart Editor** by double-clicking on a chart.

Right-click on an output element in the **PASW Statistics Viewer** to reveal a contextual menu. The contextual menu for charts includes **Edit Content ▸ In Separate Window**.

Clicking **In Separate Window** opens the PASW Statistics **Chart Editor**.

In the **Chart Editor**, graphs can be altered in numerous ways.

Shortcuts to commonly used features are arranged on the toolbars. You can hover your mouse cursor over each icon for a brief description.

This **Properties** palette is contextual, and will display different options as different parts of the chart are selected.

It can be opened by clicking the **Show Properties Window** icon ( ) on the **Chart Editor** toolbar, or by selecting **Properties** in the **Edit** menu.

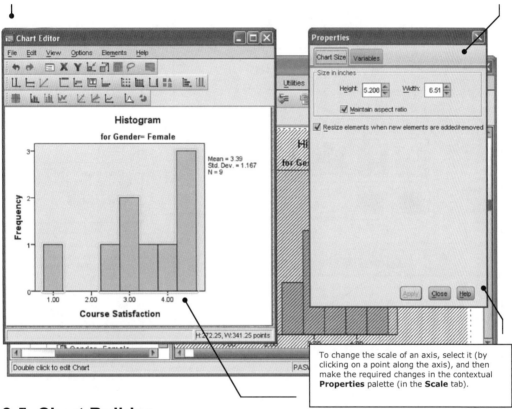

To change the scale of an axis, select it (by clicking on a point along the axis), and then make the required changes in the contextual **Properties** palette (in the **Scale** tab).

3.5. Chart Builder

The **Chart Builder** is a reasonably recent addition to PASW Statistics, and can be used to create a wide range of graphs.

3.5.1. Illustrated Example

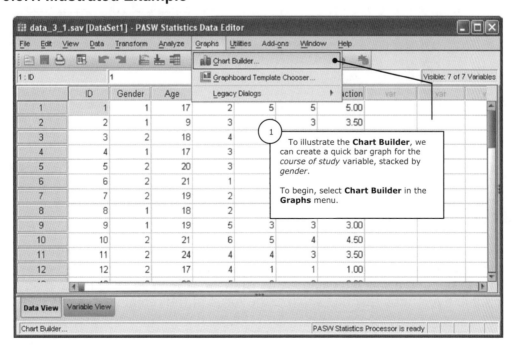

Syntax:
Run these analyses with **syntax_3_4.sps** on the companion website.

ⓘ **Tip:**
Prior to version 14 of PASW Statistics (then called SPSS Base), each type of graph had its own dialogue. These can still be accessed in the **Legacy Dialogs** menu.

To illustrate the **Chart Builder**, we can create a quick bar graph for the *course of study* variable, stacked by *gender*.

To begin, select **Chart Builder** in the **Graphs** menu.

▶▶ *Link:*
The **Chart Builder** can also be seen in action in chapter 16 (section 16.4).

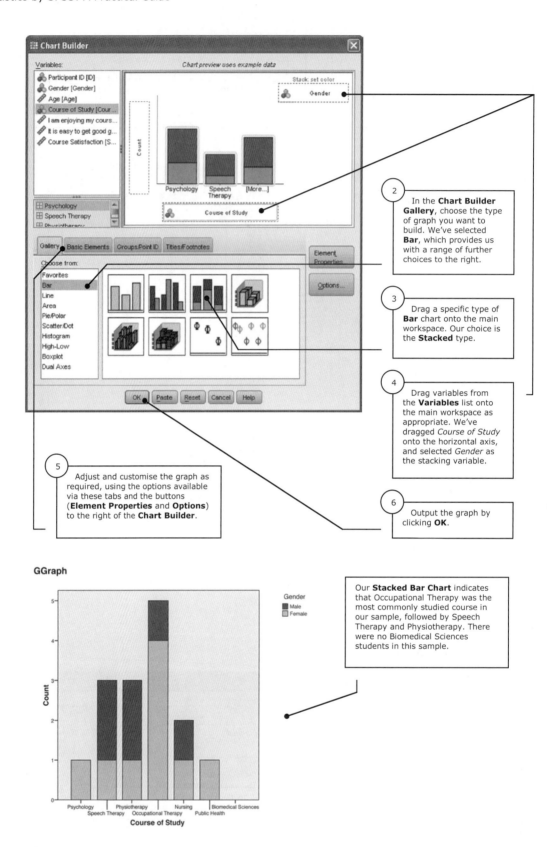

2   In the **Chart Builder Gallery**, choose the type of graph you want to build. We've selected **Bar**, which provides us with a range of further choices to the right.

3   Drag a specific type of **Bar** chart onto the main workspace. Our choice is the **Stacked** type.

4   Drag variables from the **Variables** list onto the main workspace as appropriate. We've dragged *Course of Study* onto the horizontal axis, and selected *Gender* as the stacking variable.

5   Adjust and customise the graph as required, using the options available via these tabs and the buttons (**Element Properties** and **Options**) to the right of the **Chart Builder**.

6   Output the graph by clicking **OK**.

Our **Stacked Bar Chart** indicates that Occupational Therapy was the most commonly studied course in our sample, followed by Speech Therapy and Physiotherapy. There were no Biomedical Sciences students in this sample.

3.6. Conclusion

Chapters 1 through 3 have discussed setting up and manipulating PASW Statistics data files, and some of the common ways that data can be summarised and displayed using PASW Statistics. Now, you should be ready for some hypothesis testing, which is the focus of the remainder of this book.

Chapter 4: One Sample *t* Test

💬 **AKA:**
Single sample *t* test.

Chapter Overview

4.1. Purpose of the One Sample *t* Test

To test whether a sample mean is significantly different to a predetermined/ constant value, referred to as a **Test Value** in PASW Statistics.

The **Test Value** is often, though not always, a population mean derived from prior research.

4.2. Questions We Could Answer Using the One Sample *t* Test

1. Do inner-city homeowners have higher mortgages than the national average?

To answer this question, we would need mortgage data from a (preferably random) sample of inner-city homeowners, as well as the average mortgage size throughout the entire country. This population mean (our **Test Value**) could be obtained from a government census report.

Although the **Test Value** in a one sample *t* test is often a population mean, it does not have to be. Indeed, it can be any value specified by a researcher, as illustrated in questions 2 and 3:

2. On average, do university lecturers spend more or less time at work than the 37.5 hours per week stipulated in their enterprise agreements?

3. Is there a difference between perceived quality of life today, and perceived quality of life in the 1950s?

In question 2, the test value is a condition of the lecturers' employment contracts. In question 3, it is derived from historical research.

4.3. Illustrated Example One

In an effort to improve national literacy levels (and win votes in the next election), the federal government is offering grants to any schools that can demonstrate a sufficient need for extra funding.

As part of a grant application he is preparing on behalf of his primary school, Principal Barrett wants to compare his year 7 students' performance on the National Standardised Literacy Test with that of their peers from across the country. Last year, the national year 7 average on this test was 84.6.

There were 28 year 7 students at the primary school last year. Their scores on the National Standardised Literacy Test are reproduced in Table 4.1.

Table 4.1

☐ *Data:*
This is data file
data_4_1.sav on the
companion website.

National Standardised Literacy Test Scores of the Year 7 Students (N = 28)

Student ID	Literacy Test Score
1	76
2	63
3	75
4	78
5	76
6	78
7	86
8	88
9	74
10	73
11	72
12	80
13	79
14	75
15	66
16	77
17	67
18	77
19	72
20	81
21	82
22	75
23	76
24	69
25	70
26	74
27	76
28	80

4.3.1. Setting Up the PASW Statistics Data File

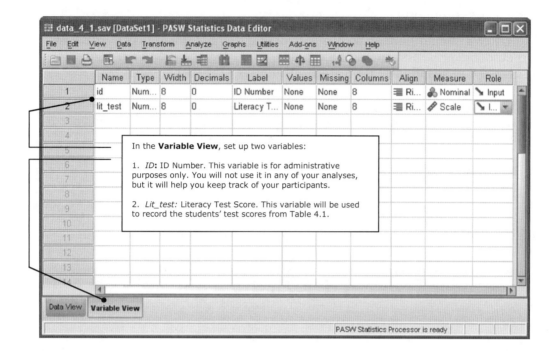

In the **Variable View**, set up two variables:

1. *ID*: ID Number. This variable is for administrative purposes only. You will not use it in any of your analyses, but it will help you keep track of your participants.

2. *Lit_test:* Literacy Test Score. This variable will be used to record the students' test scores from Table 4.1.

ⓘ *Tip:*
The **Name** you use will (by default) appear at the top of the variable column in the **Data View**. However, the **Label** will be used in your output.

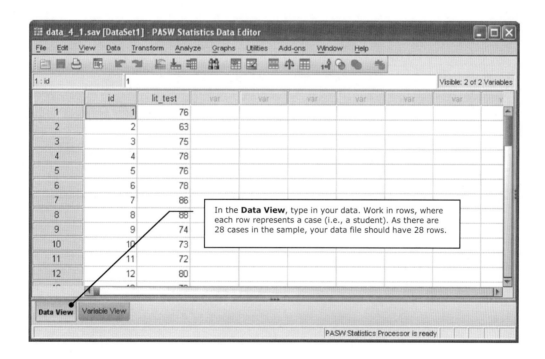

In the **Data View**, type in your data. Work in rows, where each row represents a case (i.e., a student). As there are 28 cases in the sample, your data file should have 28 rows.

4.3.2. Analysing the Data

4.3.2.1. Assumptions

Two criteria should be met before conducting a one sample *t* test. The first is methodological, and should be considered during the planning stages of research. The second will be tested with PASW Statistics.

1. **Scale of Measurement.** Interval or ratio data are required for a one sample *t* test. If your data are ordinal or nominal, you should consider a non-parametric procedure instead (see chapter 16).

2. **Normality.** The sample data should be approximately normally distributed.

4.3.2.2. PASW Statistics Procedure (Part 1: Normality)

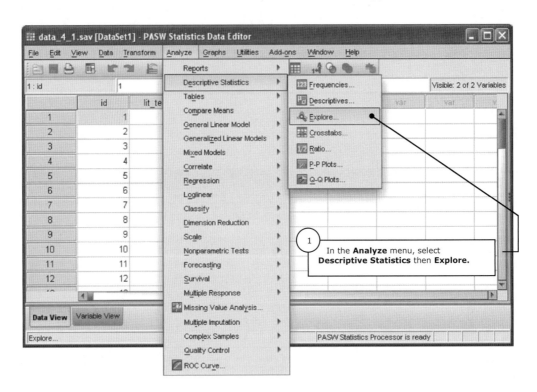

1 In the **Analyze** menu, select **Descriptive Statistics** then **Explore.**

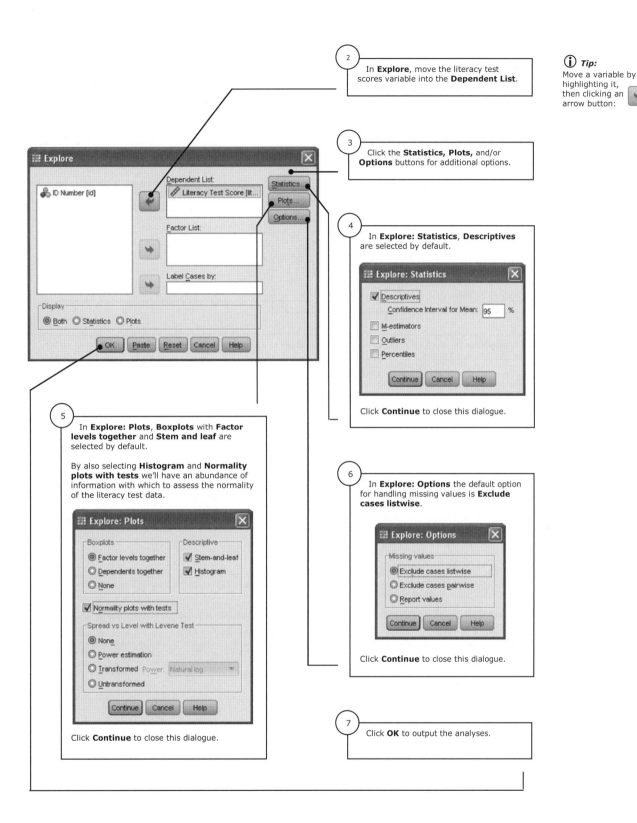

2 In **Explore**, move the literacy test scores variable into the **Dependent List**.

① *Tip:*
Move a variable by highlighting it, then clicking an arrow button:

3 Click the **Statistics, Plots,** and/or **Options** buttons for additional options.

4 In **Explore: Statistics**, **Descriptives** are selected by default.

Click **Continue** to close this dialogue.

5 In **Explore: Plots**, **Boxplots** with **Factor levels together** and **Stem and leaf** are selected by default.

By also selecting **Histogram** and **Normality plots with tests** we'll have an abundance of information with which to assess the normality of the literacy test data.

Click **Continue** to close this dialogue.

6 In **Explore: Options** the default option for handling missing values is **Exclude cases listwise**.

Click **Continue** to close this dialogue.

7 Click **OK** to output the analyses.

4.3.2.3. PASW Statistics Output (Part 1: Normality)

Explore

Case Processing Summary

	Cases					
	Valid		Missing		Total	
	N	Percent	N	Percent	N	Percent
Literacy Test Score	28	100.0%	0	.0%	28	100.0%

> The **Case Processing Summary** shows how many cases were analysed, and whether any were excluded due to missing data.

Descriptives

			Statistic	Std. Error
Literacy Test Score	Mean		75.54	1.052
	95% Confidence Interval for Mean	Lower Bound	73.38	
		Upper Bound	77.69	
	5% Trimmed Mean		75.52	
	Median		76.00	
	Variance		30.999	
	Std. Deviation		5.568	
	Minimum		63	
	Maximum		88	
	Range		25	
	Interquartile Range		7	
	Skewness		-.057	.441
	Kurtosis		.544	.858

> The table of **Descriptives** contains several useful pieces of information. For instance, the **Mean** literacy test score is 75.54; the **Standard Deviation** is 5.568; and the students' test scores ranged from 63 to 88.

The **Skewness** and **Kurtosis** statistics tell us about the shape of the distribution of literacy test scores. When both are zero, the distribution is "normal".

By dividing these values by their respective **Std. Errors** we can compute some z scores, which we can then use (with reference to the standard normal distribution) to work out the likelihood of a sample with skewness/kurtosis this extreme coming from a normally distributed population.

Dealing with **Skewness** (S) first,

$$z_S = \frac{S}{SE_S} = \frac{-0.057}{0.441} = -0.129$$

Next, we need to take the absolute value of z_S (0.129) to the standard normal distribution tables found in the appendices of most statistics texts (e.g., Howell, 2010b) to find its two-tailed probability (which will typically be the "smaller portion" or the "area beyond z" multiplied by 2). In this instance, the two-tailed probability (or likelihood) of a sample with S = -0.057 being drawn from a population with S = 0 is very high (.90 to be exact).

Repeating this process with **Kurtosis** (K),

$$z_K = \frac{K}{SE_K} = \frac{0.544}{0.858} = 0.634$$

The two-tailed probability level of z_K is .53, which is considerably higher than the .05 cut-off typically used when deciding whether or not to reject a null hypothesis. (In this instance, the null hypothesis is that the sample came from a normally distributed population. We do not want to reject this null hypothesis!)

From all of this, we can conclude that <u>this sample of data was likely drawn from a normally distributed population</u>.

If your sample is small, you can assume that your data are normally distributed if both z_S and z_K are < ±1.96 (which has a two-tailed probability of .05). For larger samples, it is recommended that you use either ±2.58 (which has a two-tailed probability of .01) or ±3.29 (which has a two-tailed probability of .001). Be aware that as N increases, SE_S and SE_K decrease, which can result in minor departures from normality appearing "statistically significant" in large samples. Consequently, these techniques for assessing normality should always be used in conjunction with the graphical methods outlined on the following page.

Tests of Normality

	Kolmogorov-Smirnov[a]			Shapiro-Wilk		
	Statistic	df	Sig.	Statistic	df	Sig.
Literacy Test Score	.106	28	.200*	.980	28	.853

a. Lilliefors Significance Correction

*. This is a lower bound of the true significance.

> PASW Statistics provides two **Tests of Normality**.
>
> Both test the null hypothesis that the data have been sampled from a normally distributed population. When *Sig.* < .05, this hypothesis is rejected, which leads one to infer that the population from which the data were drawn is probably not univariate "normal".
>
> The **Shapiro-Wilk** test is generally considered more appropriate for smaller samples, and both tests have a reputation for being over-sensitive, especially when used in larger studies (see Tabachnick & Fidell, 2007a). They should always be used in conjunction with the graphical methods of assessing normality described overleaf.
>
> Here, *W* (the **Shapiro-Wilk** test statistic) is .980 and *Sig* is .853, suggesting that the distribution of literacy test scores is normal. In other words, <u>the normality assumption is not violated</u>.

💬 *AKA:*

The *Sig* values reported by PASW Statistics are usually referred to as "*p*" in journal articles.

For example, $W(28)$ = .98, p = .853.

Literacy Test Score

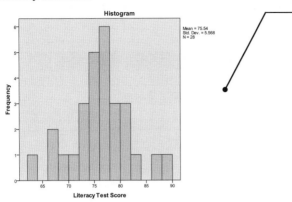

Histogram

Mean = 75.54
Std. Dev. = 5.568
N = 28

```
Literacy Test Score Stem-and-Leaf Plot

Frequency     Stem & Leaf

    1.00 Extremes     (=<63)
    3.00          6 .  679
    6.00          7 .  022344
   12.00          7 .  555666677889
    4.00          8 .  0012
    1.00          8 .  6
    1.00 Extremes     (>=88)

Stem width:         10
Each leaf:      1 case(s)
```

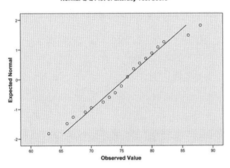

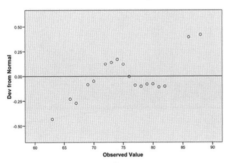

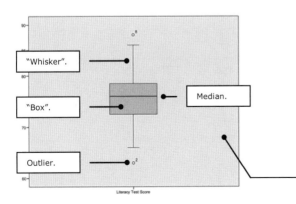

For further confirmation that the data are normally distributed, look at the **Histogram**, **Stem-and-Leaf Plot**, **Normal Q-Q Plot** and **Detrended Normal Q-Q Plot**.

1. **Histogram**. The vertical axis represents the frequency of each band (or range) of literacy test scores, which are ordered from smallest to largest along the horizontal axis. For example, one case scored between 60 and 64, and three scored between 65 and 69.

Ideally, the histogram should look "normal". That is, it should resemble an "inverted U" with greatest frequency of cases clustered around the mean, and progressively fewer cases towards the tails.

This histogram indicates that the sample data <u>are approximately normally distributed</u>.

2. **Stem-and-Leaf Plot**. This graph is very similar to the histogram (rotated 90° clockwise), although it provides more information about the specific values in the distribution. For example, the histogram indicated that three scores were between 65 and 69. The stem-and-leaf plot informs you that these sores were 66, 67, and 69.

```
    Frequency     Stem & Leaf
       3.00          6 .  679
```

In a stem-and-leaf plot, the "stem" contains the first digit of each score (6), while the "leaves" are the trailing digits (6, 7 and 9).

3. **Normal Q-Q** (Quantile-Quantile) **Plot**. This graph essentially plots the observed sample data against the values we would expect if it were normally distributed.

If the sample data are normally distributed, the points should cluster tightly around the diagonal line, as they do here.

4. **Detrended Normal Q-Q Plot**. This graph plots the deviations from the diagonal line in the **Normal Q-Q Plot**. Look carefully, and you should see the correspondence between the two.

If the sample data are normally distributed, we would expect a roughly even spread of points above and below the horizontal line, as is the case here.

In summary, these graphs further confirm that <u>the sample data are normally distributed</u>.

A roughly symmetrical **Boxplot** indicates that <u>the sample data are not skewed</u>.

The "box" contains the middle 50% of scores (i.e., the inter-quartile range). It is bounded by Tukey's hinges, which approximate the 25th and 75th percentiles.

The line in the middle is the median.

The "whiskers" extend to the highest and lowest scores in the distribution that are not outliers or extreme scores.

Scores between 1.5 and 3 box lengths above/below the box boundaries are outliers. They are denoted by a circle and data file row number.

Scores greater than 3 box lengths above/below box boundaries are extreme scores. They are denoted by an asterisk (*) and data file row number.

<u>In this data file there are two outliers</u> (on rows 2 and 8 of the data file).

4.3.2.4. PASW Statistics Procedure (Part 2: One Sample t Test)

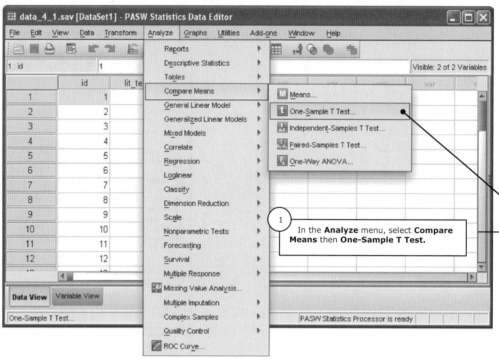

1 In the **Analyze** menu, select **Compare Means** then **One-Sample T Test**.

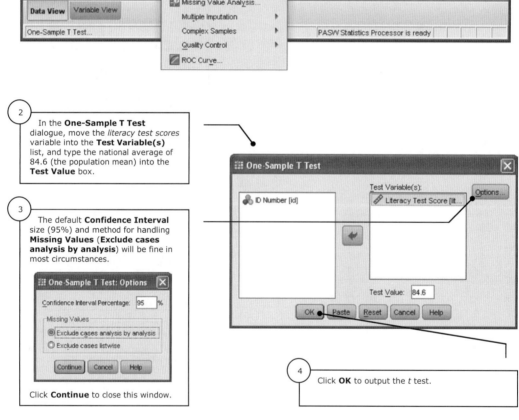

2 In the **One-Sample T Test** dialogue, move the *literacy test scores* variable into the **Test Variable(s)** list, and type the national average of 84.6 (the population mean) into the **Test Value** box.

3 The default **Confidence Interval** size (95%) and method for handling **Missing Values** (**Exclude cases analysis by analysis**) will be fine in most circumstances.

Click **Continue** to close this window.

4 Click **OK** to output the t test.

4.3.2.5. PASW Statistics Output (Part 2: One Sample t Test)

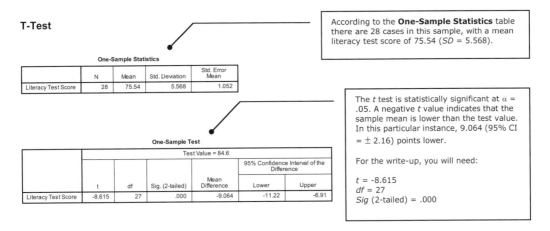

T-Test

According to the **One-Sample Statistics** table there are 28 cases in this sample, with a mean literacy test score of 75.54 (*SD* = 5.568).

One-Sample Statistics

	N	Mean	Std. Deviation	Std. Error Mean
Literacy Test Score	28	75.54	5.568	1.052

The *t* test is statistically significant at α = .05. A negative *t* value indicates that the sample mean is lower than the test value. In this particular instance, 9.064 (95% CI = ± 2.16) points lower.

For the write-up, you will need:

t = -8.615
df = 27
Sig (2-tailed) = .000

One-Sample Test

					Test Value = 84.6		
						95% Confidence Interval of the Difference	
	t	df	Sig. (2-tailed)	Mean Difference	Lower	Upper
Literacy Test Score	-8.615	27	.000	-9.064	-11.22	-6.91

4.3.3. Follow Up Analyses

4.3.3.1. Effect Size

Cohen's *d* is a useful index of the magnitude of the difference between the sample mean and **Test Value**, or population mean. It can be calculated with the following formula:

$$d = \frac{M - \mu}{s}$$

Where *M* is the sample mean, μ is the population mean (the **Test Value**), and *s* is the standard deviation of the sample data. These values are available in the *t* test output.

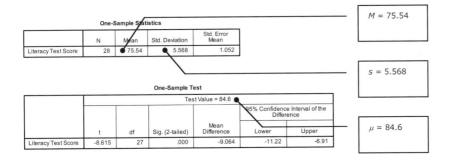

One-Sample Statistics

	N	Mean	Std. Deviation	Std. Error Mean
Literacy Test Score	28	75.54	5.568	1.052

One-Sample Test

					Test Value = 84.6		
						95% Confidence Interval of the Difference	
	t	df	Sig. (2-tailed)	Mean Difference	Lower	Upper
Literacy Test Score	-8.615	27	.000	-9.064	-11.22	-6.91

M = 75.54

s = 5.568

μ = 84.6

So,

$$d = \frac{75.54 - 84.6}{5.568} = \frac{-9.06}{5.568}$$

$$= -1.627$$

An effect size of *d* = -1.627 indicates that the sample mean is around 1.63 standard deviations lower than the population mean. According to Cohen's (1988) conventions, a difference of this magnitude is "large".

It appears that Principal Barrett has quite a strong case for arguing that the school should receive one of the literacy grants being offered.

AKA:
A sample mean can also be denoted with $\bar{X}$.

Tip:
M - μ has already been calculated by PASW Statistics and appears in the **Mean Difference** column of the **One-Sample Test** table.

Consequently, this formula can be simplified to:

$$d = \frac{MeanDifference}{s}$$

Tip:
Jacob Cohen (1988) suggests that a *d* of .20 can be considered small, a *d* of .50 is medium, and a *d* of .80 is large.

He also stresses that these values were subjectively derived with the intent that a medium effect should be "visible to the naked eye of a careful observer", a small effect should be "noticeably smaller than medium but not so small as to be trivial", and a large effect should be "the same distance above medium as small [is] below it" (Cohen, 1992, p. 156).

4.3.4. APA Style Results Write-Up

Results

A one sample *t* test was used to compare the National

Standardised Literacy Test performance of a class of 28 year-seven

students (*M* = 75.54, *SD* = 5.57) against the national average of 84.6.

The year-seven students scored 9.06 points, 95% CI [6.91, 11.22],

below the national average. This difference was found to be

statistically significant, *t*(27) = -8.62, *p* < .001, and large, *d* = 1.63.

The Shapiro-Wilk statistic and a visual inspection of the

stem-and-leaf plot confirmed that the students' test scores were

normally distributed.

(i) Tip:
When *Sig* = .000 in the PASW Statistics output, it should be reported as *p* < .001.

(i) Tip:
Report the absolute value of *d*, rather than a negative value.

A comprehensive **Results** section should provide:

- A description of each test used, and its purpose.
- Descriptive statistics for the sample, and the test value or population mean that the sample is being compared with.
- The outcome of each test, including whether or not it was statistically significant.
- The size and (if applicable) direction of each observed effect.

Additionally, a student paper will usually also describe:

- How each assumption was tested, and whether or not any were violated.
- (If applicable) how any violations were addressed.

A **Results** section should not include any interpretations of your findings. Save these for the **Discussion** section of your research report.

4.3.5. Summary

In example one, the *t* test was statistically significant, and the observed effect (or difference between the sample mean and the population mean) was large. In the following example, the opposite will occur.

4.4. Illustrated Example Two

The CEO of Acme Fashions wants to know if the unseasonably warm weather this past winter could have affected scarf sales. She knows that between 2001 and 2008 Acme sold an average of 275 scarves per store, per year.

The number of scarves sold during 2009 by 35 randomly selected Acme Fashions retail outlets is listed in Table 4.2.

Table 4.2

The Number of Scarves Sold by 35 Randomly Selected Acme Fashions Retail Outlets During 2009

Data:
This is data file **data_4_2.sav** on the companion website.

Store ID	Scarves Sold	Store ID	Scarves Sold	Store ID	Scarves Sold
1	200	13	322	25	225
2	264	14	269	26	277
3	302	15	241	27	290
4	321	16	262	28	263
5	280	17	298	29	258
6	178	18	277	30	249
7	264	19	204	31	323
8	243	20	193	32	314
9	297	21	307	33	302
10	276	22	352	34	309
11	238	23	294	35	281
12	307	24	286		

4.4.1. PASW Statistics Output (Part 1: Normality)

Explore

The procedure used to generate this output is the same as the procedure used in *Illustrated Example One*.

Syntax:
Run these analyses with **syntax_4_2.sps** on the companion website.

Case Processing Summary

	Cases					
	Valid		Missing		Total	
	N	Percent	N	Percent	N	Percent
Scarves Sold	35	100.0%	0	.0%	35	100.0%

Descriptives

		Statistic	Std. Error
Scarves Sold	Mean	273.31	6.788
	95% Confidence Interval for Mean Lower Bound	259.52	
	Upper Bound	287.11	
	5% Trimmed Mean	274.57	
	Median	277.00	
	Variance	1612.810	
	Std. Deviation	40.160	
	Minimum	178	
	Maximum	352	
	Range	174	
	Interquartile Range	53	
	Skewness	-.596	.398
	Kurtosis	.089	.778

In the **Descriptives** table we can see that, on average, each Acme store sold 273.31 scarves during 2009.

A **Standard Deviation** of 40.160 indicates that there is quite a lot of variability in this data. This is further confirmed by looking at the **Minimum** and **Maximum** statistics: 178 and 352 respectively.

The **Skewness** and **Kurtosis** statistics are both close to zero and, following the procedure described in section 4.3.2.3, z_s and z_k are both within ±1.96, <u>indicating a reasonably normal distribution</u>.

Tests of Normality

	Kolmogorov-Smirnov[a]			Shapiro-Wilk		
	Statistic	df	Sig.	Statistic	df	Sig.
Scarves Sold	.103	35	.200*	.963	35	.281

a. Lilliefors Significance Correction

*. This is a lower bound of the true significance.

A non-significant **Shapiro-Wilk** statistic (W = .963, *Sig* = .281) confirms that <u>these data are approximately normally distributed</u>.

Links:
The full output also contained a **Histogram**, a **Stem-and-Leaf Plot**, a **Normal Q-Q Plot**, a **Detrended Normal Q-Q Plot**, and a **Boxplot.** Guidelines for interpreting each of these can be found in *Illustrated Example One*.

4.4.2. PASW Statistics Output (Part 2: One Sample *t* Test)

T-Test

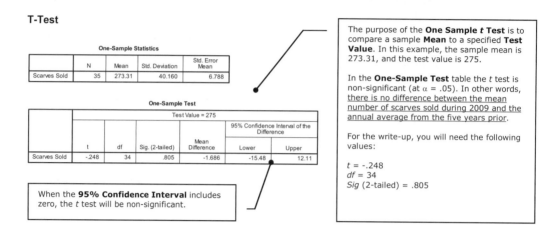

One-Sample Statistics

	N	Mean	Std. Deviation	Std. Error Mean
Scarves Sold	35	273.31	40.160	6.788

One-Sample Test

	Test Value = 275					
					95% Confidence Interval of the Difference	
	t	df	Sig. (2-tailed)	Mean Difference	Lower	Upper
Scarves Sold	-.248	34	.805	-1.686	-15.48	12.11

When the **95% Confidence Interval** includes zero, the *t* test will be non-significant.

The purpose of the **One Sample *t* Test** is to compare a sample **Mean** to a specified **Test Value**. In this example, the sample mean is 273.31, and the test value is 275.

In the **One-Sample Test** table the *t* test is non-significant (at α = .05). In other words, there is no difference between the mean number of scarves sold during 2009 and the annual average from the five years prior.

For the write-up, you will need the following values:

t = -.248
df = 34
Sig (2-tailed) = .805

4.4.3. Follow Up Analyses

4.4.3.1. Effect Size

$$d = \frac{M - \mu}{s}$$

Where *M* is the sample mean, μ is the population mean (the **Test Value**), and *s* is the standard deviation of the sample data. So,

$$d = \frac{273.31 - 275}{40.160} = \frac{-1.686}{40.160}$$

$$= -0.042$$

A *d* of 0.042 is extremely small, and could best be characterised as "no effect at all"!

Regardless of these findings however, the CEO of Acme Fashions should be extremely cautious about concluding that the weather does not influence the sale of scarves. This is because scarf sales could have been influenced by any number of factors not considered in this piece of research. For example, if scarves were "in" during 2009 (perhaps because popular celebrities started wearing them), sales may have remained buoyant despite the unseasonably warm winter.

4.4.4. APA Style Results Write-Up

Results

A one sample *t* test with an α of .05 was used to compare the average number of scarves sold by a random sample of 35 Acme Fashion retail outlets during 2009 ($M = 273.31$, $SD = 40.16$) with the company's five-year average of 275 scarves per store, per year. The Shapiro-Wilk test indicated that assumption of normality was not violated, and the *t* test was statistically non-significant, $t(34) = -0.25$, $p = .805$, $d = 0.042$, 95% CI [-15.48, 12.11].

> You will often see a non-significant result reported as *p* > .05 or simply as *ns* (for "non-significant").
>
> Although each of these is correct, it is generally advised to be as specific as possible, and to report an exact probability level (or *Sig*) wherever available.

4.5. One Sample *t* Test Checklist

Have you:

- ✔ Checked that each group of data is approximately normally distributed?
- ✔ Interpreted the results of the *t* test and taken note of the *t* value, degrees of freedom, significance, mean difference and confidence interval for your write-up?
- ✔ Calculated a measure of effect size, such as Cohen's *d*?
- ✔ Reported your results in the APA style?

Chapter 5: Independent Samples *t* Test

💬 *AKA:*
Independent groups *t* test; Between groups *t* test; Between subjects *t* test.

Chapter Overview

5.1. Purpose of the Independent Samples *t* Test

To test for a statistically significant difference between two independent sample means.

5.2. Questions We Could Answer Using the Independent Samples *t* Test

1. Is there a difference between the amount of unpaid overtime worked by male and female Occupational Therapists (OTs)?

In this example, we're asking whether two independent (or separate) groups - male OTs and female OTs – work differing amounts of unpaid overtime. Note how each participant is a member of just one group (i.e., you can be a male OT <u>or</u> a female OT, but not both). This is an essential characteristic of the independent samples *t* test.

Gender is our independent variable (IV), which has two levels: male and female. Unpaid overtime is our dependent variable (DV), which we could measure in hours per week.

Similar questions we could answer with an independent samples *t* test include:

2. Are nursing home residents with pets happier than those without?

3. Do rats injected with a growth hormone run faster than those injected with a placebo?

4. Do students who listen to classical music while studying achieve higher grades than those who listen to rock music?

5.3. Illustrated Example One

As a class project, Clare and Arash partially replicated Loftus and Palmer's (1974) classic experiment on the effects of asking leading questions on memory reconstruction. They showed video footage of a car accident to 30 participants, then asked each of them to answer one of the following questions: (1) "about how fast were the cars going when they hit each other?" or (2) "about how fast were the cars going when they smashed into each other?"

Their research investigates whether the participants who were asked the "hit" question report, on average, faster or slower speed estimates than those asked the "smashed" question.

Clare and Arash's data are reproduced in Table 5.1.

Table 5.1

☐ **Data:**
This is data file **data_5_1.sav** on the companion website.

Speed Estimates (in km/h) Given by Participants (N = 30) in Response to Either the "Hit" Question, or the "Smashed" Question

Hit Condition (Group 1)		Smashed Condition (Group 2)	
Participant ID	Est. Speed	Participant ID	Est. Speed
1	39	16	41
2	33	17	36
3	32	18	49
4	37	19	50
5	35	20	39
6	35	21	38
7	34	22	39
8	33	23	42
9	34	24	41
10	31	25	40
11	38	26	40
12	36	27	45
13	30	28	36
14	34	29	42
15	30	30	47

5.3.1. Setting Up the PASW Statistics Data File

The **Labels** entered here are what you will see during the analyses and in the output.

▶▶| *Link:*
Setting up a data file is explained in chapter 1.

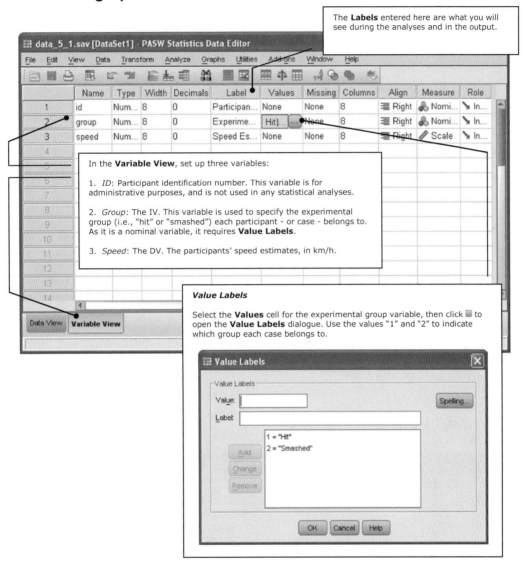

In the **Variable View**, set up three variables:

1. *ID*: Participant identification number. This variable is for administrative purposes, and is not used in any statistical analyses.

2. *Group*: The IV. This variable is used to specify the experimental group (i.e., "hit" or "smashed") each participant - or case - belongs to. As it is a nominal variable, it requires **Value Labels**.

3. *Speed*: The DV. The participants' speed estimates, in km/h.

Value Labels

Select the **Values** cell for the experimental group variable, then click ▦ to open the **Value Labels** dialogue. Use the values "1" and "2" to indicate which group each case belongs to.

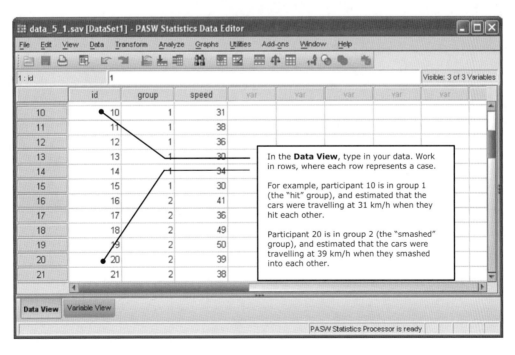

(i) *Tip:*
You can toggle between displaying *values* or *value labels* in the **Data View** by clicking the **Value Labels** button:

In the **Data View**, type in your data. Work in rows, where each row represents a case.

For example, participant 10 is in group 1 (the "hit" group), and estimated that the cars were travelling at 31 km/h when they hit each other.

Participant 20 is in group 2 (the "smashed" group), and estimated that the cars were travelling at 39 km/h when they smashed into each other.

5.3.2. Analysing the Data

📺 *Syntax:*
Run these analyses
with **syntax_5_1.sps**
on the companion
website.

5.3.2.1. Assumptions

The following criteria should be met before conducting an independent
samples *t* test. Assumptions 1 and 2 are methodological, and should have
been addressed before and during data collection. Assumptions 3 and 4 can
be tested with PASW Statistics.

💬 *AKA:*
Both interval and
ratio data are
referred to as **Scale**
data in PASW
Statistics.

1. **Scale of Measurement.** The DV should be interval or ratio data. Some
   authors (e.g., Clark-Carter, 2004) indicate that ordinal data are also
   acceptable, provided the scale has at least seven possible values.

2. **Independence.** Each participant should participate only once in the
   research, and should not influence the participation of others.

ⓘ *Tip:*
Often, we use
inferential statistics
to estimate
population
parameters from
sample data. Such
generalisations can
only be made
confidently if our
sample data has
been randomly
drawn/selected from
the population of
interest.

3. **Normality.** Each group of scores should be approximately normally
   distributed.

4. **Homogeneity of Variance.** There should be an approximately equal
   amount of variability in each set of scores.

The assumption of normality can be assessed in numerous ways (see, for
example, chapter 4). In this example, we will use the Shapiro-Wilk test, and a
visual inspection of the histograms. The homogeneity of variance assumption
is tested as part of the *t* test.

If you've used
convenience
sampling (as Clare
and Arash appear to
have done),
generalisations
should be made
cautiously.

5.3.2.2. PASW Statistics Procedure (Part 1: Normality)

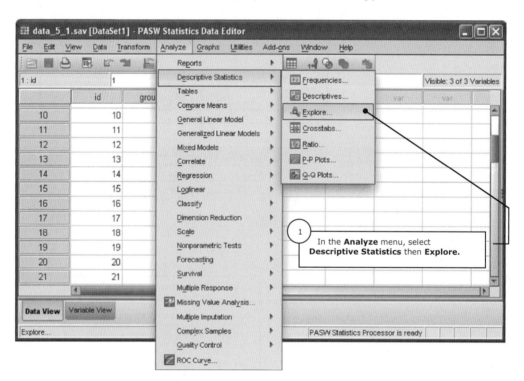

1 — In the **Analyze** menu, select
Descriptive Statistics then **Explore.**

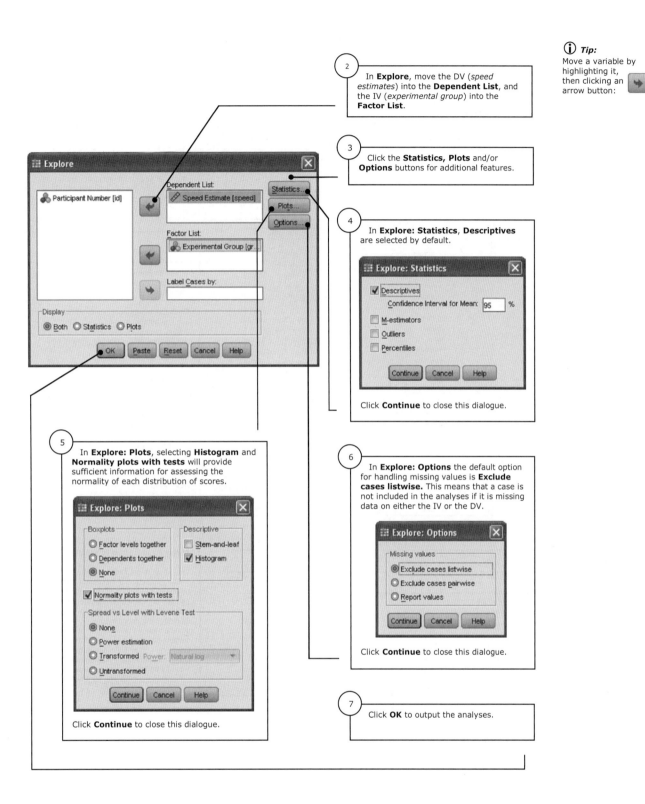

① *Tip:*
Move a variable by highlighting it, then clicking an arrow button:

2 In **Explore**, move the DV (*speed estimates*) into the **Dependent List**, and the IV (*experimental group*) into the **Factor List**.

3 Click the **Statistics, Plots** and/or **Options** buttons for additional features.

4 In **Explore: Statistics**, **Descriptives** are selected by default.

Click **Continue** to close this dialogue.

5 In **Explore: Plots**, selecting **Histogram** and **Normality plots with tests** will provide sufficient information for assessing the normality of each distribution of scores.

Click **Continue** to close this dialogue.

6 In **Explore: Options** the default option for handling missing values is **Exclude cases listwise.** This means that a case is not included in the analyses if it is missing data on either the IV or the DV.

Click **Continue** to close this dialogue.

7 Click **OK** to output the analyses.

5.3.2.3. PASW Statistics Output (Part 1: Normality)

Explore

Experimental Group

> The **Case Processing Summary** shows how many cases were analysed, and how many were dropped due to missing data.
>
> In this instance, there were 15 cases in each experimental group. No cases were excluded because of missing data.

Case Processing Summary

	Experimental Group	Cases					
		Valid		Missing		Total	
		N	Percent	N	Percent	N	Percent
Speed Estimate	Hit	15	100.0%	0	.0%	15	100.0%
	Smashed	15	100.0%	0	.0%	15	100.0%

> The table of **Descriptives** contains a range of useful information, including measures of central tendency and dispersion for each group of scores, along with skewness and kurtosis statistics.

Descriptives

	Experimental Group			Statistic	Std. Error
Speed Estimate	Hit	Mean		34.07	.700
		95% Confidence Interval for Mean	Lower Bound	32.57	
			Upper Bound	35.57	
		5% Trimmed Mean		34.02	
		Median		34.00	
		Variance		7.352	
		Std. Deviation		2.712	
		Minimum		30	
		Maximum		39	
		Range		9	
		Interquartile Range		4	
		Skewness		.167	.580
		Kurtosis		-.562	1.121
	Smashed	Mean		41.67	1.116
		95% Confidence Interval for Mean	Lower Bound	39.27	
			Upper Bound	44.06	
		5% Trimmed Mean		41.52	
		Median		41.00	
		Variance		18.667	
		Std. Deviation		4.320	
		Minimum		36	
		Maximum		50	
		Range		14	
		Interquartile Range		6	
		Skewness		.716	.580
		Kurtosis		-.298	1.121

> When **Skewness** and **Kurtosis** are both zero, the data are normally distributed.
>
> These skewness and kurtosis figures are reasonably close to zero, and z_s and z_k (see section 4.3.2.3) are within ±1.96 for both variables. This is all reassuring news!

> PASW Statistics provides two **Tests of Normality**. The **Shapiro-Wilk** test is considered more appropriate for smaller samples. A statistically significant (i.e., $Sig < .05$), W statistic is indicative of non-normality.
>
> Here, W is .967 (Sig = .804) for the "hit" data, and .920 (Sig = .193) for the "smashed" data. Thus, we can conclude <u>the assumption of normality is not violated for either group of scores</u>.

Tests of Normality

	Experimental Group	Kolmogorov-Smirnov[a]			Shapiro-Wilk		
		Statistic	df	Sig.	Statistic	df	Sig.
Speed Estimate	Hit	.110	15	.200*	.967	15	.804
	Smashed	.203	15	.099	.920	15	.193

a. Lilliefors Significance Correction

*. This is a lower bound of the true significance.

AKA:
The *Sig* figures reported by PASW Statistics are usually referred to as "*p*" in journal articles.

For example, $W(15)$ = .97, p = .804.

> A visual inspection of the **Histograms** further confirms that <u>each group of scores is approximately normally distributed</u>.

If the Assumption is Violated

The *t* test is considered robust against small to moderate violations of the normality assumption, provided the sample is reasonably large (40+), and group sizes are relatively equal. Researchers concerned about more severe violations – or non-normality combined with heterogeneity of variance – may consider data transformation (see Tabachnick & Fidell, 2007a), or a non-parametric procedure such as the Mann-Whitney *U* test (see chapter 16).

Speed Estimate

Histograms

Links:
The full PASW Statistics output included two **Normal Q-Q Plots**, and two **Detrended Normal Q-Q Plots**, which have been omitted from the current example. See chapter 3 for information about these graphs.

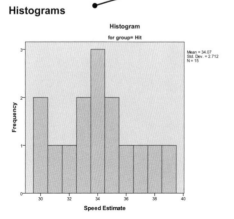

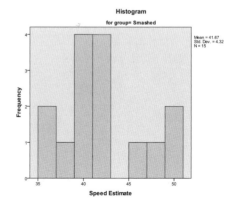

5.3.2.4. PASW Statistics Procedure (Part 2: Homogeneity of Variance & the t Test)

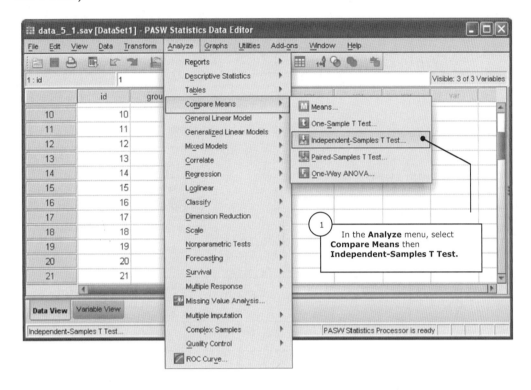

1 In the **Analyze** menu, select **Compare Means** then **Independent-Samples T Test.**

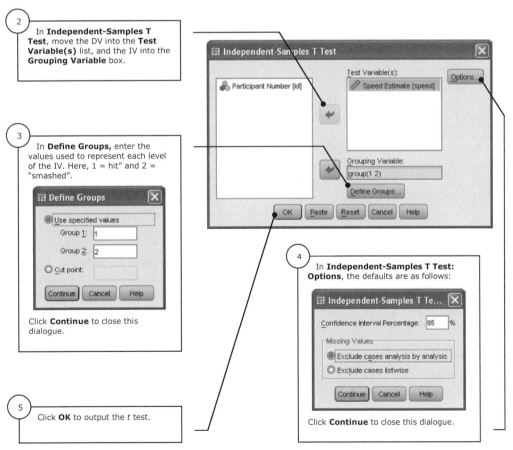

2 In **Independent-Samples T Test**, move the DV into the **Test Variable(s)** list, and the IV into the **Grouping Variable** box.

3 In **Define Groups,** enter the values used to represent each level of the IV. Here, 1 = hit" and 2 = "smashed".

Click **Continue** to close this dialogue.

5 Click **OK** to output the *t* test.

4 In **Independent-Samples T Test: Options**, the defaults are as follows:

Click **Continue** to close this dialogue.

ⓘ Tip: You can simultaneously compare two groups of participants on several factors by moving them all into the **Test Variables(s)** list.

5.3.2.5. PASW Statistics Output (Part 2: Homogeneity of Variance & the t Test)

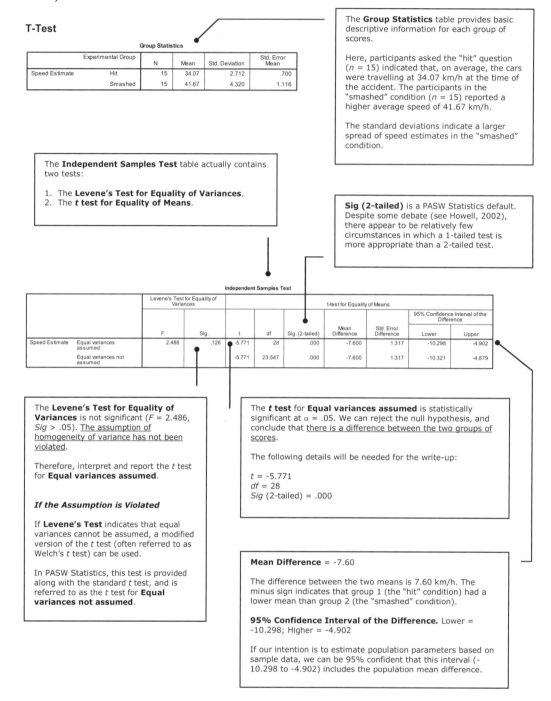

T-Test

Group Statistics

Experimental Group		N	Mean	Std. Deviation	Std. Error Mean
Speed Estimate	Hit	15	34.07	2.712	.700
	Smashed	15	41.67	4.320	1.116

The **Group Statistics** table provides basic descriptive information for each group of scores.

Here, participants asked the "hit" question (*n* = 15) indicated that, on average, the cars were travelling at 34.07 km/h at the time of the accident. The participants in the "smashed" condition (*n* = 15) reported a higher average speed of 41.67 km/h.

The standard deviations indicate a larger spread of speed estimates in the "smashed" condition.

The **Independent Samples Test** table actually contains two tests:

1. The **Levene's Test for Equality of Variances**.
2. The *t* test for Equality of Means.

Sig (2-tailed) is a PASW Statistics default. Despite some debate (see Howell, 2002), there appear to be relatively few circumstances in which a 1-tailed test is more appropriate than a 2-tailed test.

Independent Samples Test

		Levene's Test for Equality of Variances		t-test for Equality of Means					95% Confidence Interval of the Difference	
		F	Sig.	t	df	Sig. (2-tailed)	Mean Difference	Std. Error Difference	Lower	Upper
Speed Estimate	Equal variances assumed	2.486	.126	-5.771	28	.000	-7.600	1.317	-10.298	-4.902
	Equal variances not assumed			-5.771	23.547	.000	-7.600	1.317	-10.321	-4.879

The **Levene's Test for Equality of Variances** is not significant (*F* = 2.486, *Sig* > .05). The assumption of homogeneity of variance has not been violated.

Therefore, interpret and report the *t* test for **Equal variances assumed**.

If the Assumption is Violated

If **Levene's Test** indicates that equal variances cannot be assumed, a modified version of the *t* test (often referred to as Welch's *t* test) can be used.

In PASW Statistics, this test is provided along with the standard *t* test, and is referred to as the *t* test for **Equal variances not assumed**.

The *t* test for **Equal variances assumed** is statistically significant at α = .05. We can reject the null hypothesis, and conclude that there is a difference between the two groups of scores.

The following details will be needed for the write-up:

t = -5.771
df = 28
Sig (2-tailed) = .000

Mean Difference = -7.60

The difference between the two means is 7.60 km/h. The minus sign indicates that group 1 (the "hit" condition) had a lower mean than group 2 (the "smashed" condition).

95% Confidence Interval of the Difference. Lower = -10.298; Higher = -4.902

If our intention is to estimate population parameters based on sample data, we can be 95% confident that this interval (-10.298 to -4.902) includes the population mean difference.

5.3.3. Follow Up Analyses

5.3.3.1. Effect Size

The 6th edition (2010) of the *Publication Manual of the American Psychological Association* (APA) notes that:

> For the reader to appreciate the magnitude or importance of a study's findings, it is almost always necessary to include some measure of effect size in the Results section. (p. 34)

Although PASW Statistics does not automatically compute an effect size index for the *t* test, one can be easily calculated from the output PASW Statistics provides.

For example, Cohen's *d* is a scale-free measure of the separation between two group means. It provides a measure of the difference between the two group means expressed in terms of their common standard deviation. Thus, a *d* of 0.5 indicates that one-half of a standard deviation separates the two means.

d can be calculated with the following formula:

$$d = \frac{M_1 - M_2}{s_p}$$

AKA:
This formulation of *d*, which takes unequal sample sizes into account, is referred to as Hedges' *g*.

Where M_1 is the mean of group 1, M_2 is the mean of group 2, and s_p is the pooled standard deviation, calculated as:

$$s_p = \sqrt{\frac{(n_1 - 1)s_1^2 + (n_2 - 1)s_2^2}{n_1 + n_2 - 2}}$$

(i) Tip:
When both groups of scores are the same size, this formula can be simplified to:

$$s_p = \frac{s_1 + s_2}{2}$$

Where s_1 and s_2 are the standard deviations for groups 1 and 2 respectively.

Where n_1 is the size of group 1; n_2 is the size of group 2; s_1^2 is the variance of group 1; and s_2^2 is the variance of group 2.

All of these figures are available in (or easily calculated from) the **Group Statistics** table outputted with the *t* test:

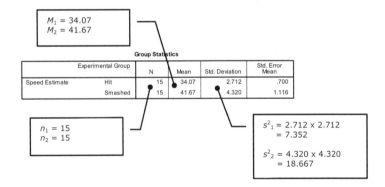

So,

$$s_p = \sqrt{\frac{(15-1)7.352 + (15-1)18.667}{15+15-2}}$$

$$= \sqrt{\frac{102.928 + 261.338}{28}}$$

$$= \sqrt{\frac{364.266}{28}} = \sqrt{13.0095}$$

$$= 3.607$$

Then,

$$d = \frac{34.07-41.67}{3.607}$$
$$= -2.11$$

① Tip:
Report the absolute value of *d*, rather than a negative value.

Cohen (1988) suggested some general conventions for effect sizes in the social sciences. According to these conventions, an effect size of *d* = .20 is considered small, *d* = .50 is medium, and *d* = .80 is large.

Cohen (1988) cautions that these general recommendations are more useful in some circumstances than others. However, the concept of small, medium, and large effect sizes can be a reasonable starting point if you do not have more precise information to work from.

5.3.4. APA Style Results Write-Up

Results

An independent samples *t* test was used to compare the

average speed estimates reported by participants in the "hit" condition

(n = 15) to the average speed estimates reported by those in the

"smashed" condition (n = 15). Neither Shapiro-Wilk statistic was

significant, indicating that the assumption of normality was not

violated. Levene's test was also non-significant, thus equal variances

can be assumed. The *t* test was statistically significant, with the "hit"

group (M = 34.07, SD = 2.71) reporting speed estimates some 7.60

km/h lower, 95% CI [-10.30, -4.90], than the "smashed" group (M =

41.67, SD = 4.32), $t(28)$ = -5.77, p < .001, two-tailed, d = 2.11.

Assumption testing is often not reported in journal articles (especially when they are not violated). It is seen much more frequently in student papers.

When the *t* test is statistically significant, both the size and direction of the effect should be reported.

In this instance, direction is indicated by the mean difference (and 95% CI), or can be inferred by comparing the two group means.

Size is indicated by Cohen's *d*.

5.3.5. Summary

In this example, the independent samples *t* test was statistically significant, and neither the normality nor homogeneity of variance assumptions were violated. Real-life research, however, is often not this "neat"!

In the second illustrated example, our *t* test is non-significant, and the homogeneity of variance assumption is violated – two issues that researchers must commonly deal with.

5.4. Illustrated Example Two

As part of a national assessment program, the students in Mrs Sommers' year 7 class recently completed an intelligence test. Their global IQ scores are tabulated in Table 5.2.

Mrs Sommers would like to know whether there is a difference between the average IQ of her male students, and that of her female students.

Table 5.2

Gender and IQ Data for Each Member of Mrs Sommers' Class (N = 35)

🖵 *Data:*
This is data file **data_5_2.sav** on the companion website.

Participant ID	Gender	IQ
1	1	100
2	1	82
3	1	109
4	2	105
5	2	110
6	1	118
7	1	97
8	1	108
9	2	112
10	1	124
11	1	104
12	2	95
13	2	106
14	2	94
15	2	109
16	1	110
17	2	105
18	2	106
19	1	105
20	1	103
21	1	112
22	1	97
23	2	99
24	1	96
25	2	108
26	2	109
27	2	104
28	1	117
29	1	114
30	2	106
31	2	99
32	2	98
33	1	89
34	1	98
35	2	100

Note. Gender is coded as 1 = male and 2 = female.

5.4.1. PASW Statistics Output (Part 1: Normality)

Syntax:
Run these analyses
with **syntax_5_2.sps**
on the companion
website.

Explore

> The procedure for generating this output is the same as that used in *Illustrated Example One*.

gender

Case Processing Summary

gender		Cases					
		Valid		Missing		Total	
		N	Percent	N	Percent	N	Percent
IQ	Male	18	100.0%	0	.0%	18	100.0%
	Female	17	100.0%	0	.0%	17	100.0%

> The **Case Processing Summary** indicates that there are 18 males in the sample, and 17 females. There is no missing data.

Descriptives

gender				Statistic	Std. Error
IQ	Male	Mean		104.61	2.508
		95% Confidence Interval for Mean	Lower Bound	99.32	
			Upper Bound	109.90	
		5% Trimmed Mean		104.79	
		Median		104.50	
		Variance		113.193	
		Std. Deviation		10.639	
		Minimum		82	
		Maximum		124	
		Range		42	
		Interquartile Range		16	
		Skewness		-.229	.536
		Kurtosis		-.050	1.038
	Female	Mean		103.82	1.304
		95% Confidence Interval for Mean	Lower Bound	101.06	
			Upper Bound	106.59	
		5% Trimmed Mean		103.92	
		Median		105.00	
		Variance		28.904	
		Std. Deviation		5.376	
		Minimum		94	
		Maximum		112	
		Range		18	
		Interquartile Range		10	
		Skewness		-.431	.550
		Kurtosis		-.871	1.063

> In the table of **Descriptives** we can see that the **Mean** male *IQ* is 104.61 (*SD* = 10.639), while the **Mean** female *IQ* is 103.82 (*SD* = 5.376).
>
> The **Variance**, **Std. Deviation**, **Minimum**, **Maximum** and **Range** for each group of scores indicate that there is more variability in the male data than the female data.
>
> The **Skewness** and **Kurtosis** statistics are all close to zero, and z_s and z_k are within ±1.96 (see section 4.3.2.3) for both genders, indicating that each group of scores is reasonably normally distributed.

Tests of Normality

gender		Kolmogorov-Smirnov[a]			Shapiro-Wilk		
		Statistic	df	Sig.	Statistic	df	Sig.
IQ	Male	.098	18	.200*	.988	18	.996
	Female	.175	17	.176	.940	17	.320

a. Lilliefors Significance Correction

*. This is a lower bound of the true significance.

> Both **Shapiro-Wilk** tests are statistically non-significant at α = .05, confirming that the normality assumption is not violated.

IQ

Histograms

> The **Histograms** also confirm that each distribution of *IQ* scores looks reasonably normal.

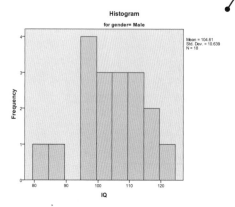

Histogram
for gender= Male

Mean = 104.61
Std. Dev. = 10.639
N = 18

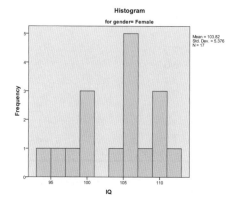

Histogram
for gender= Female

Mean = 103.82
Std. Dev. = 5.376
N = 17

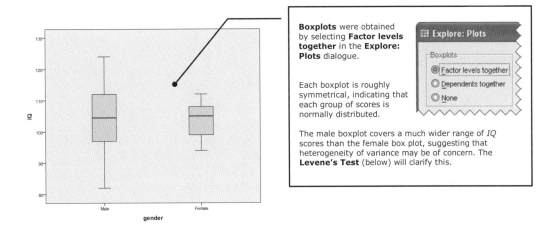

Boxplots were obtained by selecting **Factor levels together** in the **Explore: Plots** dialogue.

Each boxplot is roughly symmetrical, indicating that each group of scores is normally distributed.

The male boxplot covers a much wider range of *IQ* scores than the female box plot, suggesting that heterogeneity of variance may be of concern. The **Levene's Test** (below) will clarify this.

5.4.2. PASW Statistics Output (Part 2: Homogeneity of Variance & the *t* Test)

T-Test

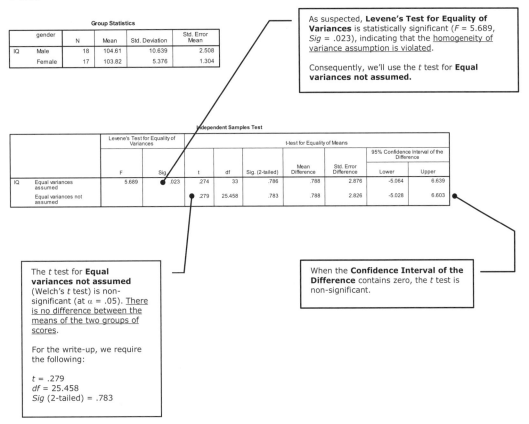

Group Statistics

	gender	N	Mean	Std. Deviation	Std. Error Mean
IQ	Male	18	104.61	10.639	2.508
	Female	17	103.82	5.376	1.304

As suspected, **Levene's Test for Equality of Variances** is statistically significant (*F* = 5.689, *Sig* = .023), indicating that the homogeneity of variance assumption is violated.

Consequently, we'll use the *t* test for **Equal variances not assumed**.

Independent Samples Test

		Levene's Test for Equality of Variances		t-test for Equality of Means					95% Confidence Interval of the Difference	
		F	Sig.	t	df	Sig. (2-tailed)	Mean Difference	Std. Error Difference	Lower	Upper
IQ	Equal variances assumed	5.689	.023	.274	33	.786	.788	2.876	-5.064	6.639
	Equal variances not assumed			.279	25.458	.783	.788	2.826	-5.028	6.603

The *t* test for **Equal variances not assumed** (Welch's *t* test) is non-significant (at α = .05). There is no difference between the means of the two groups of scores.

For the write-up, we require the following:

t = .279
df = 25.458
Sig (2-tailed) = .783

When the **Confidence Interval of the Difference** contains zero, the *t* test is non-significant.

5.4.3. Follow Up Analyses

5.4.3.1. Effect Size

$$d = \frac{M_1 - M_2}{s_p} \qquad \text{where:} \qquad s_p = \sqrt{\frac{(n_1-1)s_1^2 + (n_2-1)s_2^2}{n_1 + n_2 - 2}}$$

So,

$$s_p = \sqrt{\frac{(18-1)113.193+(17-1)28.904}{18+17-2}}$$

$$= \sqrt{\frac{1924.281+462.464}{33}}$$

$$= \sqrt{\frac{2386.745}{33}} = \sqrt{72.326} = 8.504$$

Then,

$$d = \frac{104.61-103.82}{8.504}$$

$$= 0.093$$

An effect size of this magnitude is trivial. There is virtually no difference (and certainly not a "significant" one) between the average IQs of the male and female students in Mrs Sommers' class.

5.4.4. APA Style Results Write-Up

Results

Preliminary assumption testing indicated that both the boys'

($M = 104.61$, $SD = 10.64$) and girls' ($M = 103.82$, $SD = 5.38$) IQ

scores were normally distributed, but that there was substantially more

variance in the boys' scores. Consequently, Welch's t test was used

to compare the boys' average IQ to that of the girls'. The t test was

non-significant, $t(25.46) = 0.28$, $p = .783$, two-tailed, $d = 0.09$, 95%

CI [-5.03, 6.60].

> Numbers that cannot exceed ± 1 do not need a leading zero. So, $t = 0.28$, but $p = .783$.

5.5. Independent Samples t Test Checklist

Have you:

✔   Checked that each group of data is approximately normally distributed?
✔   Checked for homogeneity of variance?
✔   Interpreted the appropriate t test (i.e., for equal variances assumed versus equal variances not assumed), and taken note of the t value, degrees of freedom, significance, mean difference and confidence interval for your write-up?
✔   Calculated a measure of effect size, such as Cohen's d?
✔   Written up your results in the APA style?

Chapter 6: Paired Samples *t* Test

💬 *AKA:*
Repeated measures *t* test; Dependent samples *t* test; Within samples *t* test; Matched samples *t* test; Correlated samples *t* test.

Additionally, "samples" is sometimes replaced by "groups", "subjects" or "participants".

Chapter Overview

6.1. Purpose of the Paired Samples *t* Test

To test for a statistically significant difference between two related sample means. Two samples are considered related when:

    a. They are both comprised of the same group of individuals, who've provided data on two separate occasions (e.g., before and after a treatment).

    b. Each individual in one sample is connected or linked with a specific individual in the other (e.g., husband and wife dyads).

Each of these situations is illustrated in section 6.2.

6.2. Questions We Could Answer Using the Paired Samples *t* Test

1. Do people report a higher sense of subjective wellbeing after 15 minutes of aerobic exercise?

To answer this question, we would need to measure the subjective wellbeing of each participant twice: before and after they've done 15 minutes of aerobic exercise. We would then use the *t* test to compare the mean of the "before" data with the mean of the "after" data.

2. Is students' short-term memory more accurate when tested in quiet or noisy surroundings?

Questions 1 and 2 are generally referred to as repeated measures designs. Question 3 is a little different, as it requires that we recruit participants in pairs, but need each to provide us with data on one occasion only. This is usually referred to as a matched design.

3. Is there a difference between fathers' and mothers' estimates of their first-born sons' IQ scores?

6.3. Illustrated Example One

A local cycling association encourages its members to wear fluorescent vests when riding after dark. It argues that doing so makes cyclists more visible to passing motorists, and thus safer.

To demonstrate this, the association's chairperson hires a driving simulator that has been specially programmed to drop a virtual cyclist into an evening driving simulation at random intervals. When a cyclist appears, the "driver" must respond – as quickly as possible – by pressing a button located on the simulator steering wheel. The simulator automatically records the time (in milliseconds) it takes the driver to react to each cyclist.

In a complete testing session (which lasts around 20 minutes) a driver will be exposed to ten cyclists: five in fluorescent vests (the experimental trials); and five in non-fluorescent attire (the control trials). These trials are presented in a random order. After all 10 trials, the driver's average reaction times for both the experimental and control conditions can be calculated. These averages, for a group of 15 participants, are reported in Table 6.1.

The chairperson thinks that reaction times will be faster during the experimental trials (i.e., when the cyclists are wearing fluorescent vests).

Table 6.1

□ *Data:*
This is data file
data_6_1.sav on the
companion website.

Average Reaction Times (in Milliseconds) to the Presentation of Virtual Cyclists in Fluorescent and Non-Fluorescent Attire During a Driving Simulation Test (N = 15)

ID	Control (Non-Fluroescent) Trials Average RT (msec)	Experimental (Fluroescent) Trials Average RT (msec)
1	345	268
2	540	340
3	430	310
4	470	322
5	420	286
6	364	320
7	388	292
8	392	388
9	378	296
10	362	304
11	420	318
12	446	312
13	452	334
14	434	346
15	498	376

6.3.1. Setting Up the PASW Statistics Data File

▶▶| *Link:*
Setting up a data
file is explained in
chapter 1.

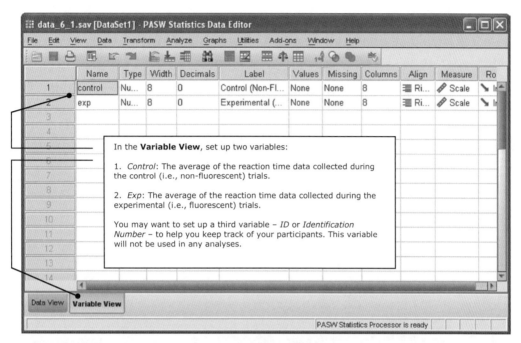

In the **Variable View**, set up two variables:

1. *Control*: The average of the reaction time data collected during the control (i.e., non-fluorescent) trials.

2. *Exp*: The average of the reaction time data collected during the experimental (i.e., fluorescent) trials.

You may want to set up a third variable – *ID* or *Identification Number* – to help you keep track of your participants. This variable will not be used in any analyses.

ⓘ *Tip:*
The **Name** you use
will appear at the top
of the variable column
in the **Data View**.
However, the **Label**
will be used in your
output.

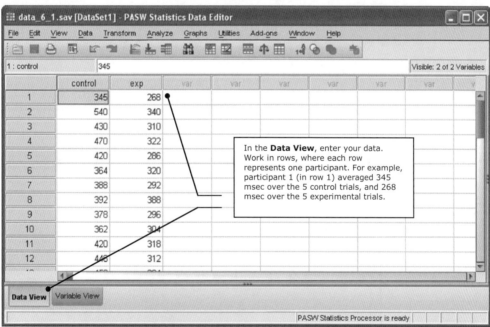

In the **Data View**, enter your data. Work in rows, where each row represents one participant. For example, participant 1 (in row 1) averaged 345 msec over the 5 control trials, and 268 msec over the 5 experimental trials.

6.3.2. Analysing the Data

6.3.2.1. Assumptions

Three assumptions should be met before conducting a paired samples *t* test. The first is methodological, and is addressed when selecting measures. The second and third can be tested with PASW Statistics.

1. **Scale of Measurement.** Interval or ratio data are required for a paired samples *t* test. If your data are ordinal or nominal, you should consider a non-parametric test instead (see chapter 16).

🖥 *Syntax:*
Run these analyses
with **syntax_6_1.sps**
on the companion
website.

2. **Normality.** Each group of scores should be approximately normally distributed.

3. **Normality of Difference Scores.** The differences between pairs of scores should be approximately normally distributed.

6.3.2.2. PASW Statistics Procedure (Part 1: Normality & Normality of Difference Scores)

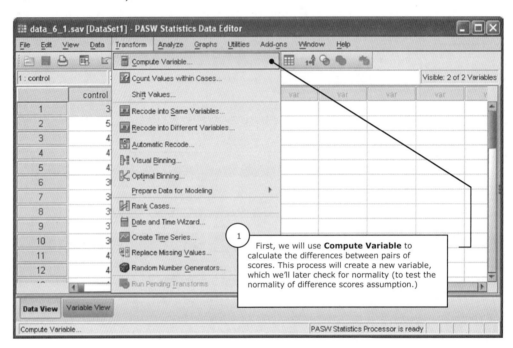

1. First, we will use **Compute Variable** to calculate the differences between pairs of scores. This process will create a new variable, which we'll later check for normality (to test the normality of difference scores assumption.)

2. Type a name for the new variable under **Target Variable**. We've used *diff* as shorthand for *difference scores*.

3. Build the **Numeric Expression** needed to calculate the difference scores using the list of available variables, the arrow button and the keypad.

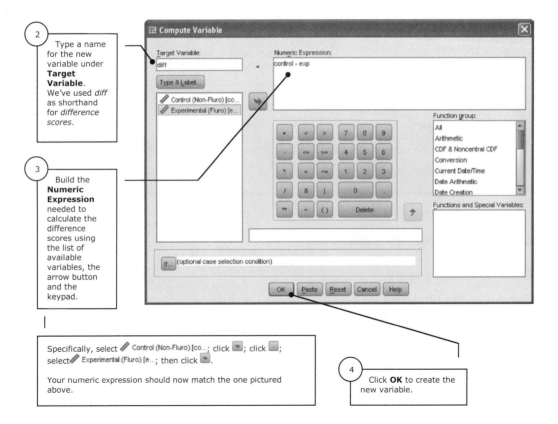

Specifically, select ✎ Control (Non-Fluro) [co...; click ▾; click ▭; select ✎ Experimental (Fluro) [e...; then click ▾.

Your numeric expression should now match the one pictured above.

4. Click **OK** to create the new variable.

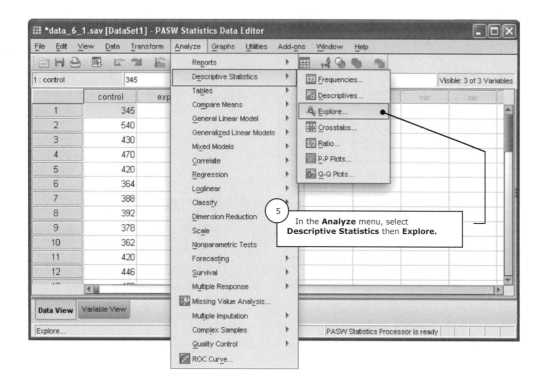

(i) *Tip:*
Although obscured by the **Analyze** menu in this illustration, there is now a third variable in the data file, called *diff*.

The first participant's score on *diff* is 77. This means that his/her average control trial reaction time of 345 msec was 77 msec slower than his/her average experimental trial reaction time of 268 msec.

6   In **Explore**, move the *control*, *experimental* and *difference scores* variables into the **Dependent List**.

7   Select **Both** to display all the options available in **Explore**, or either **Statistics** or **Plots** for a more limited range.

Here, we've limited our choices to just **Plots**.

8   Some **Histograms** will give us an adequate sense of whether or not the assumptions of normality and normality of difference scores are violated.

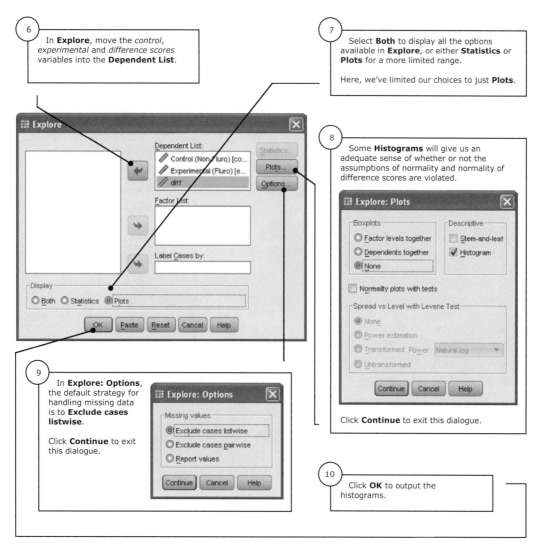

▶▶| *Link:*
There are many ways to assess normality. Several of these are illustrated in chapter 4.

9   In **Explore: Options**, the default strategy for handling missing data is to **Exclude cases listwise**.

Click **Continue** to exit this dialogue.

Click **Continue** to exit this dialogue.

10   Click **OK** to output the histograms.

6.3.2.3. PASW Statistics Output (Part 1: Normality & Normality of Difference Scores)

Explore

Case Processing Summary

	Cases					
	Valid		Missing		Total	
	N	Percent	N	Percent	N	Percent
Control (Non-Fluro)	15	100.0%	0	.0%	15	100.0%
Experimental (Fluro)	15	100.0%	0	.0%	15	100.0%
diff	15	100.0%	0	.0%	15	100.0%

The **Case Processing Summary** shows how many cases were analysed, and whether any were excluded due to missing data.

Control (Non-Fluro)

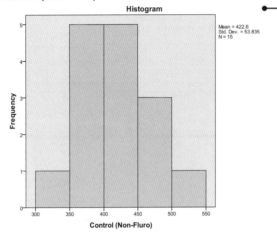

On a **Histogram**, the vertical axis represents the frequency of each band (or range) of values, which are ordered from smallest to largest along the horizontal axis.

Looking at the first histogram, one participant had an average reaction time between 300 and 350 msec during the Control trials; five had averages between 350 and 400 msec; and so on.

Ideally, a histogram should look reasonably "normal". That is, it should resemble an "inverted U" with greatest frequency of cases clustered around the mean, and progressively fewer cases towards the tails.

Even though these histograms all have slightly different appearances they are all relatively normal, thus <u>satisfying the normality and normality of difference scores assumptions</u>.

(i) Tip:
Double-click a graph to open up the **Chart Editor**. In this editor you can superimpose a normal curve on top of a histogram, as well as change its appearance in numerous ways. Turn to chapter 3 for a closer look at the PASW Statistics **Chart Editor**.

Experimental (Fluro)

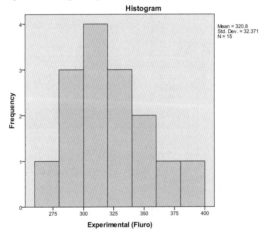

If the Assumptions are Violated

Small to moderate violations of the normality assumptions are of little concern in samples of 30+ pairs.

Severe violations may prompt the researcher to consider transformations (see Tabachnick and Fidell, 2007b) or a non-parametric alternative such as the Wilcoxon signed rank test (see chapter 16).

diff

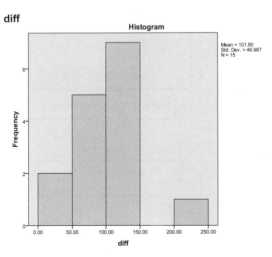

6.3.2.4. PASW Statistics Procedure (Part 2: Paired Samples t Test)

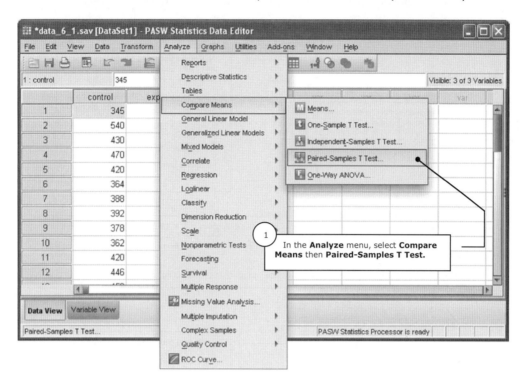

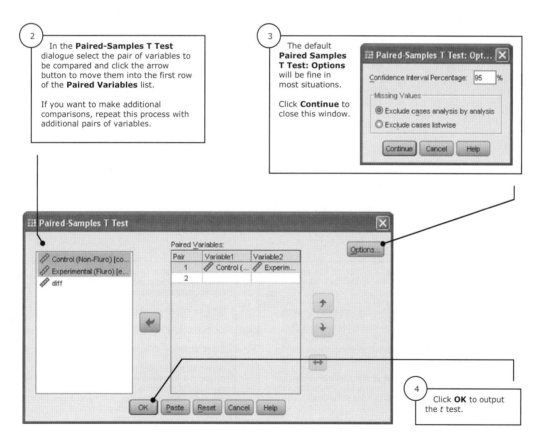

2 In the **Paired-Samples T Test** dialogue select the pair of variables to be compared and click the arrow button to move them into the first row of the **Paired Variables** list.

If you want to make additional comparisons, repeat this process with additional pairs of variables.

3 The default **Paired Samples T Test: Options** will be fine in most situations.

Click **Continue** to close this window.

(i) Tip:
To select multiple variables, hold down the Shift key on your keyboard, then click each with your mouse.

One click will select a variable; a second click will deselect it.

4 Click **OK** to output the *t* test.

6.3.2.5. PASW Statistics Output (Part 2: Paired Samples t Test)

T-Test

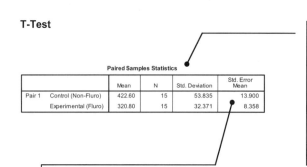

Paired Samples Statistics

		Mean	N	Std. Deviation	Std. Error Mean
Pair 1	Control (Non-Fluro)	422.60	15	53.835	13.900
	Experimental (Fluro)	320.80	15	32.371	8.358

The **Paired Samples Statistics** table includes some basic descriptive data, including the means and standard deviations for each group of scores:

Control (Non-Fluro): *M* = 422.60, *SD* = 53.835
Experimental (Fluro): *M* = 320.80, *SD* = 32.371

It certainly *appears* as though reaction times were faster during the experimental trials, but is this effect statistically significant?

The paired samples *t* test will answer this question.

If we repeated this research many hundreds of times (with samples of *N* = 15 all drawn from the same population), and calculated a pair of means for each replication, we would find that most of these means varied from those presented in the **Paired Samples Statistics** table. The **Std. Error Mean** is an estimate of this variability (expressed in standard deviation units).

A larger standard error of the mean indicates more variability in the sampling population.

▶▶ **Link:**
Correlation is covered in chapter 12.

Paired Samples Correlations

		N	Correlation	Sig.
Pair 1	Control (Non-Fluro) & Experimental (Fluro)	15	.499	.058

The **Paired Samples Correlations** table reports a Pearson's correlations coefficient, which is a commonly used index of the strength of the linear association between the two groups of scores.

Here, the correlation coefficient is .499, indicating that <u>participants who responded faster to the fluorescent cyclists tended to also respond faster to the control cyclists.</u> Although "large" by Cohen's (1988) conventions, this correlation failed to reach statistical significance (at α = .05), primarily due to the small sample size. This highlights the perils of conducting underpowered research.

Typically, you would not report this correlation when writing up a paired samples *t* test.

Paired Samples Test

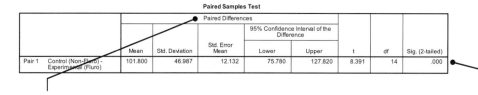

		Paired Differences							
					95% Confidence Interval of the Difference				
		Mean	Std. Deviation	Std. Error Mean	Lower	Upper	t	df	Sig. (2-tailed)
Pair 1	Control (Non-Fluro) - Experimental (Fluro)	101.800	46.987	12.132	75.780	127.820	8.391	14	.000

ⓘ **Tip:**
The difference between the two means is simply *t* x **Std. Error Mean**.

8.391 x 12.132 = 101.800

In the **Paired Differences** section of the **Paired Samples Test** table there are several useful pieces of information:

Mean. The difference between the two sample means (422.60 – 320.80 = 101.800).

Std. Deviation. The standard deviation of the difference scores. Here, the difference scores deviate from the mean of 101.800 msec by an average of 46.987 msec.

95% Confidence Interval of the Difference. We can be 95% confident that this interval (75.780 to 127.820) contains the true (population) mean difference.

The *t* test is statistically significant at α = .05. <u>Reaction times were significantly faster when the cyclists were wearing fluorescent vests.</u>

For your write-up, make note of:

t = 8.391
df = 14
Sig (2-tailed) = .000

Note. PASW Statistics rounds to three decimal places. *Sig* (2-tailed) = .000 simply means that the likelihood of observing a *t* value of ≥ 8.391 if the null hypothesis is true (i.e., if the population mean difference is zero) is less than .0005.

6.3.3. Follow Up Analyses

6.3.3.1. Effect Size

Cohen's *d* can be used to assess the size of the difference between two related sample means. It can be calculated with the following formula:

$$d = \frac{M_1 - M_2}{s_p}$$

Where M_1 and M_2 are the two sample means and s_p is the pooled standard deviation, calculated as:

$$s_p = \frac{s_1 + s_2}{2}$$

Where s_1 and s_2 are the two sample standard deviations. All of these figures can be derived from the **Paired Samples Statistics** table:

Paired Samples Statistics

		Mean	N	Std. Deviation	Std. Error Mean
Pair 1	Control (Non-Fluro)	422.60	15	53.835	13.900
	Experimental (Fluro)	320.80	15	32.371	8.358

$M_1 = 422.60$
$M_2 = 320.80$

$s_1 = 53.835$
$s_2 = 32.371$

So,

$$s_p = \frac{53.835 + 32.371}{2}$$

$$= 43.103$$

Then,

$$d = \frac{422.60 - 320.80}{43.103}$$

$$= 2.36$$

This is a very large effect. Participants in this study responded substantially faster to the fluorescent cyclists than they did to the control (non-fluorescent) cyclists.

(i) **Tip:**
Jacob Cohen (1988) suggests that a *d* of .20 can be considered small, a *d* of .50 is medium, and a *d* of .80 is large.

He also stresses that these values were subjectively derived with the intent that a medium effect should be "visible to the naked eye of a careful observer", a small effect should be "noticeably smaller than medium but not so small as to be trivial", and a large effect should be "the same distance above medium as small [is] below it" (Cohen, 1992, p. 156).

6.3.4. APA Style Results Write-Up

Results ●━━━━━━━━━━━━━━

A paired samples *t* test with an α of .05 was used to compare

mean reaction times (in milliseconds) to virtual cyclists wearing either

fluorescent ($M = 320.80$, $SD = 32.37$) or non-fluorescent ($M = 422.60$,

$SD = 53.84$) vests in an evening driving simulation. On average, the

participants reacted 101.80 msec, 95% CI [75.78, 127.82], faster

during the fluorescent trials than they did during the non-fluorescent

trials. This difference was statistically significant, $t(14) = 8.39$,

ⓘ Tip:
As *Sig/p* can never be zero, report *p* < .001, rather than *p* = .000.

$p < .001$, and large, $d = 2.36$.

It was concluded that the assumptions of normality and

normality of difference scores were not violated after outputting and

visually inspecting the relevant histograms. ●━━━━━━━━

6.3.5. Summary

The effect observed in the first illustrated example was large and significant. The effect in example two is not so clear.

6.4. Illustrated Example Two

A psychologist wishes to assess the efficacy of a new cognitive-behavioural treatment for generalised anxiety. Before beginning the treatment with a new client, she asks him/her to complete the *Beck Anxiety Inventory* (*BAI*). The pre-treatment *BAI* scores for 12 clients are listed in the first column of Table 6.2. Scores on this measure can range from 0 through to 63.

At the end of the eight-week treatment program, the psychologist asked each client to again complete the *BAI*. The post-treatment *BAI* scores for the 10 clients who completed the full eight-week program are listed in the second column of Table 6.2. Post-treatment *BAI* data are not available for two clients who dropped-out of the program after two and five weeks respectively.

Table 6.2

Beck Anxiety Inventory (BAI) Scores Before and After an Eight-Week Treatment Program for Generalised Anxiety Disorder (N = 12)

📖 *Data:*
This is data file **data_6_2.sav** on the companion website.

Client ID	Pre-Treatment BAI Score	Post-Treatment BAI Score
1	55	53
2	36	
3	39	37
4	53	49
5	46	40
6	52	
7	47	39
8	45	44
9	34	38
10	41	43
11	33	29
12	50	46

6.4.1. PASW Statistics Output (Part 1: Normality & Normality of Difference Scores)

💻 *Syntax:*
Run these analyses with **syntax_6_2.sps** on the companion website.

Explore

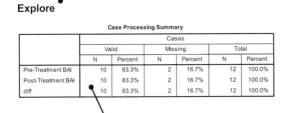

Case Processing Summary

	Cases					
	Valid		Missing		Total	
	N	Percent	N	Percent	N	Percent
Pre-Treatment BAI	10	83.3%	2	16.7%	12	100.0%
Post-Treatment BAI	10	83.3%	2	16.7%	12	100.0%
diff	10	83.3%	2	16.7%	12	100.0%

This **Explore** output was generated with the same procedures as those used in *Illustrated Example One*.

The *diff* (*difference scores*) variable was created in **Compute Variable** (in the **Transform** menu) by subtracting the *post-treatment BAI* variable from the *pre-treatment BAI* variable.

When the **Exclude cases listwise** strategy for handling missing values is selected in **Explore: Options**, only cases that have data for every variable in both the **Dependent List** and the **Factor List** are used in the analyses.

As clients 2 and 6 are missing data on some of the variables used in **Explore**, (specifically, the *post-treatment BAI* and *difference scores* variables) they are not included in any of the statistics or graphs.

If we used **Exclude cases pairwise** instead, the *pre-treatment BAI* histogram would have been based on data from 12 clients, whereas the *post-treatment BAI* and *difference scores* histograms would have been created using data from only 10 clients. This would make making comparisons difficult.

Pre-Treatment BAI

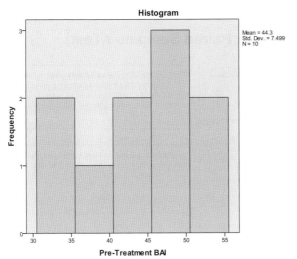

Post Treatment BAI

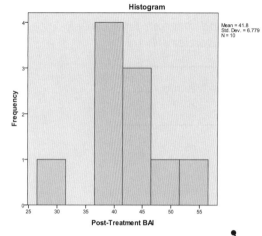

diff

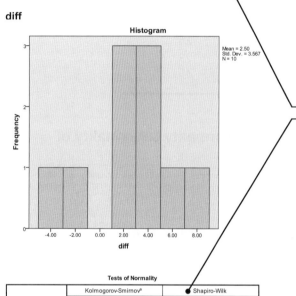

Assessing the normality of a distribution of scores using a histogram can take a bit of practice. Over time, you will develop a sense of what "normal enough" looks like, and be able to apply this mental template to each histogram you create.

In the mean time, you can easily confirm your assessments with the **Shapiro-Wilk** test. If the *Sig* value associated with the Shapiro-Wilk statistic is greater than the specified alpha level (usually .05), you can assume normality.

The Shapiro-Wilk test is a part of the **Normality plots with tests** option in **Explore: Plots**:

In this example, the three histograms look reasonably normal (although with so few cases, it can be difficult to tell). This assessment is supported by the three Shapiro-Wilk tests, which are all non-significant (at α = .05).

The normality and normality of difference scores assumptions are not violated.

►►│ Links:
Some **Normal Q-Q Plots** and **Detrended Normal Q-Q Plots,** were generated along with these normality tests. Guidelines for interpreting each can be found in chapter 4.

Tests of Normality

	Kolmogorov-Smirnov[a]			Shapiro-Wilk		
	Statistic	df	Sig.	Statistic	df	Sig.
Pre-Treatment BAI	.137	10	.200*	.957	10	.750
Post-Treatment BAI	.139	10	.200*	.983	10	.980
diff	.163	10	.200*	.960	10	.788

a. Lilliefors Significance Correction

*. This is a lower bound of the true significance.

6.4.2. PASW Statistics Output (Part 2: Paired Samples *t* Test)

T-Test

The purpose of the **Paired Samples *t* Test** is to compare two related sample means. In this example, we are comparing anxiety data collected before and after an eight-week treatment program.

Paired Samples Statistics

		Mean	N	Std. Deviation	Std. Error Mean
Pair 1	Pre-Treatment BAI	44.30	10	7.499	2.371
	Post-Treatment BAI	41.80	10	6.779	2.144

Paired Samples Correlations

		N	Correlation	Sig.
Pair 1	Pre-Treatment BAI & Post-Treatment BAI	10	.880	.001

ⓘ Tip:
You will sometimes see tests with *Sig* values only slightly above the specified alpha level (usually .05) described as "approaching significance" in the research literature.

The **Paired Samples Test** table indicates that the difference between the pre- and post-treatment means is not statistically significant (at α = .05). Make note of:

t = 2.216
df = 9
Sig (2-tailed) = .054

Paired Samples Test

		Paired Differences					t	df	Sig. (2-tailed)
		Mean	Std. Deviation	Std. Error Mean	95% Confidence Interval of the Difference				
					Lower	Upper			
Pair 1	Pre-Treatment BAI - Post-Treatment BAI	2.500	3.567	1.128	-.052	5.052	2.216	9	.054

6.4.3. Follow Up Analyses

6.4.3.1. Effect Size

$$d = \frac{M_1 - M_2}{s_p}$$

Where M_1 and M_2 are the two sample means and s_p is the pooled standard deviation (the average of the two sample standard deviations). So,

$$d = \frac{44.30 - 41.80}{7.139} = \frac{2.5}{7.139}$$

$$= 0.35$$

$$s_p = \frac{s_1 + s_2}{2} = \frac{7.499 + 6.779}{2}$$
$$= 7.139$$

Using Cohen's (1988) conventions as a guide, $d = 0.35$ is within the small to medium range.

This presents us with an interesting dilemma. We've observed a non-trivial effect size ($d = 0.35$), yet it's statistically non-significant. Why? The answer is tied to statistical power.

Statistical power refers to the likelihood of detecting an effect, where one actually exists. (By detecting, we mean finding $p < .05$.) Generally speaking, we are more likely to detect larger effects, in larger samples. Conversely, small effects in small samples can be particularly difficult to catch. As the current sample is quite small ($N = 10$), it is possible that the study was underpowered. That is, it is possible that the psychologist did not give herself a very good chance of finding a statistically significant effect.

▶▶ **Links:**
We recommend Howell (2010b) for an accessible introduction to statistical power.

Retrospective power estimates are provided as part of the output for many PASW Statistics procedures. When they are not provided, you can use the power tables included in the appendices of many good textbooks (e.g., Clark-Carter, 2009; Howell, 2010b), or one of several power calculators. The calculator that we use – G*Power – is maintained by Faul, Erdfelder, Lang, and Buchner (2007) and can be freely downloaded from http://www.psycho.uni-duesseldorf.de/abteilungen/aap/gpower3/

ⓘ **Tip:**
Retrospective power analysis can be controversial. Power is best considered (and most useful) before collecting data, when it can be used to calculate an appropriate sample size. You should avoid (mis)using it to explain away non-significant results.

Retrospective power analyses with G*Power confirmed that the *t* test was likely underpowered. It is recommended that the psychologist replicate this study with a much larger sample. She will need around 70 participants to have a decent (i.e., 80%) chance of observing an effect of around $d = 0.35$.

6.4.4. APA Style Results Write-Up

Results

A two-tailed, paired samples *t* test with an alpha level of .05 was used to compare the pre- ($M = 44.30$, $SD = 7.50$) and post-treatment ($M = 41.80$, $SD = 6.78$) *Beck Anxiety Inventory* (*BAI*) scores of 10 individuals. On average, participants' post-treatment *BAI* scores were 2.5 points lower than their pre-treatment scores, 95% CI [-0.05, 5.05]. However, this difference was not statistically significant, $t(9) = 2.22$, $p = .054$. Cohen's *d* for this test was 0.35, which can be described as small to medium.

Visual inspection of the relevant histograms indicated that neither the normality nor normality of difference scores assumptions were violated.

> **(i) Tip:**
> Numbers that cannot exceed ± 1 do not need a leading zero. Therefore, $p = .054$, but $d = 0.35$.

> In her **Method** section the author should note that 12 people began the study, but only 10 completed it. If known, the reasons for this attrition should also be reported.
>
> In her **Results** section, she is correct to report that the analyses are based on data from just 10 individuals.

> In the **Results** section, simply report what was found in the data.
>
> Interpretations, speculation about why the test did not support the hypothesis and suggestions for subsequent research should all be saved for the **Discussion**.

6.5. Paired Samples *t* Test Checklist

Have you:

- ✔ Checked that each group of data is approximately normally distributed?
- ✔ Checked that the difference scores are normally distributed?
- ✔ Interpreted the results of the *t* test and taken note of the *t* value, degrees of freedom, significance, mean difference and confidence interval for your write-up?
- ✔ Calculated a measure of effect size, such as Cohen's *d*?
- ✔ Reported your results in the APA style?

Chapter 7: One-Way Between Groups ANOVA

AKA:
One-way independent groups ANOVA; Single factor independent measures ANOVA; One-way between subjects ANOVA.

Chapter Overview

7.1. Purpose of the One-Way Between Groups ANOVA

To test for statistically significant differences between three or more independent sample means.

7.2. Questions We Could Answer Using the One-Way Between Groups ANOVA

1. Are there differences between the results of students who take an exam at different times of the day (9am, 11am or 3pm)?

In this example we're asking whether the three independent (or separate) groups – students taking the exam at 9am, students taking the exam at 11am, and students taking the exam at 3pm – produce different exam results. Each participant is a member of just one group (i.e., participants can take the exam at 9am or 11am or 3pm, but they cannot take it more than once).

Time of the day is the independent variable (IV), which has three levels: 9am, 11am and 3pm. Exam result is the dependent variable (DV).

Similar questions we could answer with a one-way between groups ANOVA include:

2. Does the amount of television watched by Australian children vary from state to state?

3. Are rats injected with growth hormone A more or less active than rats injected with either growth hormone B or a placebo?

4. Is employment contract type (e.g., casual, fixed-term or continuing) related to scores on a life satisfaction survey?

7.3. Illustrated Example One

A researcher wants to investigate the impact of education support services on the attitudes of teachers towards the mainstreaming (i.e., inclusion) of children with disabilities into their classrooms. She selects 30 teachers to participate in the study and divides them into three groups. The first group consists of 10 teachers who receive no additional classroom support; the second group consists of 10 teachers who receive three hours of in-class teacher aide time per day; and the third group consists of 10 teachers who participate in a weekly peer support program.

At the end of the school term each teacher is asked to complete the *Attitude Towards Mainstreaming Scale* which is an 18-item questionnaire containing statements about the inclusion of children with disabilities into schools (Berryman, Neal, & Robinson, 1980). Scores on this scale can range from 0 to 18, with higher scores reflecting more positive or favourable attitudes towards mainstreaming.

The researcher would like to know whether the teachers who received support held more positive attitudes towards mainstreaming and, if so, which types of support resulted in the most positive attitudes.

The 30 teachers' scores on the *Attitude Towards Mainstreaming Scale* are reproduced in Table 7.1.

📮 *Data:*
This is data file **data_7_1.sav** on the companion website.

Table 7.1

"Attitude Towards Mainstreaming Scale" Scores of Teachers Receiving Different Types of Support in the Classroom (N = 30)

No Support Condition (Group 1)	Teacher Aide Condition (Group 2)	Peer Support Condition (Group 3)
14	16	10
11	14	14
9	15	16
6	11	13
7	17	11
13	18	8
14	9	14
10	14	9
8	11	12
12	15	11

7.3.1. Setting Up the PASW Statistics Data File

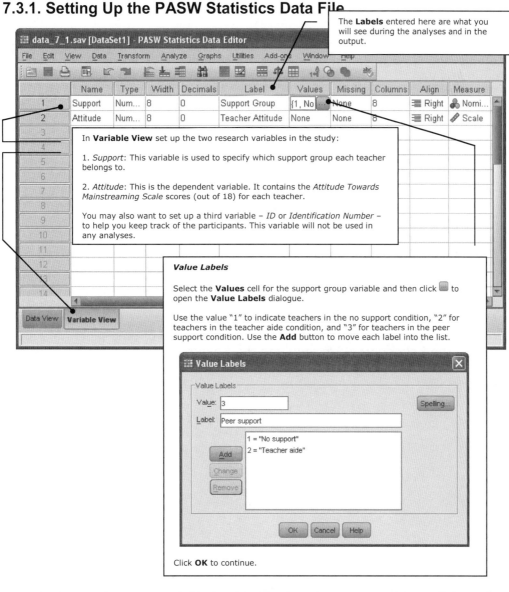

The **Labels** entered here are what you will see during the analyses and in the output.

▶▶ *Link:*
Setting up a data file is illustrated in greater detail in chapter 1.

In **Variable View** set up the two research variables in the study:

1. *Support*: This variable is used to specify which support group each teacher belongs to.

2. *Attitude*: This is the dependent variable. It contains the *Attitude Towards Mainstreaming Scale* scores (out of 18) for each teacher.

You may also want to set up a third variable – *ID* or *Identification Number* – to help you keep track of the participants. This variable will not be used in any analyses.

Value Labels

Select the **Values** cell for the support group variable and then click ▦ to open the **Value Labels** dialogue.

Use the value "1" to indicate teachers in the no support condition, "2" for teachers in the teacher aide condition, and "3" for teachers in the peer support condition. Use the **Add** button to move each label into the list.

Click **OK** to continue.

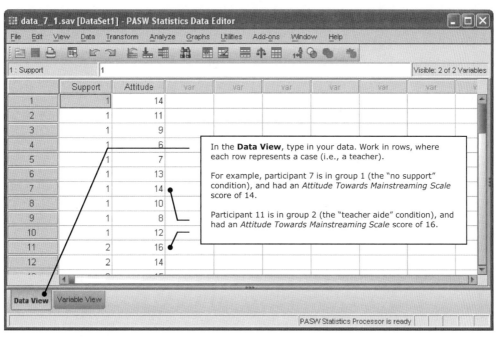

In the **Data View**, type in your data. Work in rows, where each row represents a case (i.e., a teacher).

For example, participant 7 is in group 1 (the "no support" condition), and had an *Attitude Towards Mainstreaming Scale* score of 14.

Participant 11 is in group 2 (the "teacher aide" condition), and had an *Attitude Towards Mainstreaming Scale* score of 16.

Syntax:
Run these analyses
with **syntax_7_1.sps**
on the companion
website.

7.3.2. Analysing the Data

7.3.2.1. Assumptions

The following criteria should be met before conducting a one-way between groups ANOVA.

1. **Scale of Measurement.** The DV should be interval or ratio data.

2. **Independence.** Each participant should participate only once in the research, and should not influence the participation of others.

3. **Normality.** Each group of scores should be approximately normally distributed. Although ANOVA is quite robust with respect to moderate violations of this assumption.

4. **Homogeneity of Variance.** There should be an approximately equal amount of variability in each set of scores.

Assumptions 1 and 2 are methodological, and should have been addressed before and during data collection. Assumptions 3 and 4 can be addressed in a number of ways, including visual inspection of a range of data plots, or with tests like the Kolmogorov-Smirnov and the Shapiro-Wilk.

7.3.2.2. PASW Statistics Procedure (Part 1: Normality)

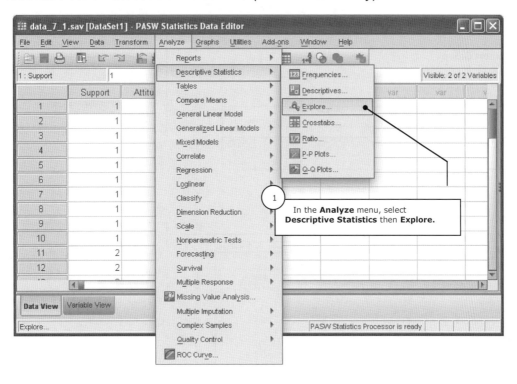

1. In the **Analyze** menu, select **Descriptive Statistics** then **Explore**.

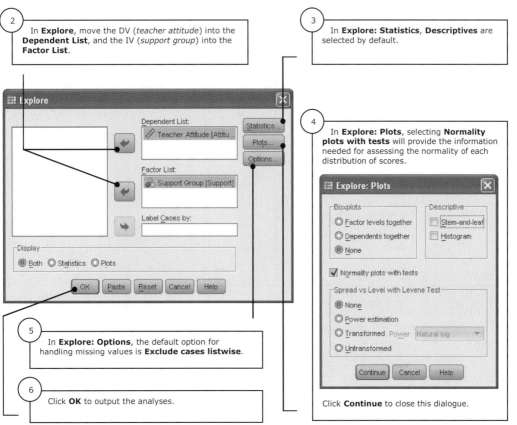

2. In **Explore**, move the DV (*teacher attitude*) into the **Dependent List**, and the IV (*support group*) into the **Factor List**.

3. In **Explore: Statistics**, **Descriptives** are selected by default.

ⓘ Tip:
Move a variable by highlighting it, then clicking an arrow button:

4. In **Explore: Plots**, selecting **Normality plots with tests** will provide the information needed for assessing the normality of each distribution of scores.

5. In **Explore: Options**, the default option for handling missing values is **Exclude cases listwise**.

6. Click **OK** to output the analyses.

Click **Continue** to close this dialogue.

7.3.2.3. PASW Statistics Output (Part 1: Normality)

Explore

Support Group

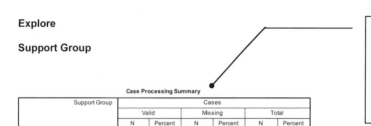

The **Case Processing Summary** shows how many cases were analysed, and how many were dropped due to missing data.

In this instance, there were 10 cases in each group. No cases were excluded from the analyses, as there was no missing data.

Case Processing Summary

Support Group		Cases					
		Valid		Missing		Total	
		N	Percent	N	Percent	N	Percent
Teacher Attitude	No support	10	100.0%	0	.0%	10	100.0%
	Teacher aide	10	100.0%	0	.0%	10	100.0%
	Peer support	10	100.0%	0	.0%	10	100.0%

Descriptives

The table of **Descriptives** contains a range of useful information, including measures of central tendency, dispersion, skewness and kurtosis for each group of scores.

Support Group			Statistic	Std. Error
Teacher Attitude	No support	Mean	10.40	.909
		95% Confidence Interval for Mean — Lower Bound	8.34	
		Upper Bound	12.46	
		5% Trimmed Mean	10.44	
		Median	10.50	
		Variance	8.267	
		Std. Deviation	2.875	
		Minimum	6	
		Maximum	14	
		Range	8	
		Interquartile Range	6	
		Skewness	-.151	.687
		Kurtosis	-1.357	1.334
	Teacher aide	Mean	14.00	.907
		95% Confidence Interval for Mean — Lower Bound	11.95	
		Upper Bound	16.05	
		5% Trimmed Mean	14.06	
		Median	14.50	
		Variance	8.222	
		Std. Deviation	2.867	
		Minimum	9	
		Maximum	18	
		Range	9	
		Interquartile Range	5	
		Skewness	-.459	.687
		Kurtosis	-.652	1.334
	Peer support	Mean	11.80	.786
		95% Confidence Interval for Mean — Lower Bound	10.02	
		Upper Bound	13.58	
		5% Trimmed Mean	11.78	
		Median	11.50	
		Variance	6.178	
		Std. Deviation	2.486	
		Minimum	8	
		Maximum	16	
		Range	8	
		Interquartile Range	4	
		Skewness	.122	.687
		Kurtosis	-.671	1.334

When the **Skewness** and **Kurtosis** statistics for a distribution of scores are both zero, it is perfectly normal.

Here, the skewness and kurtosis statistics for the three distributions are all fairly close to zero. Also, z_s and z_k are within ±1.96 for all three groups (see section 4.3.2.3).

This suggests that <u>all three groups of data are approximately normal</u>.

▶▶ Links:
The full PASW Statistics output also included three **Normal Q-Q Plots** and three **Detrended Normal Q-Q Plots**.

These have been described elsewhere in this text (for example, chapter 4).

Tests of Normality

Support Group		Kolmogorov-Smirnov[a]			Shapiro-Wilk		
		Statistic	df	Sig.	Statistic	df	Sig.
Teacher Attitude	No support	.117	10	.200*	.942	10	.578
	Teacher aide	.200	10	.200*	.947	10	.635
	Peer support	.126	10	.200*	.979	10	.959

a. Lilliefors Significance Correction

*. This is a lower bound of the true significance.

If the **Shapiro-Wilk** statistic (W) is significant (i.e., $Sig < .05$), the distribution is not normal.

Here, W ranges from .942 to .979, and all three tests are non-significant (all $Sig > .05$).

Therefore, we can conclude that <u>all three groups of data are normally distributed</u>.

7.3.2.4. PASW Statistics Procedure (Part 2: Homogeneity of Variance & the ANOVA)

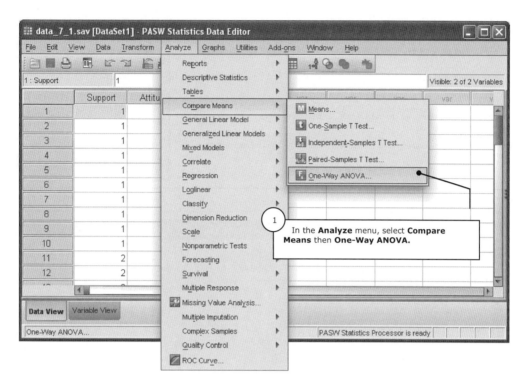

1 In the **Analyze** menu, select **Compare Means** then **One-Way ANOVA.**

(i) *Tip:*
Univariate (in the **General Linear Model** submenu) can also be used to perform a one-way between groups ANOVA.

2 In **One-Way ANOVA**, move the DV (*teacher attitude*) into the **Dependent List**, and the IV (*support group*) into the **Factor** box.

4 To identify specifically which group(s) differ from the others, we can conduct either planned **Contrasts** or **Post Hoc** analyses.

Planned **Contrasts** are preferred when the researcher has a strong theory or hypothesis about where differences will lie.

When a researcher is less certain about the sources of significance, he or she can cast a wider net with one of several **Post Hoc** tests.

Both procedures are available in PASW Statistics, and are illustrated overleaf. In reality, however, a researcher would only ever need to use one of these approaches.

3 In **One-Way ANOVA: Options**, select **Descriptive** and **Homogeneity of variance test.**

Descriptive will, as the name suggests, provide some basic summary statistics for each level of the IV.

Homogeneity of variance test will provide Levene's test for equality of variance (which is necessary for testing the homogeneity of variance assumption).

Click **Continue** to close this dialogue.

One-Way ANOVA: Options

Statistics
☑ Descriptive
☐ Fixed and random effects
☑ Homogeneity of variance test
☐ Brown-Forsythe
☐ Welch

☐ Means plot

Missing Values
⦿ Exclude cases analysis by analysis
◯ Exclude cases listwise

Continue   Cancel   Help

One-Way ANOVA

Dependent List:
✎ Teacher Attitude [Attitu...

Factor:
🔒 Support Group [Support]

Contrasts...
Post Hoc...
Options...

OK   Paste   Reset   Cancel   Help

5 Once you have finished, click **OK** to output the analyses.

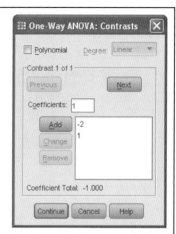

AKA:

Planned contrasts are often referred to as planned comparisons or *a priori* contrasts.

(i) Tip:

When developing weights, the following should be observed:

1. Contrasts are made between groups with negative weights and groups with positive weights. Here, the teachers who did not receive any support (assigned a negative weight) were contrasted with those who did (assigned positive weights).

2. The weights for a planned contrast must sum to zero. In our example, -2 + 1 + 1 = 0.

See Field (2009) or any good behavioural science statistics text for further details.

AKA:

Post hoc comparisons are often referred to as unplanned or *a posteriori* contrasts.

Planned Contrasts

In the current example we could predict that the attitudes of teachers who do not receive any support will differ from the attitudes of teachers who receive some support (in the form of a regular teacher aide or peer support). This hypothesis can be tested with a planned contrast.

Each planned contrast performed requires a set of weights. The weights in this contrast are as follows:

 -2 = No support
 +1 = Teacher aide
 +1 = Peer support

In **One-Way ANOVA: Contrasts**, enter the weight of the first group (-2) into the **Coefficients** box, then click the **Add** button. Repeat this process for the remaining two groups.

To specify a second contrast, click the **Next** button, and enter the appropriate weights.

Post Hoc Comparisons

Post hoc tests are used to detect differences between all possible combinations of groups.

In **One-Way ANOVA: Post Hoc Multiple Comparisons,** select **Tukey**. The **Tukey** test is appropriate when the groups being compared are all the same size, and have similar variances.

Other post hoc tests you may encounter include:

- **Gabriel's** procedure, which has greater statistical power when group sizes are unequal.
- **Hochberg's GT2**, which is appropriate when group sizes are extremely unequal.
- **Games-Howell** procedure, which can be used when equal variances cannot be assumed.

Click **Continue** to close this dialogue box.

7.3.2.5. PASW Statistics Output (Part 2: Homogeneity of Variance & the ANOVA)

Oneway

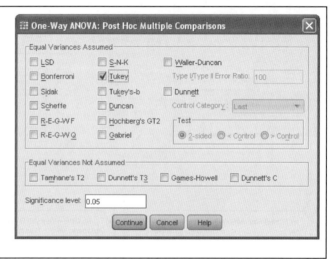

Descriptives

Teacher Attitude

	N	Mean	Std. Deviation	Std. Error	95% Confidence Interval for Mean Lower Bound	95% Confidence Interval for Mean Upper Bound	Minimum	Maximum
No support	10	10.40	2.875	.909	8.34	12.46	6	14
Teacher aide	10	14.00	2.867	.907	11.95	16.05	9	18
Peer support	10	11.80	2.486	.786	10.02	13.58	8	16
Total	30	12.07	3.051	.557	10.93	13.21	6	18

The **Descriptives** table contains basic descriptive information about teacher attitudes in each of the three conditions.

Teachers without any support had a **Mean** score on the *Attitude Towards Mainstreaming Scale* of 10.40. The other two groups had somewhat higher means.

The **Std.** (Standard) **Deviations** indicate a slightly larger spread of scores in the "no support" group (*SD* = 2.875) relative to the other two groups.

Levene's **Test of Homogeneity of Variances** will indicate whether or not this should concern us.

The **Test of Homogeneity of Variances** table shows that Levene's statistic is not significant at α = .05 (*F* = .186, *Sig* = .831). We can conclude that the assumption of homogeneity of variance has not been violated.

If the Assumption is Violated

ANOVA is not sensitive to violations of the equality of variances assumption when group sizes are moderate to large and approximately equal.

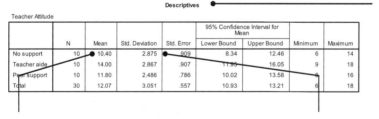

Test of Homogeneity of Variances

Teacher Attitude

Levene Statistic	df1	df2	Sig.
.186	2	27	.831

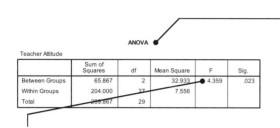

ANOVA

Teacher Attitude

	Sum of Squares	df	Mean Square	F	Sig.
Between Groups	65.867	2	32.933	4.359	.023
Within Groups	204.000	27	7.556		
Total	269.867	29			

The **ANOVA** table contains the main output for the ANOVA test.

This table reports both **Between Groups** variance (the variability in the data that can be attributed to the IV plus chance factors), and **Within Groups** variance (the variability in the data attributable to chance factors alone).

(Chance factors include both individual differences and experimental error.)

F (also referred to as the *F*-ratio) is the essential test statistic for the ANOVA. This value (4.359) will be included in the results write-up, along with the following:

df (Between Groups) = 2
df (Within Groups) = 27
Sig = .023

Sig indicates the likelihood of an *F*-ratio this large occurring by chance. Generally, a cut-off of .05 (referred to as alpha or α) is used in the health and behavioural sciences to determine whether or not an effect is occurring due to chance alone. As our research *Sig* is less than .05 we can conclude that there is a statistically significant difference between the attitudes of at least two of the teacher groups.

However, we do not yet know exactly which groups are different. To find out exactly which groups differ we need to interpret the output from either the **Multiple Comparisons** (i.e., post hoc comparisons) or **Contrast Tests** (i.e., planned contrasts) tables.

Planned contrasts are appropriate when the researcher has a specific hypothesis about which group(s) will differ from the others.

The **Contrast Coefficients** table shows the weights that we assigned to the different teacher groups in the previous section. This table will have one row for every contrast specified.

Contrast Coefficients

Contrast	Support Group		
	No support	Teacher aide	Peer support
1	-2	1	1

Contrast Tests

		Contrast	Value of Contrast	Std. Error	t	df	Sig. (2-tailed)
Teacher Attitude	Assume equal variances	1	5.00	2.129	2.348	27	.026
	Does not assume equal variances	1	5.00	2.179	2.295	16.910	.035

As Levene's **Test of Homogeneity of Variances** was not significant (*F* = .186, *Sig* = .831), and thus the assumption of homogeneity of variance has not been violated, we should read the **Assume equal variances** row(s) of the **Contrast Tests** table.

If Levene's test was statistically significant (i.e., *Sig* < .05), we would simply read the **Does not assume equal** variances row(s) instead.

Reading the **Assume equal variances** row of the **Contrast Tests** table, we can see that *t* = 2.348 and *Sig* = .026.

This finding supports the hypothesis that the attitudes of teachers without support are significantly different to the attitudes of those receiving support (i.e., *Sig* < .05).

Further contrasts can be conducted to investigate hypothesised differences between the remaining groups if desired.

Post hoc tests are conducted when the researcher has no specific hypotheses about which group(s) will differ from the others.

The **Multiple Comparisons** table shows the output for the **Tukey HSD** (Honestly Significant Difference) tests we requested.

Post Hoc Tests

Multiple Comparisons

Teacher Attitude
Tukey HSD

(I) Support Group	(J) Support Group	Mean Difference (I-J)	Std. Error	Sig.	95% Confidence Interval	
					Lower Bound	Upper Bound
No support	Teacher aide	-3.600*	1.229	.018	-6.65	-.55
	Peer support	-1.400	1.229	.499	-4.45	1.65
Teacher aide	No support	3.600*	1.229	.018	.55	6.65
	Peer support	2.200	1.229	.192	-.85	5.25
Peer support	No support	1.400	1.229	.499	-1.65	4.45
	Teacher aide	-2.200	1.229	.192	-5.25	.85

*. The mean difference is significant at the 0.05 level.

Read this table in rows, where each row compares one pair of means.

For example, there is a -3.600 difference between the mean of the no support group and the mean of the teacher aide group. This difference is statistically significant at α = .05 (*Sig* = .018).

Scanning down the table indicates that no other pairs of means differ significantly.

Teacher Attitude

Tukey HSD[a]

Support Group	N	Subset for alpha = 0.05	
		1	2
No support	10	10.40	
Peer support	10	11.80	11.80
Teacher aide	10		14.00
Sig.		.499	.192

Means for groups in homogeneous subsets are displayed.

a. Uses Harmonic Mean Sample Size = 10.000.

The **Homogenous Subsets** table provides information similar to that in the **Multiple Comparisons** table.

Here, there is no difference between the mean of the no support group and the mean of the peer support group (*p* = .499). Likewise, there is no difference between the mean of the peer support group and the mean of the teacher aide group (*p* = .192).

7.3.3. Follow Up Analyses

7.3.3.1. Effect Size

The 6th edition (2010) of the *Publication Manual of the American Psychological Association* (APA) notes that:

> For the reader to appreciate the magnitude or importance of a study's findings, it is almost always necessary to include some measure of effect size in the Results section. (p. 34)

Although PASW Statistics does not automatically compute effect size indices for the omnibus (or overall) one-way between groups ANOVA, planned contrasts or post hoc comparisons, they can be calculated from output that the program does provide. Here, we will use eta-squared (η^2) as an index of the omnibus effect size, and Cohen's *d* for the contrasts and comparisons.

7.3.3.1.1. Effect Size Calculations for the Omnibus ANOVA

Eta-squared (η^2) is interpreted as the proportion of variability in the data that can be attributed to the IV. It can be calculated with the following formula:

$$\eta^2 = \frac{SS_{Between}}{SS_{Total}}$$

Where $SS_{Between}$ is the between groups sum of squares, and SS_{Total} is the total sum of squares. Both are included in the **ANOVA** summary table:

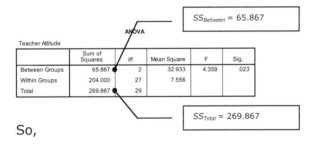

$SS_{Between} = 65.867$

ANOVA

Teacher Attitude

	Sum of Squares	df	Mean Square	F	Sig.
Between Groups	65.867	2	32.933	4.359	.023
Within Groups	204.000	27	7.556		
Total	269.867	29			

$SS_{Total} = 269.867$

So,

$$\eta^2 = \frac{65.867}{269.867}$$

$$= .244$$

① *Tip:*
Cohen (1988) suggested that $\eta^2 = .01$ could be considered small, $\eta^2 = .059$ could be considered medium, and $\eta^2 = .138$ could be considered large.

Here, we can attribute 24.4% (.244 x 100) of the variability in teachers' attitudes to the amount and type of support they received. According to Cohen's (1988) conventions, this is a large effect.

To convert η^2 to f, Cohen's measure of effect size for ANOVA, the following formula can be used:

$$f = \sqrt{\frac{\eta^2}{1-\eta^2}}$$

In the current example,

$$f = \sqrt{\frac{.244}{1-.244}} = .568$$

Once again, this effect can be considered large.

7.3.3.1.2. Effect Size Calculations for the Contrasts and Comparisons

Cohen's *d* is a measure of the size of the difference between two means, expressed in standard deviations. It can be calculated using the following formula:

$$d = \frac{2t}{\sqrt{df_{Within}}}$$

t and *df_{Within}* can be read off (or easily calculated from) the **Contrast Tests** and **Multiple Comparisons** tables.

Dealing with our planned contrast first:

$t = 2.348$

Contrast Tests

		Contrast	Value of Contrast	Std. Error	t	df	Sig. (2-tailed)
Teacher Attitude	Assume equal variances	1	5.00	2.129	2.348	27	.026
	Does not assume equal variances	1	5.00	2.179	2.295	16.910	.035

df_{Within} = 27

Note that this value can also be located in the **ANOVA** summary table.

So,

$$d = \frac{2(2.348)}{\sqrt{27}} = \frac{4.696}{5.196} = 0.903$$

The difference between the attitudes towards mainstreaming of teachers who did and did not receive support is substantial.

Next, we can calculate *d* for each of the post hoc comparisons in the **Multiple Comparisons** table.

Looking first at the difference between the group of teachers without any support, and the group with teacher aides:

Multiple Comparisons

Teacher Attitude
Tukey HSD

(I) Support Group	(J) Support Group	Mean Difference (I-J)	Std. Error	Sig.	95% Confidence Interval Lower Bound	95% Confidence Interval Upper Bound
No support	Teacher aide	-3.600*	1.229	.018	-6.65	-.55
	Peer support	-1.400	1.229	.499	-4.45	1.65
Teacher aide	No support	3.600*	1.229	.018	.55	6.65
	Peer support	2.200	1.229	.192	-.85	5.25
Peer support	No support	1.400	1.229	.499	-1.65	4.45
	Teacher aide	-2.200	1.229	.192	-5.25	.85

*. The mean difference is significant at the 0.05 level.

t = Mean Difference / Std. Error

So, *t* = -3.600 / 1.229 = -2.929

So,

df_{Within} = 27

This value can be derived from the **ANOVA** summary table.

$$d = \frac{2(-2.929)}{\sqrt{27}} = \frac{-5.858}{5.196} = -1.127$$

Predictably, this effect is large.

Following the same procedure, the difference between the attitudes of teachers without support and teachers with peer support was medium sized (d = 0.438), whilst the difference between the attitudes of the peer and teacher aide supported teachers was medium-to-large (d = 0.689).

7.3.4. APA Style Results Write-Up

Results

A one-way between groups analysis of variance (ANOVA) was used to investigate the impact that support had on the attitudes of teachers towards the mainstreaming of children with disabilities into their classrooms.

Inspection of the skewness, kurtosis and Shapiro-Wilk statistics indicated that the assumption of normality was supported for each of the three conditions. Levene's statistic was non-significant, $F(2, 27) = .186$, $p = .831$, and thus the assumption of homogeneity of variance was not violated.

> Even though assumption testing is generally not reported in journal articles, it is often required in student papers.

The ANOVA was statistically significant, indicating that the teachers' attitudes towards mainstreaming were influenced by the support they received, $F(2, 27) = 4.36$, $p = .023$, $\eta^2 = .244$.

Post hoc analyses with Tukey's HSD (using an α of .05) revealed that the teachers without support ($M = 10.40$, $SD = 2.88$) had significantly lower attitude scores than teachers receiving regular teacher aide time ($M = 14.00$, $SD = 2.87$). However, there was no significant difference between the attitude scores of the teachers without support and the teachers receiving peer support ($M = 11.80$, $SD = 2.49$), nor between the attitude scores of the teachers receiving peer support and those receiving regular aide time. Effect sizes for these three comparisons were $d = 1.13$, 0.44 and 0.69 respectively.

> If the ANOVA is statistically significant, follow-up with post hoc analyses.
>
> Post hoc comparisons are not necessary if the ANOVA is non-significant.

7.3.5. Summary

In the previous example, the ANOVA was statistically significant, and assumptions of normality and homogeneity of variance were not violated. In the following example we will discuss how to report non-significant findings.

7.4. Illustrated Example Two

A researcher wants to evaluate the effectiveness of four different diets. She divides her sample of 33 participants into four groups, and assigns a different diet (plus regular exercise) to each:

> Group 1: Fruit diet plus exercise (11 participants).
> Group 2: Restricted calories diet plus exercise (8 participants).
> Group 3: Low carbohydrates diet plus exercise (8 participants).
> Group 4: Low sugar diet plus exercise (6 participants).

The amount of weight lost (in kilograms) by each participant is recorded at completion of the research. These data are reproduced in Table 7.2.

The researcher asks, "Are there differences between the average amounts of weight lost by participants on the different diet and exercise programs?"

Table 7.2

Weight Lost (in kgs) by Participants on Four Different Diet and Exercise Programs (N = 33)

Data:
This is data file **data_7_2.sav** on the companion website.

Fruit Diet & Exercise (Group 1)	Restricted Calories Diet & Exercise (Group 2)	Low Carbohydrates Diet & Exercise (Group 3)	Low Sugar Diet & Exercise (Group 4)
-4	-5	-7	-5
-2	-3	-1	-3
-5	-4	-5	-1
-1	-2	-3	-2
-4	-4	-5	-2
-3	-5	-4	-5
-3	-2	-5	
-1	-5	-1	
-4			
-2			
-1			

Syntax:
Run these analyses
with **syntax_7_2.sps**
on the companion
website.

7.4.1. PASW Statistics Output (Part 1: Normality)

The assumption of normality can be assessed in numerous ways (see chapter 4). In this example, Shapiro Wilk tests and boxplots are used.

Explore

Diet & Exercise Group

> The procedure for generating this output is the same as that used in *Illustrated Example One*.

Case Processing Summary

Diet & Exercise Group		Cases					
		Valid		Missing		Total	
		N	Percent	N	Percent	N	Percent
Weight Lost (kg)	Fruit and exercise	11	100.0%	0	.0%	11	100.0%
	Restricted calories and exercise	8	100.0%	0	.0%	8	100.0%
	Low carbohydrates and exercise	8	100.0%	0	.0%	8	100.0%
	Low sugar and exercise	6	100.0%	0	.0%	6	100.0%

> The **Case Processing Summary** indicates that 11 people participated in the fruit diet and exercise group, 8 participated in the restricted calories diet and exercise group, and so on.
>
> There is no missing data.

Links:
The full PASW
Statistics output also
contained a table of
Descriptives, four
Normal Q-Q Plots
and four **Detrended
Normal Q-Q Plots**,
which we've omitted
for the sake of brevity.
Examples of these
tables and graphs, and
guidelines for their
interpretation can be
found elsewhere in
this book.

Tests of Normality

Diet & Exercise Group		Kolmogorov-Smirnov[a]			Shapiro-Wilk		
		Statistic	df	Sig.	Statistic	df	Sig.
Weight Lost (kg)	Fruit and exercise	.178	11	.200*	.903	11	.201
	Restricted calories and exercise	.210	8	.200*	.843	8	.082
	Low carbohydrates and exercise	.204	8	.200*	.911	8	.358
	Low sugar and exercise	.225	6	.200*	.876	6	.252

a. Lilliefors Significance Correction

*. This is a lower bound of the true significance.

> The **Shapiro-Wilk** is the more appropriate **Test of Normality** for research with small samples.
>
> A significant (i.e., *Sig* < .05) Shapiro-Wilk statistic indicates a violation of the normality assumption.
>
> In the current data set, <u>each group of weight loss scores is approximately normally distributed</u>.

Boxplots were obtained by selecting **Factor levels together** in the **Explore: Plots** dialogue.

These plots can be used to visually assess the normality assumption.

> A **Boxplot** illustrates how a variable is spread. The "box" contains the middle 50% of scores (i.e., the interquartile range), and is bounded by Tukey's hinges, which approximate the 25th and 75th percentiles.
>
> The line in the middle of the box is the median, and the "whiskers" extend to the highest and lowest scores (excluding outliers and extreme scores).
>
> Outliers and extreme scores are denoted by a small circle or asterisk (*) respectively. <u>There are no outliers or extreme scores in this data set</u>.

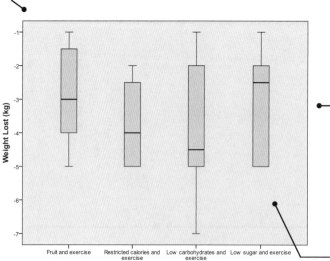

> These four **Boxplots** are all roughly symmetrical, indicating that <u>each group of scores is approximately normally distributed</u>.
>
> When a group of scores are normally distributed, the box should be in the middle of the range, and the median line should be near the middle of the box.

7.4.2. PASW Statistics Output (Part 2: Homogeneity of Variance & the ANOVA)

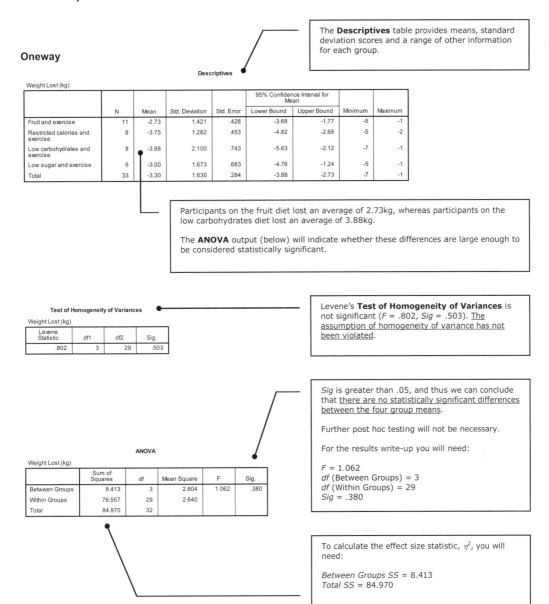

Oneway

The **Descriptives** table provides means, standard deviation scores and a range of other information for each group.

Descriptives

Weight Lost (kg)

	N	Mean	Std. Deviation	Std. Error	95% Confidence Interval for Mean Lower Bound	95% Confidence Interval for Mean Upper Bound	Minimum	Maximum
Fruit and exercise	11	-2.73	1.421	.428	-3.68	-1.77	-5	-1
Restricted calories and exercise	8	-3.75	1.282	.453	-4.82	-2.68	-5	-2
Low carbohydrates and exercise	8	-3.88	2.100	.743	-5.63	-2.12	-7	-1
Low sugar and exercise	6	-3.00	1.673	.683	-4.76	-1.24	-5	-1
Total	33	-3.30	1.630	.284	-3.88	-2.73	-7	-1

Participants on the fruit diet lost an average of 2.73kg, whereas participants on the low carbohydrates diet lost an average of 3.88kg.

The **ANOVA** output (below) will indicate whether these differences are large enough to be considered statistically significant.

Test of Homogeneity of Variances

Weight Lost (kg)

Levene Statistic	df1	df2	Sig.
.802	3	29	.503

Levene's **Test of Homogeneity of Variances** is not significant (F = .802, Sig = .503). The assumption of homogeneity of variance has not been violated.

Sig is greater than .05, and thus we can conclude that there are no statistically significant differences between the four group means.

Further post hoc testing will not be necessary.

For the results write-up you will need:

F = 1.062
df (Between Groups) = 3
df (Within Groups) = 29
Sig = .380

ANOVA

Weight Lost (kg)

	Sum of Squares	df	Mean Square	F	Sig.
Between Groups	8.413	3	2.804	1.062	.380
Within Groups	76.557	29	2.640		
Total	84.970	32			

To calculate the effect size statistic, η^2, you will need:

Between Groups SS = 8.413
Total SS = 84.970

7.4.3. Follow Up Analyses

7.4.3.1. Effect Size

$$\eta^2 = \frac{SS_{Between}}{SS_{Total}} = \frac{8.413}{84.970} = .099$$

(i) Tip:
An index of effect size should be reported regardless of whether or not the ANVOA is statistically significant.

An η^2 of .099 indicates that 10% of the variability in the weight loss data can be attributed to the IV (the different diets).

7.4.4. APA Style Results Write-Up

Results

To evaluate the effectiveness of four different diets on the amount of weight participants lost a one-way between groups analysis of variance (ANOVA) was conducted.

The ANOVA assumptions of normality and homogeneity of variance were not violated, and the F test was not significant, $F(3, 29) = 1.06$, $p = .380$. It should be noted, however, that $\eta^2 = 0.10$, which can be characterised as medium.

Reserve speculation about why the ANOVA was not statistically significant until the **Discussion** section of your report.

In the **Results** section, simply state the facts.

7.5. One-Way Between Groups ANOVA Checklist

Have you:

- ✔ Checked that each group of data is approximately normally distributed?
- ✔ Checked for homogeneity of variance?
- ✔ Taken note of the F-ratio, degrees of freedom, and significance for your write-up?
- ✔ Calculated a measure of effect size such as η^2?
- ✔ Written up your results in the APA style?

Chapter 8: Factorial Between Groups ANOVA

Chapter Overview

8.1. Purpose of the Factorial Between Groups ANOVA

The factorial between groups ANOVA is used to test hypotheses about means when there are two or more independent variables. As the participants are only tested once, this is called a between groups design.

The most common type of factorial ANOVA is the 2x2 design, where there are two independent variables, each with two levels.

Factorial designs are useful because they provide:

1. A main effect for each independent variable in isolation.
2. An interaction effect, which examines the influence of the independent variables in combination.

8.2. Questions We Could Answer Using the Factorial Between Groups ANOVA

1. Do the class evaluation scores of male and female students differ if the lecturer is male versus female?

In this example, we're asking whether the two independent (or separate) groups – male and female students – give different evaluations of a class. In the factorial between groups ANOVA we are also interested in examining the additional variable of lecturer gender.

Student gender and lecturer gender are our independent variables (IVs), and each variable has two levels (male or female). Class evaluation score is our dependent variable (DV).

Other questions we could answer with a factorial between groups ANOVA include:

2. Does career performance improve for individuals enrolled in two different types of career support programs (mentoring versus coaching) over time (pre, post and six months after support program)?

3. Does age (12 years, 15 years and 18 years) influence the attitudes of adolescents (male versus female) towards physical exercise?

8.3. Illustrated Example One

A researcher wanted to examine the impact that room temperature has on exam scores of students sitting a 1st year Physics exam. The researcher believed that there would be a difference in exam scores between students in hotter (29°C) rooms and students in cooler rooms (23°C). Additionally, the researcher believed that exam difficulty would influence the impact of room temperature on exam scores (easy versus hard exams).

The researcher asked 40 participants to take part in study. The participants were divided into four evenly sized groups:

    a. 10 participants were asked to complete an easy exam in a cool (23°C) room.
    b. 10 participants were asked to complete a hard exam in a cool (23°C) room.
    c. 10 participants were asked to complete an easy exam in a hot (29°C) room.
    d. 10 participants were asked to complete a hard exam in a hot (29°C) room.

The exam scores from the participants in each group were then compared to answer three basic questions:

1. Does room temperature influence exam score? (That is, is there a main effect for room temperature?)
2. Does exam difficulty influence exam score? (That is, is there a main effect for exam difficulty?)
3. Does the influence of room temperature on exam score depend on the difficulty of the exam? (That is, is there an interaction effect?)

The data from this study are presented in Table 8.1.

Table 8.1

Exam Scores of Participants (N = 40) in Each of Four Experimental Conditions

□ *Data:*
This is data file
data_8_1.sav on the
companion website.

Group 1[a]	Group 2[b]	Group 3[c]	Group 4[d]
45	36	43	24
46	35	42	29
45	29	44	30
37	34	40	33
47	32	39	38
43	31	38	29
48	34	45	25
36	34	39	21
46	23	33	20
45	21	36	26

Note. [a] = Easy exam in a cool room; [b] = Hard exam in a cool room; [c] = Easy exam in a hot room; [d] = Hard exam in a hot room.

8.3.1. Setting Up the PASW Statistics Data File

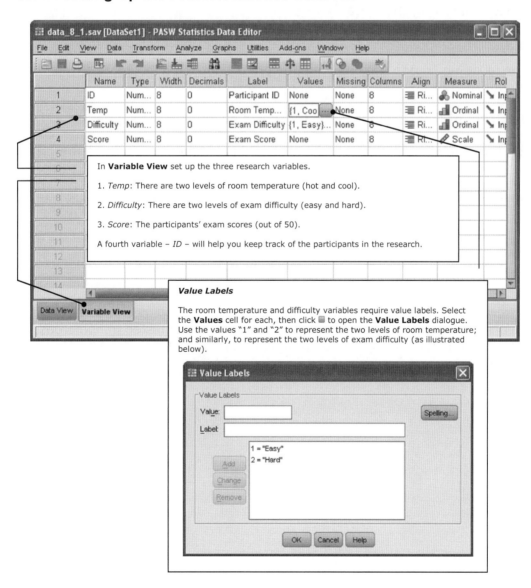

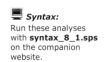
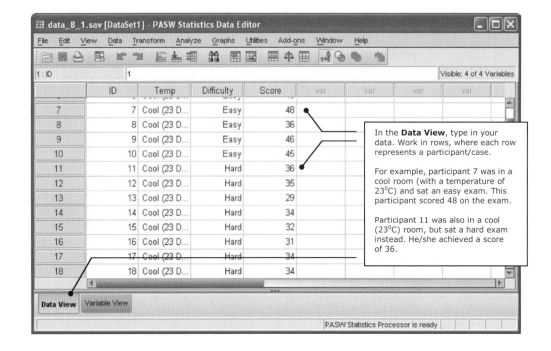

Syntax:
Run these analyses with **syntax_8_1.sps** on the companion website.

8.3.2. Analysing the Data

8.3.2.1. Assumptions

The following criteria should be met before conducting a factorial between groups ANOVA.

1. **Scale of Measurement.** The DV should be interval or ratio data.

2. **Independence.** Each participant should participate only once in the research, and should not influence the participation of others.

3. **Normality.** Each group of scores should be approximately normally distributed. ANOVA is quite robust against moderate violations of this assumption.

4. **Homogeneity of Variance.** The variance of data in groups should be approximately the same.

Assumptions 1 and 2 are methodological and should have been addressed before and during data collection. Assumption 3 can be tested in numerous ways, as illustrated below. Assumption 4 is tested as part of the ANOVA.

8.3.2.2. PASW Statistics Procedure (Part 1: Normality)

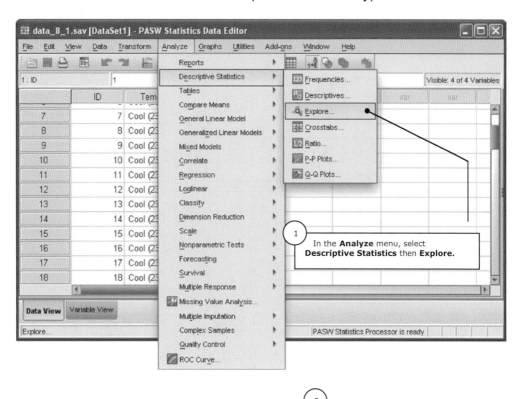

1 In the **Analyze** menu, select **Descriptive Statistics** then **Explore**.

2 In **Explore**, move the DV (*exam score*) into the **Dependent List**, and the IV's (*room temperature* and *exam difficulty*) into the **Factor List**.

3 In **Explore: Plots**, select **Normality plots with tests**.

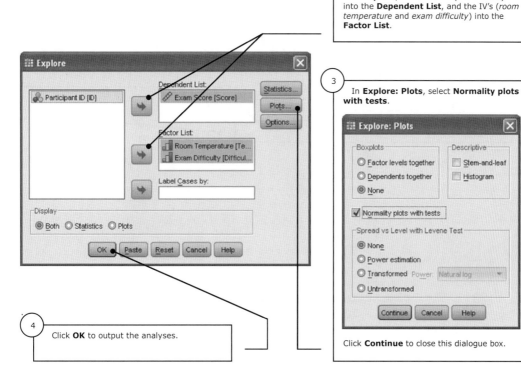

4 Click **OK** to output the analyses.

Click **Continue** to close this dialogue box.

8.3.2.3. PASW Statistics Output (Part 1: Normality)

Explore ●————————————————————

Room Temperature

The **Explore** output for each IV is presented separately. The output for *room temperature* is first.

Case Processing Summary

Room Temperature		Cases					
		Valid		Missing		Total	
		N	Percent	N	Percent	N	Percent
Exam Score	Cool (23 Degrees)	20	100.0%	0	.0%	20	100.0%
	Hot (29 Degrees)	20	100.0%	0	.0%	20	100.0%

Descriptives

Room Temperature				Statistic	Std. Error
Exam Score	Cool (23 Degrees)	Mean		37.35	1.791
		95% Confidence Interval for Mean	Lower Bound	33.60	
			Upper Bound	41.10	
		5% Trimmed Mean		37.67	
		Median		36.00	
		Variance		64.134	
		Std. Deviation		8.008	
		Minimum		21	
		Maximum		48	
		Range		27	
		Interquartile Range		13	
		Skewness		-.401	.512
		Kurtosis		-.631	.992
	Hot (29 Degrees)	Mean		33.70	1.750
		95% Confidence Interval for Mean	Lower Bound	30.04	
			Upper Bound	37.36	
		5% Trimmed Mean		33.83	
		Median		34.50	
		Variance		61.274	
		Std. Deviation		7.828	
		Minimum		20	
		Maximum		45	
		Range		25	
		Interquartile Range		13	
		Skewness		-.277	.512
		Kurtosis		-1.150	.992

When the **Skewness** and **Kurtosis** statistics for a distribution of scores are both zero, it is perfectly normal.

Here, the skewness and kurtosis statistics for both sets of scores are fairly close to zero. Also, z_s and z_k are within ±1.96 for both (see section 4.3.2.3).

This suggests that <u>both groups of data are approximately normal</u>.

PASW Statistics provides two separate **Tests of Normality**.

The **Shapiro-Wilk** test is considered more appropriate for small to medium sized samples. If the **Shapiro-Wilk** statistic (*W*) is significant (i.e., *Sig* < .05), the group of scores is not normally distributed.

Here, *W* is .920 (*Sig* = .098) for the cool room and .945 (*Sig* = .297) for the hot room.

We can conclude that <u>the assumption of normality has not been violated for the room temperature variable</u>.

Tests of Normality

| Room Temperature | | Kolmogorov-Smirnov[a] | | | Shapiro-Wilk | | |
		Statistic	df	Sig.	Statistic	df	Sig.
Exam Score	Cool (23 Degrees)	.180	20	.088	.920	20	.098
	Hot (29 Degrees)	.159	20	.200[*]	.945	20	.297

a. Lilliefors Significance Correction

*. This is a lower bound of the true significance.

Exam Difficulty

Case Processing Summary

Exam Difficulty		Cases					
		Valid		Missing		Total	
		N	Percent	N	Percent	N	Percent
Exam Score	Easy	20	100.0%	0	.0%	20	100.0%
	Hard	20	100.0%	0	.0%	20	100.0%

Descriptives

Exam Difficulty				Statistic	Std. Error
Exam Score	Easy	Mean		41.85	.960
		95% Confidence Interval for Mean	Lower Bound	39.84	
			Upper Bound	43.86	
		5% Trimmed Mean		42.00	
		Median		43.00	
		Variance		18.450	
		Std. Deviation		4.295	
		Minimum		33	
		Maximum		48	
		Range		15	
		Interquartile Range		7	
		Skewness		-.492	.512
		Kurtosis		-.878	.992
	Hard	Mean		29.20	1.220
		95% Confidence Interval for Mean	Lower Bound	26.65	
			Upper Bound	31.75	
		5% Trimmed Mean		29.22	
		Median		29.50	
		Variance		29.747	
		Std. Deviation		5.454	
		Minimum		20	
		Maximum		38	
		Range		18	
		Interquartile Range		10	
		Skewness		-.274	.512
		Kurtosis		-1.093	.992

> Looking at the relevant **Skewness**, **Kurtosis** and **Shapiro-Wilk** statistics, the exam scores on both the easy and hard papers also appear to be approximately normally distributed.

▶▶| *Links:*
The full output also included **Normal Q-Q Plots** and **Detrended Normal Q-Q Plots**.

These have been described elsewhere in this text (e.g., ch 4).

Tests of Normality

Exam Difficulty		Kolmogorov-Smirnov[a]			Shapiro-Wilk		
		Statistic	df	Sig.	Statistic	df	Sig.
Exam Score	Easy	.168	20	.140	.935	20	.191
	Hard	.135	20	.200*	.946	20	.317

a. Lilliefors Significance Correction

*. This is a lower bound of the true significance.

8.3.2.4. PASW Statistics Procedure (Part 2: Homogeneity of Variance & the ANOVA)

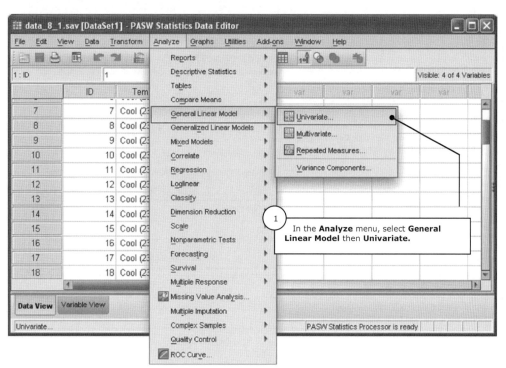

1. In the **Analyze** menu, select **General Linear Model** then **Univariate**.

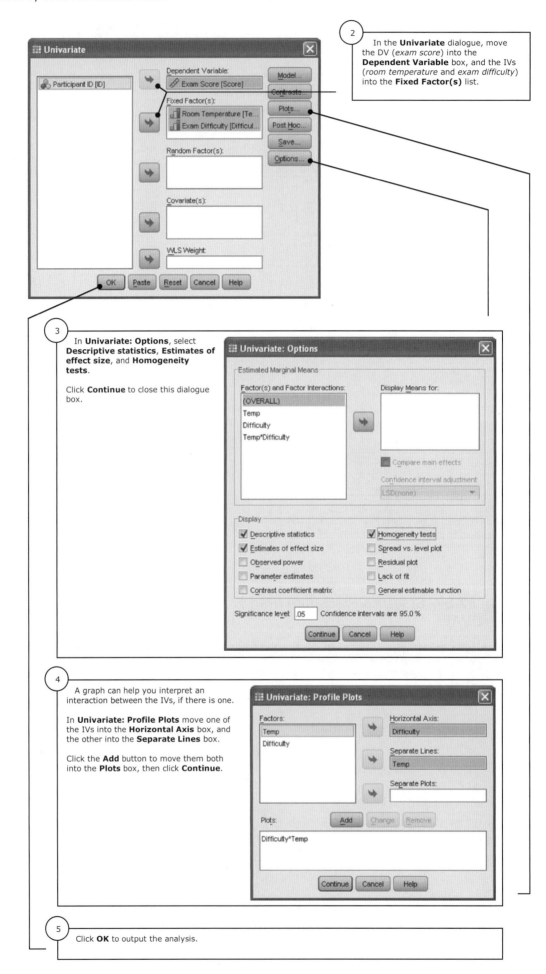

2 In the **Univariate** dialogue, move the DV (*exam score*) into the **Dependent Variable** box, and the IVs (*room temperature* and *exam difficulty*) into the **Fixed Factor(s)** list.

3 In **Univariate: Options**, select **Descriptive statistics**, **Estimates of effect size**, and **Homogeneity tests**.

Click **Continue** to close this dialogue box.

4 A graph can help you interpret an interaction between the IVs, if there is one.

In **Univariate: Profile Plots** move one of the IVs into the **Horizontal Axis** box, and the other into the **Separate Lines** box.

Click the **Add** button to move them both into the **Plots** box, then click **Continue**.

5 Click **OK** to output the analysis.

8.3.2.5. PASW Statistics Output (Part 2: Homogeneity of Variance and the ANOVA)

Multivariate Analysis of Variance

Between-Subjects Factors

		Value Label	N
Room Temperature	1	Cool (23 Degrees)	20
	2	Hot (29 Degrees)	20
Exam Difficulty	1	Easy	20
	2	Hard	20

The **Between-Subjects Factors** table provides basic information about the two IVs, and the number of participants who experienced each.

Here, we can see that 20 participants sat one of the exams in the cool room, and so on.

With the **Descriptive Statistics** table, we can get an initial sense of the main effects (if any) of each IV.

For example, the 20 participants who sat the easy exam (regardless of whether they sat it in a cool or hot room) had a mean score of 41.85. The 20 participants who sat the hard exam averaged just 29.20.

The factorial ANOVA will be used to determine whether or not this apparent main effect of exam difficulty is statistically significant.

Similarly, the ANOVA will also test the main effect of room temperature. (That is, whether or not the difference between the cool room mean of 37.35 and the hot room mean of 33.70 is statistically significant.)

The **Descriptive Statistics** table can also be used to gain an initial sense of any interaction between the two IVs. That is, whether the effects of one variable (e.g., room temperature) on exam scores are dependent on the other variable (e.g., exam difficulty).

Looking first at the participants who sat the easy exam. Those who did so in the cool room ($n = 10$) scored an average of 43.80, which is 3.9 points higher than those participants who sat this exam in the hot room ($n = 10$, $M = 39.90$).

Next, we can compare the participants who sat the hard exam in the cool room ($n = 10$, $M = 30.90$) with the participants who sat it in the hot room ($n = 10$, $M = 27.50$). The difference here is 3.4 points.

So, on the face of things, it appears as though room temperature has a similar effect on exam performance regardless of whether the exam is easy or difficult. That is, increasing the room temperature from 23°C to 29°C results in a similar decrease in performance on both the easy (3.9 points) and hard (3.4 points) exams.

The factorial ANOVA will confirm the absence of a statistically significant interaction between the room temperature and exam difficulty variables.

Descriptive Statistics

Dependent Variable: Exam Score

Room Temperature	Exam Difficulty	Mean	Std. Deviation	N
Cool (23 Degrees)	Easy	43.80	4.077	10
	Hard	30.90	5.131	10
	Total	37.35	8.008	20
Hot (29 Degrees)	Easy	39.90	3.725	10
	Hard	27.50	5.482	10
	Total	33.70	7.828	20
Total	Easy	41.85	4.299	20
	Hard	29.20	5.454	20
	Total	35.53	8.032	40

Levene's Test of Equality of Error Variances[a]

Dependent Variable: Exam Score

F	df1	df2	Sig.
.620	3	36	.607

Tests the null hypothesis that the error variance of the dependent variable is equal across groups.

a. Design: Intercept + Temp + Difficulty + Temp * Difficulty

The **Levene's Test for Equality of Error Variances** is not significant at $\alpha = .05$, $F(3, 36) = .62$, $p = .607$. The assumption of homogeneity of variance has not been violated.

If the Assumption is Violated

ANOVA is not sensitive to violations of the equal variances assumption when samples are moderate to large and approximately equally sized.

Therefore, when designing a study you should always try to ensure that your group sizes are equal.

Tests of Between-Subjects Effects

Dependent Variable: Exam Score

Source	Type III Sum of Squares	df	Mean Square	F	Sig.	Partial Eta Squared
Corrected Model	1734.075[a]	3	578.025	26.613	.000	.689
Intercept	50481.025	1	50481.025	2324.232	.000	.985
Temp	133.225	1	133.225	6.134	.018	.146
Difficulty	1600.225	1	1600.225	73.677	.000	.672
Temp * Difficulty	.625	1	.625	.029	.866	.001
Error	781.900	36	21.719			
Total	52997.000	40				
Corrected Total	2515.975	39				

a. R Squared = .689 (Adjusted R Squared = .663)

The **Tests of Between-Subjects Effects** table reports the outcome of the ANOVA.

The **Temp** row indicates that there is a significant main effect for temperature, $F(1, 36) = 6.13$, $p = .018$.

The **Difficulty** row indicates that there is a significant main effect for exam difficulty, $F(1, 36) = 73.68$, $p < .001$.

The **Temp*Difficulty** row indicates that the two IVs do not interact, $F(1, 36) = .03$, $p = .866$.

Each of these findings should be included in your written results section.

Profile Plots

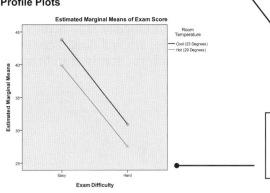

Estimated Marginal Means of Exam Score

Room Temperature
— Cool (23 Degrees)
— Hot (29 Degrees)

Estimated Marginal Means

Exam Difficulty
Easy   Hard

As can be seen in this plot of marginal means, the impact of room temperature on exam scores is similar, regardless of exam difficulty. (Or, if you prefer, the impact of exam difficulty on exam scores is similar, regardless of room temperature.)

8.3.3. APA Style Results Write-Up

Results

A factorial between groups analysis of variance (ANOVA) was used to compare the average exam scores of four groups of participants: (a) participants completing an easy exam in a cool (23°C) room, (b) participants completing an easy exam in a hot (29°C) room, (c) participants completing a hard exam in a cool (23°C) room, and (d) participants completing a hard exam in a hot (29°C) room. Shapiro-Wilk and Levene's tests were used to evaluate the assumptions of normality and homogeneity of variance respectively. Neither was violated.

> Assumption testing is often not reported in journal articles (particularly when the assumptions are not violated). However, it is more commonly expected in student reports.

The main effect of room temperature on exam scores was statistically significant, $F(1, 36) = 6.14$, $p = .018$, with participants in the cool room ($M = 37.35$, $SD = 8.01$) achieving significantly higher exam scores than participants in the warm room ($M = 33.70$, $SD = 7.83$). Partial eta-squared (η^2) for this effect was .146.

> Partial eta-squared for each main effect and for the interaction effect(s) can be read directly from the **Tests of Between-Subjects Effects** table in the PASW Statistics output.
>
> Although commonly reported, its use can be problematic due to its tendency to overestimate true (population) effect sizes.
>
> In *Illustrated Example Two* we will describe how to calculate omega-squared as an alternative (and often preferred) measure of effect size for factorial ANOVA.

The main effect of exam difficulty was also statistically significant, $F(1, 36) = 73.68$, $p < .001$, and large, partial $\eta^2 = .672$. The participants in the easy exam condition achieved significantly higher exam scores ($M = 41.85$, $SD = 4.30$) than those in the hard exam condition ($M = 29.20$, $SD = 5.45$).

> When PASW Statistics reports *Sig* as .000, use *p* < .001 in your results.

There was no interaction between room temperature and exam difficulty, $F(1, 36) = .029$, $p = .866$, partial $\eta^2 = .001$.

8.3.4. Summary

In *Illustrated Example One*, both main effects were statistically significant, while the interaction was not. In the following example, we will illustrate the follow-up testing that should be performed when an interaction is significant.

8.4. Illustrated Example Two

To investigate the impact of both background noise and practice on adults' reading comprehension, a researcher divided a group of 30 volunteers into three evenly sized groups. Members of the first group were individually taken to a silent room, where half were tested immediately on a simple reading comprehension task ($n = 5$), while the other half ($n = 5$) were given 10 minutes of practice time before testing commenced. Members of the second and third groups experienced similar treatment, in rooms with low and high levels of background noise respectively.

The researcher wants to know if:

1. Background noise impacts on reading comprehension?
2. Practice impacts on reading comprehension?
3. There is an interaction between background noise and practice on reading comprehension?

The participants' reading comprehension scores are reproduced in Table 8.2.

Table 8.2

Reading Comprehension Scores of Participants Under Two Levels of Practice and Three Levels of Background Noise (N = 30)

📁 *Data:*
This is **data_8_2.sav** on the companion website.

	No Practice			**10 Minutes Practice**	
ID	Background Noise	Reading Comprehension	ID	Background Noise	Reading Comprehension
1	No Noise	12	16	No Noise	17
2	No Noise	13	17	No Noise	17
3	No Noise	14	18	No Noise	18
4	No Noise	14	19	No Noise	18
5	No Noise	16	20	No Noise	20
6	Low Noise	11	21	Low Noise	15
7	Low Noise	12	22	Low Noise	16
8	Low Noise	12	23	Low Noise	17
9	Low Noise	14	24	Low Noise	18
10	Low Noise	14	25	Low Noise	18
11	High Noise	9	26	High Noise	10
12	High Noise	10	27	High Noise	10
13	High Noise	12	28	High Noise	13
14	High Noise	12	29	High Noise	14
15	High Noise	14	30	High Noise	14

8.4.1. PASW Statistics Output

Univariate Analysis of Variance

Between-Subjects Factors

		Value Label	N
Level of background noise	.00	No noise	10
	1.00	Low noise	10
	2.00	High noise	10
Practice	.00	No practice	15
	1.00	Practice	15

Descriptive Statistics

Dependent Variable:Reading Comprehension Level

Level of backgrou...	Practice	Mean	Std. Deviation	N
No noise	No practice	13.8000	1.48324	5
	Practice	18.0000	1.22474	5
	Total	15.9000	2.55821	10
Low noise	No practice	12.6000	1.34164	5
	Practice	16.8000	1.30384	5
	Total	14.7000	2.54078	10
High noise	No practice	11.4000	1.94936	5
	Practice	12.2000	2.04939	5
	Total	11.8000	1.93218	10
Total	No practice	12.6000	1.80476	15
	Practice	15.6667	2.96808	15
	Total	14.1333	2.87358	30

The table of **Descriptive Statistics** seems to suggest that, generally speaking:

1. Participants given the opportunity to practice (*M* = 15.67) scored higher on the reading comprehension task than participants who were denied this opportunity (*M* = 12.60).

2. Participants exposed to high background noise (*M* = 11.80) performed worse on the comprehension task than participants in the low noise and no noise conditions (*M* = 14.70 and *M* = 15.90 respectively).

3. The effects of practice on reading comprehension seem to be influenced by background noise levels. (That is, practice seems more beneficial when background noise is low or absent, than it does when background noise is high.)

Whether or not these effects are statistically significant will be determined by examining the ANOVA output reported in the **Tests of Between-Subjects Effects** table.

Levene's Test of Equality of Error Variances[a]

Dependent Variable:Reading Comprehension Level

F	df1	df2	Sig.
1.127	5	24	.373

Tests the null hypothesis that the error variance of the dependent variable is equal across groups.

a. Design: Intercept + Noise + Practice + Noise * Practice

Levene's Test of Equality of Error Variances is not significant at α = .05. Therefore, homogeneity of variance assumption is not violated.

Tests of Between-Subjects Effects

Dependent Variable:Reading Comprehension Level

Source	Type III Sum of Squares	df	Mean Square	F	Sig.	Partial Eta Squared
Corrected Model	178.667[a]	5	35.733	14.105	.000	.746
Intercept	5992.533	1	5992.533	2365.474	.000	.990
Noise	88.867	2	44.433	17.539	.000	.594
Practice	70.533	1	70.533	27.842	.000	.537
Noise * Practice	19.267	2	9.633	3.803	.037	.241
Error	60.800	24	2.533			
Total	6232.000	30				
Corrected Total	239.467	29				

a. R Squared = .746 (Adjusted R Squared = .693)

The **Tests of Between-Subjects Effects** table reports the outcome of the factorial ANOVA.

The **Noise** row indicates a statistically significant background noise main effect, *F*(2, 24) = 17.54, *p* < .001.

The **Practice** row indicates a statistically significant practice main effect, *F*(1, 24) = 27.84, *p* < .001.

The **Practice*Noise** row indicates a significant interaction effect, *F*(2, 24) = 3.80, *p* = .037. This means that the effects of practice on comprehension depend on background noise levels.

Profile Plots

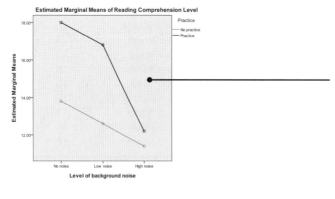

A graph of the interaction can aid with interpretation, and can be a useful addition to a written results section.

8.4.2. Follow Up Analyses

8.4.2.1. Simple Effects Analyses

When an interaction is statistically significant, it can be further explored with simple effects analyses. This simply involves looking at the effects of one IV on the DV at each level of the second IV. For example, we can look at the effects of background noise at each level of practice; or look at the effects of practice at each level of background noise. We will do the latter.

8.4.2.1.1. PASW Statistics Procedure

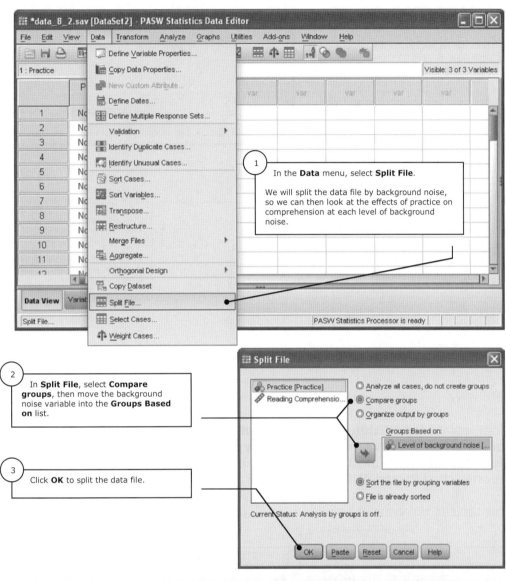

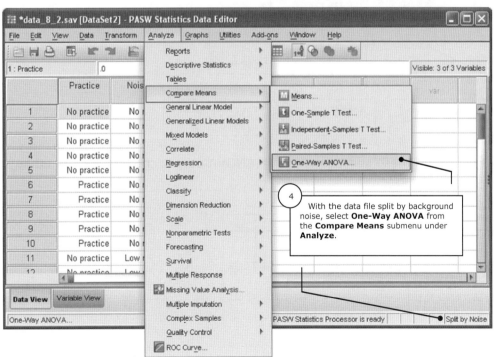

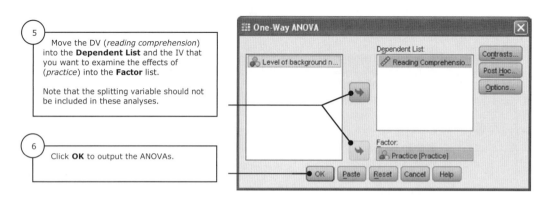

5 Move the DV (*reading comprehension*) into the **Dependent List** and the IV that you want to examine the effects of (*practice*) into the **Factor** list.

Note that the splitting variable should not be included in these analyses.

6 Click **OK** to output the ANOVAs.

8.4.2.1.2. PASW Statistics Output

Oneway

PASW Statistics has produced three ANOVAs, which we can use to examine the effects of practice on comprehension at each level of background noise.

However, before we can interpret these ANOVAs, we need to make a few small adjustments.

ANOVA

Reading Comprehension Level

Level of background noise		Sum of Squares	df	Mean Square	F	Sig.
No noise	Between Groups	44.100	1	44.100	23.838	.001
	Within Groups	14.800	8	1.850		
	Total	58.900	9			
Low noise	Between Groups	44.100	1	44.100	25.200	.001
	Within Groups	14.000	8	1.750		
	Total	58.100	9			
High noise	Between Groups	1.600	1	1.600	.400	.545
	Within Groups	32.000	8	4.000		
	Total	33.600	9			

Before we can interpret these ANOVAs, we need to make some adjustments.

First, replace the **Within Groups Sum of Squares**, **df** and **Mean Square** values for each ANOVA with the **Error Type III Sum of Squares**, **df** and **Mean Square** from the **Tests of Between Subjects Effects** table that reported the factorial ANOVA in section 8.4.1.

This table is reproduced below:

Tests of Between-Subjects Effects

Dependent Variable:Reading Comprehension Level

Source	Type III Sum of Squares	df	Mean Square	F	Sig.	Partial Eta Squared
Corrected Model	178.667[a]	5	35.733	14.105	.000	.746
Intercept	5992.533	1	5992.533	2365.474	.000	.990
Noise	88.867	2	44.433	17.539	.000	.594
Practice	70.533	1	70.533	27.842	.000	.537
Noise * Practice	19.267	2	9.633	3.803	.037	.241
Error	60.800	24	2.533			
Total	6232.000	30				
Corrected Total	239.467	29				

a. R Squared = .746 (Adjusted R Squared = .693)

Error Type III Sum of Squares = 60.800

Error df = 24.

Error Mean Square = 2.533.

After making these substitutions, the ANOVA output should look something like:

ANOVA

Reading Comprehension Level

Level of background noise		Sum of Squares	df	Mean Square	F	Sig.
No noise	Between Groups	44.100 60.80	1 24	44.100 2.53	23.838	.001
	Within Groups	~~14.800~~	~~8~~	~~1.850~~		
	Total	58.900	9			
Low noise	Between Groups	44.100 60.80	1 24	44.100 2.53	25.200	.001
	Within Groups	~~14.000~~	~~8~~	~~1.750~~		
	Total	58.100	9			
High noise	Between Groups	1.600 60.80	1 24	1.600 2.53	.400	.545
	Within Groups	~~32.000~~	~~8~~	~~4.000~~		
	Total	33.600	9			

Next, recalculate F for each ANOVA with:

$$F = \frac{MS_{Between}}{MS_{Within}}$$

So,

ANOVA

Reading Comprehension Level

Level of background noise		Sum of Squares	df	Mean Square	F	Sig.
No noise	Between Groups	44.100	1	44.100	17.41	.001
	Within Groups	14.800	8	1.850		
	Total	58.900	9			
Low noise	Between Groups	44.100	1	44.100	17.41	.001
	Within Groups	14.000	8	1.750		
	Total	58.100	9			
High noise	Between Groups	1.600	1	1.600	0.68	.545
	Within Groups	32.000	8	4.000		
	Total	33.600	9			

Finally, turn to the tables of critical F-values that are available in any good statistics textbook, and find the $F_{critical}$ for $(df_{Between}, df_{Within})$ at $\alpha = .01$.

Howell (2010b, p. 674) reports $F_{critical}$ for $df = (1, 24)$ at $\alpha = .01$ as 7.82.

When $F_{observed} > F_{critical}$, the test is statistically significant.

Consequently, we can report that practice has a significant effect on reading comprehension in the absence of any background noise, $F(1, 24) = 17.41$, $p < .01$, and when background noise levels are low, $F(1, 24) = 17.41$, $p < .01$. However, practice does not influence reading comprehension when background noise levels are high, $F(1, 24) = 0.68$, ns.

8.4.2.2. Effect Size (Omega-Squared)

As discussed in Illustrated Example One, the use of partial eta-squared (which is provided by PASW Statistics) is not generally recommended. Research has shown the partial eta-squared has a positive bias and will often result in the calculation of large effect sizes when they are not justified. Instead, omega-squared (ω^2) can be used to provide an unbiased estimate of effect size. Omega-squared is not provided by PASW Statistics but can be calculated by hand.

To calculate ω^2 we first need to determine the variance component ($\hat{\sigma}^2$) for each of the effects in our analysis using the formulas below. In our example there are two main effects (noise and practice) plus an interaction effect.

$$\hat{\sigma}^2_{Noise} = \frac{(nl-1)(MS_{Noise} - MS_{Error})}{(Number\ of\ People\ Per\ Condition)(nl)(pl)}$$

$$\hat{\sigma}^2_{Practice} = \frac{(pl-1)(MS_{Practice} - MS_{Error})}{(Number\ of\ People\ Per\ Condition)(nl)(pl)}$$

$$\hat{\sigma}^2_{Interaction} = \frac{(nl-1)(pl-1)(MS_{Interaction} - MS_{Error})}{(Number\ of\ People\ Per\ Condition)(nl)(pl)}$$

Where nl is Number of levels of the noise variable, pl is the number of the levels of the practice variable, and MS_{Noise}, $MS_{Practice}$, $MS_{Interaction}$ and MS_{Error} can be read off the **Tests of Between Subjects Effects** table.

Tests of Between-Subjects Effects

Dependent Variable:Reading Comprehension Level

Source	Type III Sum of Squares	df	Mean Square	F	Sig.	Partial Eta Squared
Corrected Model	178.667ᵃ	5	35.733	14.105	.000	.746
Intercept	5992.533	1	5992.533	2365.474	.000	.990
Noise	88.867	2	44.433	17.539	.000	.594
Practice	70.533	1	70.533	27.842	.000	.537
Noise * Practice	19.267	2	9.633	3.803	.037	.241
Error	60.800	24	2.533			
Total	6232.000	30				
Corrected Total	239.467	29				

a. R Squared = .746 (Adjusted R Squared = .693)

$MS_{Noise} = 44.433$

$MS_{Practice} = 70.533$

$MS_{Interaction} = 9.633$

$MS_{Error} = 2.533$

So,

$$\hat{\sigma}^2_{Noise} = \frac{(3-1)(44.433-2.533)}{(5)(3)(2)} = 2.793$$

$$\hat{\sigma}^2_{Practice} = \frac{(2-1)(70.533-2.533)}{(5)(3)(2)} = 2.267$$

$$\hat{\sigma}^2_{Interaction} = \frac{(3-1)(2-1)(9.633-2.533)}{(5)(3)(2)} = 0.473$$

We also need to calculate the total variance estimate:

$$\hat{\sigma}^2_{Total} = \hat{\sigma}^2_{Noise} + \hat{\sigma}^2_{Practice} + \hat{\sigma}^2_{Interaction} + MS_{Error}$$

So,

$$\hat{\sigma}^2_{Total} = 2.793 + 2.267 + 0.473 + 2.533$$
$$= 8.066$$

Omega-squared for each effect can now be calculated by dividing the relevant variance estimates by the total variance estimate:

$$\omega^2_{Noise} = \frac{2.793}{8.066} = 0.346$$

$$\omega^2_{Practice} = \frac{2.267}{8.066} = 0.281$$

$$\omega^2_{Interaction} = \frac{0.473}{8.066} = 0.059$$

To convert ω^2 to f, Cohen's measure of effect size for ANOVA, the following formula can be used:

$$f = \sqrt{\frac{\omega^2}{1-\omega^2}}$$

In the current example, $f_{Noise} = .727$, $f_{Practice} = .625$ and $f_{Interaction} = .251$.

Cohen (1988) suggested that $f = .10$ could be considered small, $f = .25$ could be considered medium and $f = .40$ could be considered large.

8.4.3. APA Style Results Write-Up

Results

A factorial between groups analysis of variance (ANOVA) was used to investigate the effects of background noise and practice on reading comprehension.

> Don't forget to report the results of your assumption testing (as we did in the previous example).

The ANOVA revealed a statistically significant main effect for background noise, $F(2, 24) = 17.54$, $p < .001$, $\omega^2 = .346$, as well as a statistically significant main effect for practice, $F(1, 24) = 27.84$, $p < .001$, $\omega^2 = .281$.

Furthermore, a statistically significant interaction indicated that the effects of practice on reading comprehension depend on background noise levels, $F(2, 24) = 3.80$, $p = .037$, $\omega^2 = .059$. The nature of this interaction is illustrated in Figure 1.

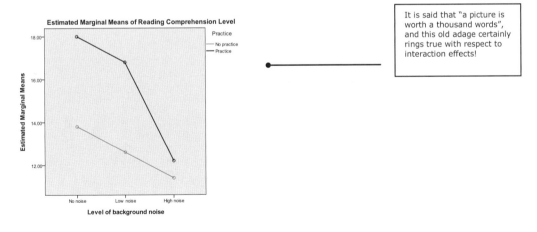

> It is said that "a picture is worth a thousand words", and this old adage certainly rings true with respect to interaction effects!

Figure 1. The effects of background noise and practice on reading comprehension.

Simple effects analyses were used to further examine the interaction between practice and background noise. These analyses indicated that practice has a statistically significant (positive) effect on reading comprehension in the absence of any background noise,

$F(1, 24) = 17.41, p < .01$, and when background noise levels are low, $F(1, 24) = 17.41, p < .01$. However, practice does not influence reading comprehension when background noise levels are high, $F(1, 24) = 0.68$, *ns*.

We could also look at the effects of noise at each level of practice, which would involve both simple effects analyses and (if the simple effects were significant) simple comparisons.

Basically, we would split the data file by practice, and then run a one-way ANOVA with reading comprehension as the DV and background noise as the IV.

This would produce two one-way ANOVAs, one for each level of practice. If either (or both) of these were significant, we would need to locate the source of this significance with either post-hoc testing, or a series of planned comparisons.

8.5. Factorial Between Groups ANOVA Checklist

Have you:

- ✔ Checked that each group of data is approximately normally distributed?
- ✔ Checked for homogeneity of variance?
- ✔ Identified the main effects for all independent variables and taken note of the *F* value, degrees of freedom, and significance level for each?
- ✔ Recorded partial eta-squared or calculated omega-squared?
- ✔ Identified the interaction effect and taken note of its *F* value, degrees of freedom, and significance level?
- ✔ Performed simple effects analyses (and/or simple comparisons) if the interaction is significant?
- ✔ Written up your results in the APA style?

Chapter 9: One-Way Repeated Measures ANOVA

AKA:
Single factor analysis of variance, Single factor ANOVA, One-way within-subjects ANOVA.

Chapter Overview

Included With:
Repeated measures ANOVA is a part of the *PASW Advanced Statistics* add-on module.

It is not included in *PASW Statistics*, and is consequently not available in *PASW Statistics Student Version*.

9.1. Purpose of the One-Way Repeated Measures ANOVA

To test for statistically significant differences between three or more related sample means. Samples are considered related when:
    a. They are comprised of the same group of individuals, who've provided data on multiple occasions (e.g., in spring, summer, autumn and winter). This is a repeated measures research design.
    b. Each individual in one sample is connected or linked with one specific individual in each other sample. This is a matched design.

9.2. Questions We Could Answer Using the One-Way Repeated Measures ANOVA

1. Does counselling alter teachers' attitudes towards their jobs over time (pre-counselling, post-counselling, two month follow up, one year follow up)?

To answer this question, we would need to measure the attitudes of the same group of teachers (the dependent variable) at four separate points in time (the independent variable). Hence, we are using a repeated measures design.

2. Do children prefer some brands of chocolate to others?

3. Does relaxation training improve the behaviour of boys with Attention Deficit Hyperactivity Disorder (ADHD)? Are these improvements still evident six months after the completion of the relaxation training program?

9.3. Illustrated Example One

As part of an ongoing market research program, a local toyshop invited 20 children to play with four new toys (in a counterbalanced order), and then rate how much they "liked" each toy on a scale from "1" to "10". The manager of the toyshop plans to order extra quantities of the most popular toy(s) in the lead-up to the busy Christmas shopping period.

The children's ratings of the four toys are reproduced in Table 9.1.

Table 9.1

Ratings of Four New Toys by a Sample of 20 Children

☐ *Data:*
This is data file
data_9_1.sav on the
companion website.

Participant ID	Robotic Allosaurus	Remote Controlled Space Shuttle	Indoor Light-up Kite	Interactive Teddy Bear
1	7	2	7	4
2	8	2	8	5
3	5	3	7	6
4	8	4	8	5
5	7	5	8	3
6	6	4	6	2
7	9	2	9	1
8	8	3	7	4
9	4	5	8	6
10	3	6	10	5
11	7	5	9	6
12	7	2	6	2
13	4	6	9	4
14	5	4	9	8
15	6	3	10	4
16	4	5	9	7
17	9	7	8	9
18	8	5	7	7
19	9	4	9	6
20	7	7	10	8

9.3.1. Setting Up The PASW Statistics Data File

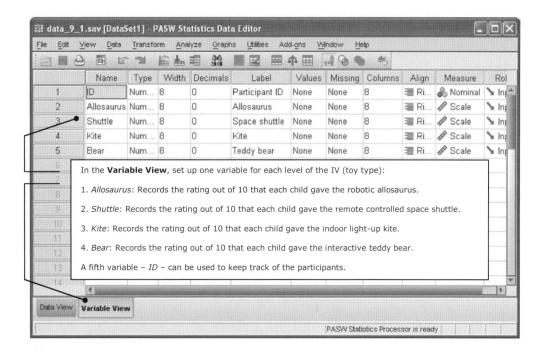

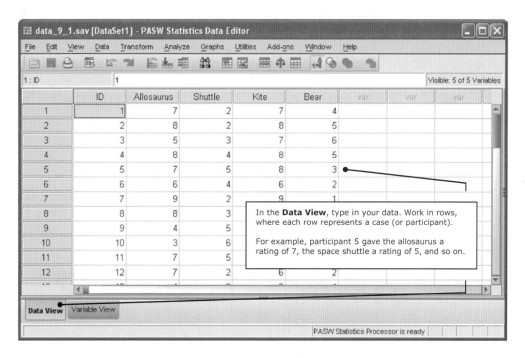

9.3.2. Analysing the Data

9.3.2.1. Assumptions

The following criteria should be met before conducting a one-way repeated measures ANOVA. Assumption 1 is methodological and should have been addressed before and during data collection. Assumptions 2, 3 and 4 can be tested with PASW Statistics and PASW Advanced Statistics.

1. **Scale of Measurement.** The DV should be interval or ratio data.

Syntax:
Run these analyses with **syntax_9_1.sps** on the companion website.

2. **Normality.** Each group of scores should be approximately normally distributed.

3. **Homogeneity of Variance.** There should be an approximately equal amount of variability in each set of scores.

4. **Sphericity.** The variability in the differences between any pair of groups should be the same as the variability in the differences between any other pair of groups (Garson, 2009). PASW Advanced Statistics automatically tests the sphericity assumption.

9.3.2.2. PASW Statistics Procedure (Part 1: Normality)

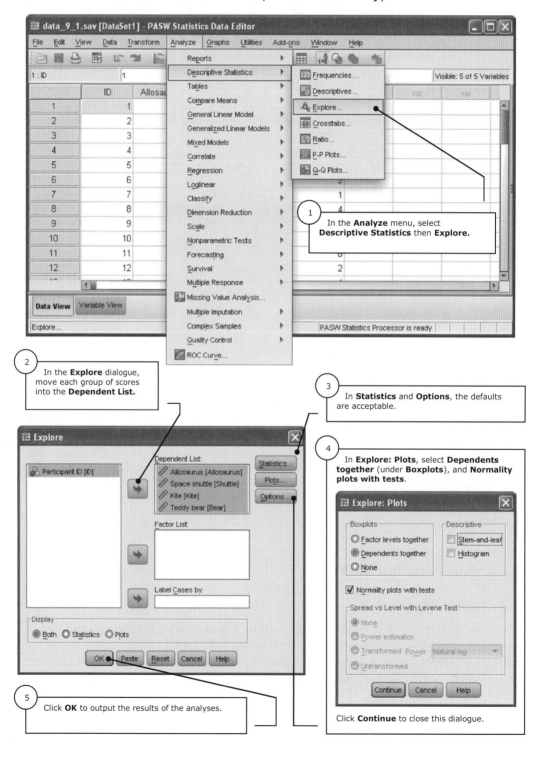

9.3.2.3. PASW Statistics Output (Part 1: Normality)

Explore

Case Processing Summary

	Cases					
	Valid		Missing		Total	
	N	Percent	N	Percent	N	Percent
Allosaurus	20	100.0%	0	.0%	20	100.0%
Space shuttle	20	100.0%	0	.0%	20	100.0%
Kite	20	100.0%	0	.0%	20	100.0%
Teddy bear	20	100.0%	0	.0%	20	100.0%

The **Case Processing Summary** shows how many cases were analysed, and how many were dropped due to missing data.

In this instance, there were 20 participants in the study. No cases were excluded from the analyses because of missing data.

Descriptives

			Statistic	Std. Error
Allosaurus	Mean		6.55	.413
	95% Confidence Interval for Mean	Lower Bound	5.68	
		Upper Bound	7.42	
	5% Trimmed Mean		6.61	
	Median		7.00	
	Variance		3.418	
	Std. Deviation		1.849	
	Minimum		3	
	Maximum		9	
	Range		6	
	Interquartile Range		3	
	Skewness		-.421	.512
	Kurtosis		-.922	.992
Space shuttle	Mean		4.20	.360
	95% Confidence Interval for Mean	Lower Bound	3.45	
		Upper Bound	4.95	
	5% Trimmed Mean		4.17	
	Median		4.00	
	Variance		2.589	
	Std. Deviation		1.609	
	Minimum		2	
	Maximum		7	
	Range		5	
	Interquartile Range		2	
	Skewness		.145	.512
	Kurtosis		-.890	.992
Kite	Mean		8.20	.277
	95% Confidence Interval for Mean	Lower Bound	7.62	
		Upper Bound	8.78	
	5% Trimmed Mean		8.22	
	Median		8.00	
	Variance		1.537	
	Std. Deviation		1.240	
	Minimum		6	
	Maximum		10	
	Range		4	
	Interquartile Range		2	
	Skewness		-.236	.512
	Kurtosis		-.814	.992
Teddy bear	Mean		5.10	.481
	95% Confidence Interval for Mean	Lower Bound	4.09	
		Upper Bound	6.11	
	5% Trimmed Mean		5.11	
	Median		5.00	
	Variance		4.621	
	Std. Deviation		2.150	
	Minimum		1	
	Maximum		9	
	Range		8	
	Interquartile Range		3	
	Skewness		-.108	.512
	Kurtosis		-.515	.992

The table of **Descriptives** contains a range of useful information, including measures of central tendency, dispersion, skewness and kurtosis for each group of scores.

When the **Skewness** and **Kurtosis** statistics for a distribution of scores are both zero, it is perfectly normal.

Here, the skewness and kurtosis statistics for the four distributions are all fairly close to zero. Also, z_s and z_k are within ±1.96 for all four groups (see section 4.3.2.3).

This suggests that <u>all four groups of data are approximately normal</u>.

Tests of Normality

	Kolmogorov-Smirnov[a]			Shapiro-Wilk		
	Statistic	df	Sig.	Statistic	df	Sig.
Allosaurus	.196	20	.042	.923	20	.112
Space shuttle	.140	20	.200*	.927	20	.134
Kite	.191	20	.055	.920	20	.100
Teddy bear	.112	20	.200*	.974	20	.828

a. Lilliefors Significance Correction

*. This is a lower bound of the true significance.

Due to the small size of the sample (*N*=20), the **Shapiro-Wilk** test is a more appropriate **Test of Normality** than the **Kolmogorov-Smirnov** test.

If the **Shapiro-Wilk** statistic (*W*) is significant (i.e., *Sig* < .05), the group of scores is not normally distributed.

Here, *W* is non-significant (i.e., *Sig* > .05) for all four groups of ratings.

We can conclude that <u>the assumption of normality is not violated</u>.

If the Assumption is Violated

Small-to-moderate departures from normality are usually not of concern. Researchers faced with severe departures from normality should consider using the non-parametric Friedman two-way ANOVA instead.

▶▶| *Links:*
When **Normality plots with tests** is selected in **Explore: Plots**, PASW Statistics produces a **Normal Q-Q Plot** and a **Detrended Normal Q-Q Plot** for each group of scores, in addition to the **Tests of Normality** pictured to the left.

We've omitted these plots from the current example, but have discussed them in detail elsewhere in this book.

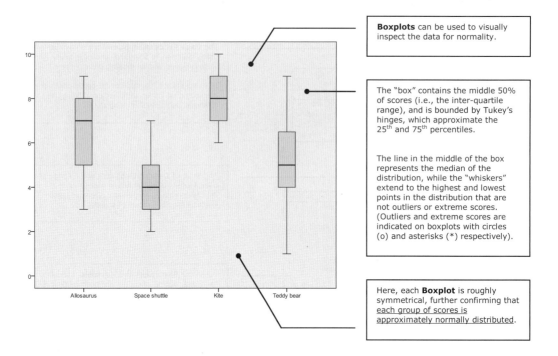

Boxplots can be used to visually inspect the data for normality.

The "box" contains the middle 50% of scores (i.e., the inter-quartile range), and is bounded by Tukey's hinges, which approximate the 25th and 75th percentiles.

The line in the middle of the box represents the median of the distribution, while the "whiskers" extend to the highest and lowest points in the distribution that are not outliers or extreme scores. (Outliers and extreme scores are indicated on boxplots with circles (o) and asterisks (*) respectively).

Here, each **Boxplot** is roughly symmetrical, further confirming that each group of scores is approximately normally distributed.

9.3.2.4. PASW Advanced Statistics Procedure (Part 2: Homogeneity of Variance, Sphericity, & the ANOVA)

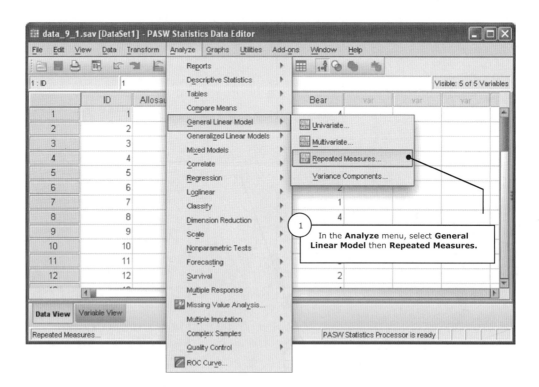

In the **Analyze** menu, select **General Linear Model** then **Repeated Measures**.

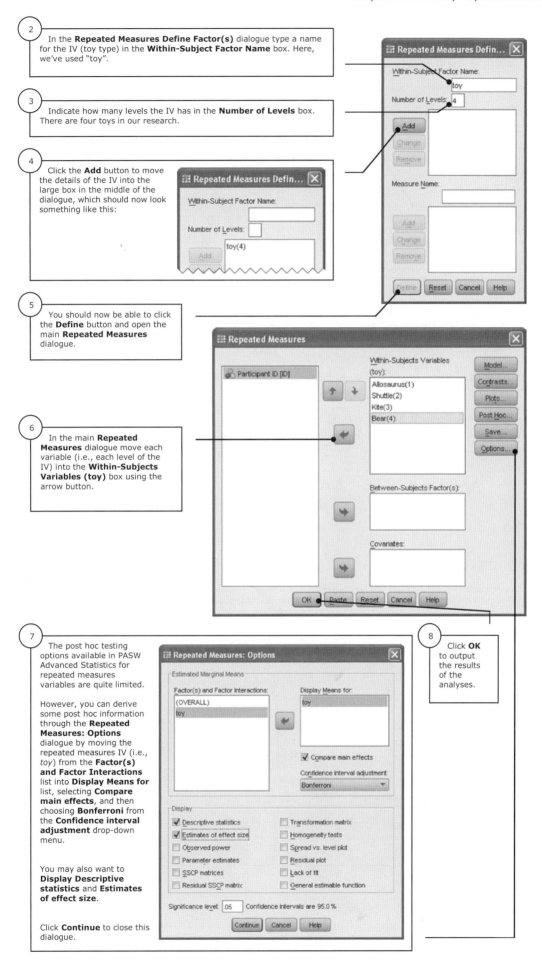

2

In the **Repeated Measures Define Factor(s)** dialogue type a name for the IV (toy type) in the **Within-Subject Factor Name** box. Here, we've used "toy".

3

Indicate how many levels the IV has in the **Number of Levels** box. There are four toys in our research.

4

Click the **Add** button to move the details of the IV into the large box in the middle of the dialogue, which should now look something like this:

5

You should now be able to click the **Define** button and open the main **Repeated Measures** dialogue.

6

In the main **Repeated Measures** dialogue move each variable (i.e., each level of the IV) into the **Within-Subjects Variables (toy)** box using the arrow button.

7

The post hoc testing options available in PASW Advanced Statistics for repeated measures variables are quite limited.

However, you can derive some post hoc information through the **Repeated Measures: Options** dialogue by moving the repeated measures IV (i.e., *toy*) from the **Factor(s) and Factor Interactions** list into **Display Means for** list, selecting **Compare main effects**, and then choosing **Bonferroni** from the **Confidence interval adjustment** drop-down menu.

You may also want to **Display Descriptive statistics** and **Estimates of effect size**.

Click **Continue** to close this dialogue.

8

Click **OK** to output the results of the analyses.

9.3.2.5. PASW Advanced Statistics Output (Part 2: Homogeneity of Variance, Sphericity, & the ANOVA)

General Linear Model

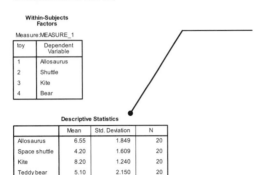

Within-Subjects Factors

Measure:MEASURE_1

toy	Dependent Variable
1	Allosaurus
2	Shuttle
3	Kite
4	Bear

Descriptive Statistics

	Mean	Std. Deviation	N
Allosaurus	6.55	1.849	20
Space shuttle	4.20	1.609	20
Kite	8.20	1.240	20
Teddy bear	5.10	2.150	20

The **Descriptive Statistics** table provides **Means** and **Std.** (standard) **Deviations** for each level of the IV (i.e., each group of scores).

With a **Mean** rating of 8.20 ($SD = 1.24$), the kite appears to be substantially more popular than the other three toys, which averaged ratings of between 4.20 and 6.55.

The standard deviations indicate greater variability in the participants' ratings of the teddy bear than in their ratings of the other three toys.

PASW Advanced Statistics does not provide a test of homogeneity of variance for repeated measures variables. Consequently, we will use the F_{max} test, which can be easily calculated by hand. The formula for F_{max} is:

$$F_{max} = \frac{\text{Largest sample variance}}{\text{Smallest sample variance}}$$

where, the largest sample variance is simply the largest SD squared, and the smallest sample variance is the smallest SD squared.

So,

$$\text{Largest sample variance} = 2.150 * 2.150$$
$$= 4.6225$$

and,

$$\text{Smallest sample variance} = 1.240 * 1.240$$
$$= 1.5376$$

then,

$$F_{max} = \frac{4.6225}{1.5376} = 3.006$$

According to Tabachnick and Fidell (2007a), homogeneity of variance can be assumed when F_{max} is less than 10. Here, F_{max} is 3.006, indicating that <u>the homogeneity of variance assumption has not been violated</u>.

ⓘ Tip:
Some authors suggest a more conservative F_{max} cut-off of 3 when examining repeated measures data for homogeneity of variance.

If the Assumption is Violated

The outcome of a repeated measures ANOVA is not sensitive to small-to-moderate violations of the homogeneity of variance assumption.

The **Multivariate Tests** can be interpreted in much the same way as the univariate tests (overleaf).

For our purposes, these can be ignored.

Multivariate Tests[b]

Effect		Value	F	Hypothesis df	Error df	Sig.	Partial Eta Squared
toy	Pillai's Trace	.877	40.362[a]	3.000	17.000	.000	.877
	Wilks' Lambda	.123	40.362[a]	3.000	17.000	.000	.877
	Hotelling's Trace	7.123	40.362[a]	3.000	17.000	.000	.877
	Roy's Largest Root	7.123	40.362[a]	3.000	17.000	.000	.877

a. Exact statistic

b. Design: Intercept
Within Subjects Design: toy

Mauchly's Test of Sphericity[b]

Measure:MEASURE_1

Within Subjects Effect	Mauchly's W	Approx. Chi-Square	df	Sig.	Epsilon[a]		
					Greenhouse-Geisser	Huynh-Feldt	Lower-bound
toy	.537	11.013	5	.051	.719	.814	.333

Tests the null hypothesis that the error covariance matrix of the orthonormalized transformed dependent variables is proportional to an identity matrix.

a. May be used to adjust the degrees of freedom for the averaged tests of significance. Corrected tests are displayed in the Tests of Within-Subjects Effects table.

b. Design: Intercept
Within Subjects Design: toy

If **Mauchly's Test of Sphericity** is non-significant (i.e., *Sig* > .05) the assumption of sphericity has been met.

In the current example Mauchly's *W* is .537 (*Sig* = .051). As the *Sig* value associated with *W* is (just) above .05, the sphericity assumption is not violated. Consequently, we read the **Sphericity Assumed** rows of the **Tests of Within-Subjects Effects** table.

If the Assumption is Violated

If the sphericity assumption is violated, the degrees of freedom for the repeated measures ANOVA should be adjusted by multiplying them by one of the three **Epsilons** provided in the **Mauchly's Test of Sphericity** table. The observed *F*-ratio should then be evaluated against the critical *F* for the newly adjusted degrees of freedom to determine whether or not the test is statistically significant. Tables of critical *F* values can be found in the appendices of most good statistics texts.

Fortunately, PASW Advanced Statistics does all this hard work for us! So, in practical terms, if the sphericity assumption is violated, read and report the results of one of the **Epsilon** adjusted tests, rather than the **Sphericity Assumed** test.

Tabachnick and Fidell (2007a) note that the **Huynh-Feldt Epsilon** is preferred over the more conservative **Greenhouse-Geisser Epsilon**.

Tests of Within-Subjects Effects

Measure:MEASURE_1

Source		Type III Sum of Squares	df	Mean Square	F	Sig.	Partial Eta Squared
toy	Sphericity Assumed	183.837	3	61.279	22.843	.000	.546
	Greenhouse-Geisser	183.837	2.158	85.206	22.843	.000	.546
	Huynh-Feldt	183.837	2.443	75.241	22.843	.000	.546
	Lower-bound	183.837	1.000	183.837	22.843	.000	.546
Error(toy)	Sphericity Assumed	152.912	57	2.683			
	Greenhouse-Geisser	152.912	40.994	3.730			
	Huynh-Feldt	152.912	46.423	3.294			
	Lower-bound	152.912	19.000	8.048			

The **Tests of Within-Subjects Effects** table reports the outcome of the repeated measures ANOVA. As the sphericity assumption is not violated, we should read and interpret the **Sphericity Assumed** rows of this table.

The *F*-ratio for the ANOVA is 22.843, and the significance is .000 (which is less than our α level of .05). Consequently, we can conclude that some toys are "liked" significantly more than others.

The following details will be needed for the write-up:

F = 22.843
df = (3, 57)
Sig = .000

Partial Eta Squared (η^2) is a measure of effect size. It can be interpreted as the proportion of variance in the children's ratings that can be attributed to the IV (the different toys).

Here, partial η^2 is .546, which is quite large.

Tests of Within-Subjects Contrasts

Measure:MEASURE_1

Source	toy	Type III Sum of Squares	df	Mean Square	F	Sig.	Partial Eta Squared
toy	Linear	.122	1	.122	.034	.855	.002
	Quadratic	2.813	1	2.813	1.126	.302	.056
	Cubic	180.902	1	180.902	92.032	.000	.829
Error(toy)	Linear	68.127	19	3.586			
	Quadratic	47.438	19	2.497			
	Cubic	37.348	19	1.966			

As the levels of the IV in the current study do not form an increasing or decreasing continuum, these tests for significant trends in the data cannot be interpreted.

Tests of Between-Subjects Effects

Measure:MEASURE_1
Transformed Variable:Average

Source	Type III Sum of Squares	df	Mean Square	F	Sig.	Partial Eta Squared
Intercept	2892.013	1	2892.013	702.326	.000	.974
Error	78.238	19	4.118			

The **Tests of Between-Subjects Effects** are a default part of the PASW Advanced Statistics output. They can be used to determine if the overall mean of the data is significantly different from zero. These results are not usually interpreted or reported.

Estimated Marginal Means

toy

Estimates

Measure:MEASURE_1

toy	Mean	Std. Error	95% Confidence Interval	
			Lower Bound	Upper Bound
1	6.550	.413	5.685	7.415
2	4.200	.360	3.447	4.953
3	8.200	.277	7.620	8.780
4	5.100	.481	4.094	6.106

> As the repeated measures ANOVA was statistically significant, it is useful to determine specifically which toys proved most popular with the children.
>
> **Pairwise Comparisons** indicate that the mean rating for toy 1 (the Allosaurus) was significantly higher than the mean rating for toy 2 (the shuttle; $Sig = .007$), but significantly lower than the mean rating for toy 3 (the kite; $Sig = .044$).
>
> Furthermore, the kite was also liked significantly more than both the shuttle ($Sig. = .000$) and the bear ($Sig. = .000$).

Pairwise Comparisons

Measure:MEASURE_1

(I) toy	(J) toy	Mean Difference (I-J)	Std. Error	Sig.ᵃ	95% Confidence Interval for Differenceᵃ	
					Lower Bound	Upper Bound
1	2	2.350*	.617	.007	.534	4.166
	3	-1.650*	.549	.044	-3.266	-.034
	4	1.450	.659	.242	-.490	3.390
2	1	-2.350*	.617	.007	-4.166	-.534
	3	-4.000*	.348	.000	-5.024	-2.976
	4	-.900	.390	.195	-2.048	.248
3	1	1.650*	.549	.044	.034	3.266
	2	4.000*	.348	.000	2.976	5.024
	4	3.100*	.470	.000	1.718	4.482
4	1	-1.450	.659	.242	-3.390	.490
	2	.900	.390	.195	-.248	2.048
	3	-3.100*	.470	.000	-4.482	-1.718

Based on estimated marginal means

*. The mean difference is significant at the .05 level.

a. Adjustment for multiple comparisons: Bonferroni.

> (i) **Tip:**
> You may need to refer back to the **Within-Subjects Factors** table (reproduced below) to determine which toys are being compared in each pairwise comparison.

Within-Subjects Factors

Measure:MEASURE_1

toy	Dependent Variable
1	Allosaurus
2	Shuttle
3	Kite
4	Bear

9.3.3. APA Style Results Write-Up

Results

A one-way repeated measures analysis of variance (ANOVA) was used to compare 20 children's ratings of four different toys.

Boxplots and Shapiro-Wilk statistics indicated that the assumption of normality was supported; F_{max} was 3.006, demonstrating homogeneity of variances; and Mauchly's test indicated that the assumption of sphericity was not violated.

The ANOVA results show that the sample preferred some toys to others, $F(3, 57) = 22.84$, $p < .001$, partial $\eta^2 = .55$. Pairwise comparisons further revealed that the Indoor Light-up Kite ($M = 8.20$, $SD = 1.24$) was "liked" significantly more than the Robotic Allosaurus ($M = 6.55$, $SD = 1.85$), the Interactive Teddy Bear ($M = 5.10$, $SD = 2.15$) and the Remote Controlled Space Shuttle ($M = 4.20$, $SD = 1.61$). Finally, the Allosaurus was rated significantly higher than the Shuttle, but not the Teddy Bear.

> When PASW Advanced Statistics reports Sig as .000, used $p < .001$ in your results.

9.3.4. Summary

In this example, the one-way repeated measures ANOVA was statistically significant and assumptions of normality, homogeneity of variance, and sphericity were not violated. However, when conducting real-life research it is not uncommon to find data that violates one or more of the assumptions, or that does not lead to the significant results one is hoping for. In the second illustrated example, some of these issues will be dealt with.

9.4. Illustrated Example Two

A researcher decided to partially replicate a study by Krawitz (2004), which investigated the effectiveness of a training workshop on the confidence of clinicians working with people diagnosed with borderline personality disorder (BPD).

The researcher invited 40 clinicians to participate in the two-day training workshop, which had been designed to promote knowledge and understanding of BPD. Each clinician was asked to rate their confidence (using a 10-point rating scale) on three occasions: before the workshop; immediately after the workshop; and again four months later.

The data from this study are presented in Table 9.2.

Table 9.2

Confidence Ratings Given by Clinicians (N = 40) On Three Occasions

> 🖵 *Data:*
> This is data file **data_9_2.sav** on the companion website.

ID	Pre-Training	Post-Training	4-Month Follow-Up	ID	Pre-Training	Post-Training	4-Month Follow-Up
1	1	4	5	21	5	6	5
2	3	4	7	22	5	5	5
3	1	2	2	23	5	7	7
4	2	4	3	24	5	6	4
5	7	8	6	25	2	3	8
6	2	2	3	26	4	6	3
7	4	6	6	27	5	7	7
8	2	3	8	28	5	8	3
9	6	9	8	29	5	8	4
10	3	9	9	30	3	3	6
11	6	6	7	31	6	7	6
12	3	8	5	32	6	7	5
13	3	7	4	33	6	7	7
14	3	4	5	34	6	6	6
15	1	5	9	35	2	3	7
16	4	5	5	36	2	4	4
17	5	5	6	37	7	8	7
18	8	9	4	38	7	9	7
19	2	2	4	39	7	10	3
20	4	9	2	40	4	5	4

9.4.1. PASW Statistics Output (Part 1: Normality)

Syntax:
Run these analyses
with **syntax_9_2.sps**
on the companion
website.

Explore

Case Processing Summary

	Cases					
	Valid		Missing		Total	
	N	Percent	N	Percent	N	Percent
Pre-Training Rating of Confidence	40	100.0%	0	.0%	40	100.0%
Post-Training Rating of Confidence	40	100.0%	0	.0%	40	100.0%
Follow-up Rating of Confidence	40	100.0%	0	.0%	40	100.0%

> The procedure for generating this output is the same as that used in *Illustrated Example One*.

Descriptives

			Statistic	Std. Error
Pre-Training Rating of Confidence	Mean		4.18	.304
	95% Confidence Interval for Mean	Lower Bound	3.56	
		Upper Bound	4.79	
	5% Trimmed Mean		4.17	
	Median		4.00	
	Variance		3.687	
	Std. Deviation		1.920	
	Minimum		1	
	Maximum		8	
	Range		7	
	Interquartile Range		4	
	Skewness		.035	.374
	Kurtosis		-1.018	.733
Post-Training Rating of Confidence	Mean		5.90	.354
	95% Confidence Interval for Mean	Lower Bound	5.18	
		Upper Bound	6.62	
	5% Trimmed Mean		5.92	
	Median		6.00	
	Variance		5.015	
	Std. Deviation		2.240	
	Minimum		2	
	Maximum		10	
	Range		8	
	Interquartile Range		4	
	Skewness		-.099	.374
	Kurtosis		-.991	.733
Follow-up Rating of Confidence	Mean		5.40	.295
	95% Confidence Interval for Mean	Lower Bound	4.80	
		Upper Bound	6.00	
	5% Trimmed Mean		5.39	
	Median		5.00	
	Variance		3.477	
	Std. Deviation		1.865	
	Minimum		2	
	Maximum		9	
	Range		7	
	Interquartile Range		3	
	Skewness		.053	.374
	Kurtosis		-.805	.733

Tests of Normality

	Kolmogorov-Smirnov[a]			Shapiro-Wilk		
	Statistic	df	Sig.	Statistic	df	Sig.
Pre-Training Rating of Confidence	.141	40	.043	.948	40	.063
Post-Training Rating of Confidence	.113	40	.200*	.954	40	.105
Follow-up Rating of Confidence	.130	40	.089	.958	40	.146

a. Lilliefors Significance Correction

*. This is a lower bound of the true significance.

> All three **Shapiro-Wilk** statistics are non-significant (as *Sig* > .05). Therefore, <u>the assumption of normality is not violated.</u>

9.4.2. PASW Advanced Statistics Output (Part 2: Homogeneity of Variance, Sphericity, & the ANOVA)

General Linear Model

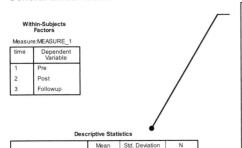

Within-Subjects Factors

Measure:MEASURE_1

time	Dependent Variable
1	Pre
2	Post
3	Followup

Descriptive Statistics

	Mean	Std. Deviation	N
Pre-Training Rating of Confidence	4.18	1.920	40
Post-Training Rating of Confidence	5.90	2.240	40
Follow-up Rating of Confidence	5.40	1.865	40

> With the **Std.** (standard) **Deviations** reported in the **Descriptive Statistics** table we can calculate F_{max}, which is used to assess the homogeneity of variance assumption.
>
> $$F_{max} = \frac{\text{Largest sample variance}}{\text{Smallest sample variance}}$$
>
> So,
>
> $$F_{max} = \frac{2.240^2}{1.865^2} = \frac{5.0176}{3.4782} = 1.4425$$
>
> According to Tabachnick and Fidell (2007a), homogeneity of variance can be assumed when F_{max} is less than 10. Here, F_{max} is 1.4425, indicating that <u>the homogeneity of variance assumption has not been violated</u>.

Multivariate Tests[b]

Effect		Value	F	Hypothesis df	Error df	Sig.	Partial Eta Squared
time	Pillai's Trace	.594	27.795[a]	2.000	38.000	.000	.594
	Wilks' Lambda	.406	27.795[a]	2.000	38.000	.000	.594
	Hotelling's Trace	1.463	27.795[a]	2.000	38.000	.000	.594
	Roy's Largest Root	1.463	27.795[a]	2.000	38.000	.000	.594

a. Exact statistic

b. Design: Intercept
 Within Subjects Design: time

> We're going to ignore these **Multivariate Tests** in favour of the univariate tests reported below.

Mauchly's Test of Sphericity[b]

Measure:MEASURE_1

Within Subjects Effect	Mauchly's W	Approx. Chi-Square	df	Sig.	Epsilon[a] Greenhouse-Geisser	Huynh-Feldt	Lower-bound
time	.621	18.085	2	.000	.725	.746	.500

Tests the null hypothesis that the error covariance matrix of the orthonormalized transformed dependent variables is proportional to an identity matrix.

a. May be used to adjust the degrees of freedom for the averaged tests of significance. Corrected tests are displayed in the Tests of Within-Subjects Effects table.

b. Design: Intercept
 Within Subjects Design: time

> **Mauchly's Test of Sphericity** is statistically significant (W = .621, Sig = .000), indicating that <u>the sphericity assumption has been violated</u>.
>
> Consequently, we will read and interpret the **Huynh-Feldt** rows of the **Tests of Within-Subjects Effects** below.

Tests of Within-Subjects Effects

Measure:MEASURE_1

Source		Type III Sum of Squares	df	Mean Square	F	Sig.	Partial Eta Squared
time	Sphericity Assumed	63.017	2	31.508	11.121	.000	.222
	Greenhouse-Geisser	63.017	1.451	43.440	11.121	.000	.222
	Huynh-Feldt	63.017	1.492	42.234	11.121	.000	.222
	Lower-bound	63.017	1.000	63.017	11.121	.002	.222
Error(time)	Sphericity Assumed	220.983	78	2.833			
	Greenhouse-Geisser	220.983	56.576	3.906			
	Huynh-Feldt	220.983	58.191	3.798			
	Lower-bound	220.983	39.000	5.666			

> Tabachnick and Fidell (2007a) advise against using partial η^2 when sphericity cannot be assumed.
>
> They recommend calculating a lower-bound estimate of η^2 instead, which we will do later.

> The **Tests of Within-Subjects Effects** table reports the outcome of the repeated measures ANOVA. As the sphericity assumption is violated, the **Huynh-Feldt** rows should be read and interpreted, rather than the **Sphericity Assumed** rows.
>
> From the ANOVA we can conclude that <u>there is a significant difference between at least two of the mean confidence ratings</u>. **Pairwise Comparisons** (overleaf) will be needed to locate the source(s) of this significance.
>
> For your write-up, take note of:
>
> F = 11.121
> df = (1.492, 58.191)
> Sig = .000

Tests of Within-Subjects Contrasts

Measure:MEASURE_1

Source	time	Type III Sum of Squares	df	Mean Square	F	Sig.	Partial Eta Squared
time	Linear	30.012	1	30.012	8.902	.005	.186
	Quadratic	33.004	1	33.004	14.382	.001	.269
Error(time)	Linear	131.488	39	3.371			
	Quadratic	89.496	39	2.295			

As the levels of the IV in the current study do not form an increasing or decreasing continuum, these tests for significant trends in the data cannot be interpreted.

Tests of Between-Subjects Effects

Measure:MEASURE_1
Transformed Variable:Average

Source	Type III Sum of Squares	df	Mean Square	F	Sig.	Partial Eta Squared
Intercept	3193.008	1	3193.008	490.281	.000	.926
Error	253.992	39	6.513			

In most circumstances, the **Tests of Between-Subjects Effects** can be ignored.

However, when the sphericity assumption is violated the **Error Type III Sum of Squares** is required for the calculation of a lower-bound estimate of η^2.

Take note of this figure (253.992), as we will use it later.

Estimated Marginal Means

time

Estimates

Measure:MEASURE_1

time	Mean	Std. Error	95% Confidence Interval	
			Lower Bound	Upper Bound
1	4.175	.304	3.561	4.789
2	5.900	.354	5.184	6.616
3	5.400	.295	4.804	5.996

Pairwise Comparisons

Measure:MEASURE_1

(I) time	(J) time	Mean Difference (I-J)	Std. Error	Sig.a	95% Confidence Interval for Difference a	
					Lower Bound	Upper Bound
1	2	-1.725*	.237	.000	-2.319	-1.131
	3	-1.225*	.411	.015	-2.252	-.198
2	1	1.725*	.237	.000	1.131	2.319
	3	.500	.447	.811	-.619	1.619
3	1	1.225*	.411	.015	.198	2.252
	2	-.500	.447	.811	-1.619	.619

Based on estimated marginal means

*. The mean difference is significant at the .05 level.

a. Adjustment for multiple comparisons: Bonferroni.

The **Pairwise Comparisons** table indicates that there is a significant difference between the mean confidence of the clinicians at time 1 (before the training) and time 2 (after the training), and also between time 1 and time 3 (the 4-month follow up).

The difference between the confidence ratings at time 2 and time 3 is not significant.

9.4.3. Follow Up Analyses

9.4.3.1. Effect Size

Tabachnick and Fidell (2007a) advise against using partial η^2 (as reported in the **Tests of Within-Subjects Effects** table) as a measure of effect size when the sphericity assumption has been violated. Instead, they recommend calculating a lower-bound estimate of η^2, using the following formula:

$$\eta_L^2 = \frac{SS_A}{SS_A + SS_S + SS_{AS}}$$

where SS_A is the sum of squares attributable to the IV (available in the **Tests of Within-Subjects Effects** table), SS_s is the sum of squares attributable to the cases (available in the **Tests of Between-Subjects Effects** table), and SS_{AS} is the sum of squares attributable to error (available in the **Tests of Within-Subjects Effects** table). So,

$$\eta_L^2 = \frac{63.017}{63.017 + 253.992 + 220.983}$$

$$= .117$$

9.4.4. APA Style Results Write-Up

Results

A one-way repeated measures ANOVA was used to partially replicate a study by Krawitz (2004), which investigated the impact of a training workshop on the confidence of clinicians working with people diagnosed with borderline personality disorder (BPD). In the current study, 40 clinicians were asked to rate their confidence using a 10-point rating scale. The rating scale was administered pre- and post-training, and again during a four month follow-up.

The Shapiro-Wilk and F_{max} statistics were used to test the assumptions of normality and homogeneity of variance respectively. Neither was violated. However, Mauchly's test indicated that the sphericity assumption was violated. Consequently, the Huynh-Feldt correction was employed.

> As the sphericity assumption was violated, **Huynh-Feldt** adjusted degrees of freedom have been reported, and used to evaluate the statistical significance of F.

The repeated measures ANOVA indicated that the clinicians' confidence ratings did change significantly over time, $F (1.49, 58.19) = 11.12, p < .001, \eta^2_L = .12$.

A series of pairwise comparisons revealed that the average pre-training confidence level ($M = 4.18, SD = 1.92$) was significantly lower than the average post-training ($M = 5.90, SD = 2.24$), and average four-month follow-up ($M = 5.40, SD = 1.86$) confidence levels. There was no difference between the post-training and four-month follow-up confidence levels.

> Only report the results of pairwise comparisons if the omnibus ANOVA is statistically significant.
>
> If the ANOVA is not significant, pairwise comparisons are redundant.

9.5. One-Way Repeated Measures ANOVA Checklist

Have you:

- ✔ Checked that each group of data is approximately normally distributed?
- ✔ Checked for homogeneity of variance?
- ✔ Checked Mauchly's test of sphericity?
- ✔ Taken note of the F-ratio, (appropriate) degrees of freedom and (appropriate) significance level for your write-up?
- ✔ Taken note of the partial η^2 (or calculated a lower-bound η^2 if the sphericity assumption is violated)?
- ✔ Taken note of the results of each pairwise comparison (if the ANOVA was significant)?
- ✔ Written up your results in the APA style?

Chapter 10: One-Way Analysis of Covariance (ANCOVA)

Chapter Overview

10.1. Purpose of the One-Way ANCOVA

To test for a statistically significant difference between two or more independent samples (or levels of an independent variable) after statistically controlling for the effects of a "third variable", referred to as a covariate.

10.2. Questions We Could Answer Using the One-Way ANCOVA

1. After statistically controlling for the amount of time they spend practicing their musical instruments each week, is there a difference in overall musical ability between children taught with three different methods of music instruction?

The amount of time that children spend practicing musical instruments can vary from just minutes per week through to literally hours per day. Any researcher trying to test the hypothesis that children taught with instruction method A have greater musical ability than children taught with methods B and C may well find that the variability in musical ability within these three groups is so large (mainly due to individual differences in practice) that a simple between-groups ANOVA fails to reach statistical significance.

By using practice time as a covariate, we can partial out (or remove) its effects on the DV (musical ability), leaving us with a far more powerful test of the original research hypothesis.

A similar question that we could address with a one-way ANCOVA is:

2. After statistically controlling for individual differences in marriage satisfaction, are there differences between the scores of young,

middle-aged and elderly adult groups on a scale measuring healthy lifestyle behaviours?

10.3. Illustrated Example One

To investigate the hypothesis that students in some university courses experience significantly higher levels of stress than others, students studying medicine, dentistry and podiatry were asked to complete a stress measure at the beginning of their respective degrees, and again 12 months later.

Here, course of study is the IV and stress at the end of first year is the DV. To control for (or partial out) any pre-existing individual differences in stress, the stress data that was collected when the students started their degrees will be used as a covariate.

The data from this study are presented in Table 10.1.

Table 10.1

□ *Data:*
This is data file
data_10_1.sav
on the companion
website.

Student Stress Scores at the Beginning and End of First Year (N = 30)

Medical students		Dentistry students		Podiatry students	
Start	End	Start	End	Start	End
14	29	8	14	21	22
17	19	12	16	23	29
17	27	12	14	16	14
20	22	15	19	11	17
12	19	12	14	3	16
14	26	11	16	23	15
14	20	10	17	12	18
15	23	9	13	20	19
17	22	15	15	17	21
12	18	12	18	23	27

Note. Start = Stress score at the beginning of first year; End = Stress score at the end of first year.

10.3.1. Setting Up the PASW Statistics Data File

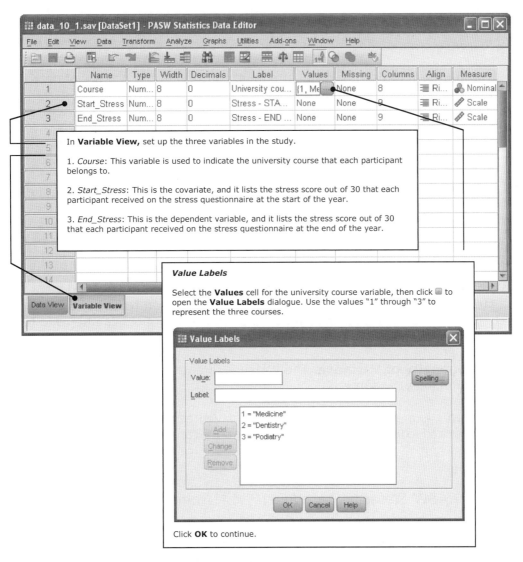

In **Variable View,** set up the three variables in the study.

1. *Course*: This variable is used to indicate the university course that each participant belongs to.

2. *Start_Stress*: This is the covariate, and it lists the stress score out of 30 that each participant received on the stress questionnaire at the start of the year.

3. *End_Stress*: This is the dependent variable, and it lists the stress score out of 30 that each participant received on the stress questionnaire at the end of the year.

Value Labels

Select the **Values** cell for the university course variable, then click ▣ to open the **Value Labels** dialogue. Use the values "1" through "3" to represent the three courses.

1 = "Medicine"
2 = "Dentistry"
3 = "Podiatry"

Click **OK** to continue.

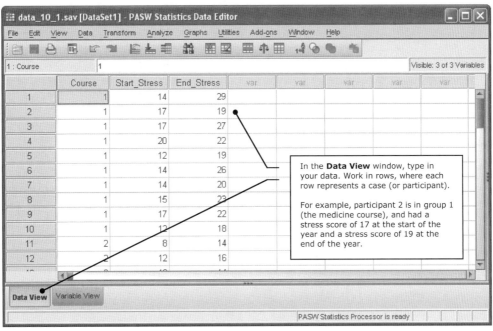

In the **Data View** window, type in your data. Work in rows, where each row represents a case (or participant).

For example, participant 2 is in group 1 (the medicine course), and had a stress score of 17 at the start of the year and a stress score of 19 at the end of the year.

■ *Syntax:*
Run these
analyses with
syntax_10_1.sps
on the companion
website.

10.3.2. Analysing the Data

10.3.2.1. Assumptions

The following criteria should be met before conducting a one-way ANCOVA.

1. **Independence.** Each participant should contribute just one set of data to the research, and should not influence the participation of others.

2. **Normality.** Each group of scores should be approximately normally distributed. ANCOVA is quite robust over moderate violations of this assumption.

3. **Homogeneity of Regression Slopes**. The slopes of the regression lines should be the same for each group formed by the independent variable and measured on the dependent variable. The more this assumption is violated, the more likely it is that incorrect statistical conclusions will be drawn.

4. **Linearity.** The relationship between the covariate and the dependent variable should be linear. Violations of this assumption reduce the power of the ANCOVA to find significant differences. Scatterplots can be used to evaluate linearity.

5. **Homogeneity of Variance.** The variance of data in the groups should be approximately the same.

Assumption 1 is methodological and should have been addressed before and during data collection. Assumptions 2, 3, 4 and 5 can be tested with PASW Statistics.

10.3.2.2. PASW Statistics Procedure (Part 1: Normality)

The assumption of normality can be assessed in numerous ways. In this example, we will use the Shapiro-Wilk test and a set of histograms.

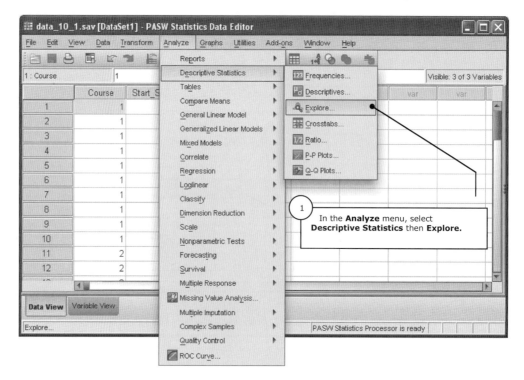

In the **Analyze** menu, select **Descriptive Statistics** then **Explore**.

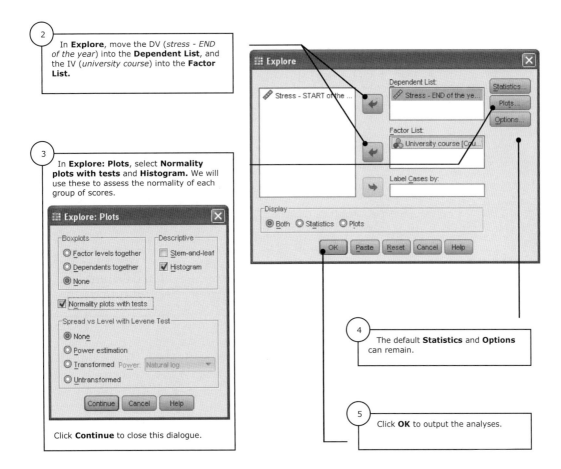

2 In **Explore**, move the DV (*stress - END of the year*) into the **Dependent List**, and the IV (*university course*) into the **Factor List.**

3 In **Explore: Plots**, select **Normality plots with tests** and **Histogram.** We will use these to assess the normality of each group of scores.

Click **Continue** to close this dialogue.

4 The default **Statistics** and **Options** can remain.

5 Click **OK** to output the analyses.

10.3.2.3. PASW Statistics Output (Part 1: Normality)

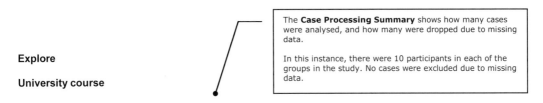

Explore

University course

The **Case Processing Summary** shows how many cases were analysed, and how many were dropped due to missing data.

In this instance, there were 10 participants in each of the groups in the study. No cases were excluded due to missing data.

Case Processing Summary

University course		Cases					
		Valid		Missing		Total	
		N	Percent	N	Percent	N	Percent
Stress - END of the year	Medicine	10	100.0%	0	.0%	10	100.0%
	Dentistry	10	100.0%	0	.0%	10	100.0%
	Podiatry	10	100.0%	0	.0%	10	100.0%

Descriptives

University course			Statistic	Std. Error
Stress - END of the year	Medicine	Mean	22.50	1.186
		95% Confidence Interval for Mean Lower Bound	19.82	
		Upper Bound	25.18	
		5% Trimmed Mean	22.39	
		Median	22.00	
		Variance	14.056	
		Std. Deviation	3.749	
		Minimum	18	
		Maximum	29	
		Range	11	
		Interquartile Range	7	
		Skewness	.569	.687
		Kurtosis	-.920	1.334
	Dentistry	Mean	15.60	.618
		95% Confidence Interval for Mean Lower Bound	14.20	
		Upper Bound	17.00	
		5% Trimmed Mean	15.56	
		Median	15.50	
		Variance	3.822	
		Std. Deviation	1.955	
		Minimum	13	
		Maximum	19	
		Range	6	
		Interquartile Range	3	
		Skewness	.482	.687
		Kurtosis	-.811	1.334
	Podiatry	Mean	19.80	1.583
		95% Confidence Interval for Mean Lower Bound	16.22	
		Upper Bound	23.38	
		5% Trimmed Mean	19.61	
		Median	18.50	
		Variance	25.067	
		Std. Deviation	5.007	
		Minimum	14	
		Maximum	29	
		Range	15	
		Interquartile Range	8	
		Skewness	.858	.687
		Kurtosis	-.234	1.334

> The table of **Descriptives** contains a range of useful information, including measures of central tendency and dispersion for each group of scores, along with skewness and kurtosis statistics.

> Both z_s and z_k are within within ±1.96 for all three groups of stress scores (see section 4.3.2.3), indicating that <u>normality can be assumed</u>.

Tests of Normality

University course	Kolmogorov-Smirnov[a]			Shapiro-Wilk		
	Statistic	df	Sig.	Statistic	df	Sig.
Stress - END of the year Medicine	.153	10	.200[*]	.924	10	.392
Dentistry	.193	10	.200[*]	.940	10	.555
Podiatry	.163	10	.200[*]	.915	10	.320

a. Lilliefors Significance Correction

*. This is a lower bound of the true significance.

> Due to the small sample size (N=30) the **Shapiro-Wilk** test is considered more appropriate than the **Kolmogorov-Smirnov** test.
>
> If the Shapiro-Wilk statistic (W) is significant (i.e., Sig < .05), the group of scores is not normally distributed.
>
> Here, W ranges from .915 to .940, and are all non-significant (i.e. Sig > .05). Thus, we can conclude <u>the assumption of normality is not violated.</u>

> ***If the Assumption is Violated***
>
> The ANCOVA is considered robust against small to moderate violations of the normality assumption, provided the scores for the covariate alone are normally distributed.

▶▶ *Links:*
The full PASW Statistics output also included three **Normal Q-Q Plots** and three **Detrended Normal Q-Q Plots**.

These have been described elsewhere in this text (for example, chapter 4).

Stress – END of the year

Histograms

> A visual inspection of the **Histograms** further confirms that <u>each group of scores is approximately normally distributed</u>.

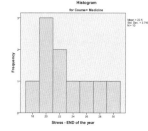

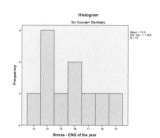

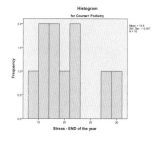

10.3.2.4. PASW Statistics Procedure (Part 2: Homogeneity of Regression Slopes)

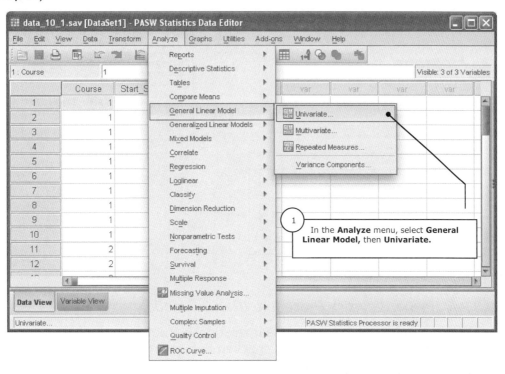

1 In the **Analyze** menu, select **General Linear Model,** then **Univariate.**

2 Move the DV (*stress - END of the year*) into the **Dependent Variable** box, and the IV (*university course*) into the **Fixed Factor(s)** list.

3 Move the covariate (*stress - START of the year*) into the **Covariates(s)** list.

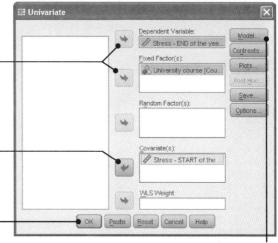

4 In the **Univariate: Model** dialogue select **Custom**, then:

1. Use the arrow button to move the IV (*course*) from the **Factors & Covariates** list into the **Model** list.

2. Repeat this process for the covariate (stress at the start of the year, or *Start_Stress*).

2. Highlight both the IV and the covariate in the **Factors & Covariates** list (by clicking one, and then the other) then click the arrow button to add an *IV * covariate* **Interaction** term to the **Model** list.

Click **Continue** to close this dialogue.

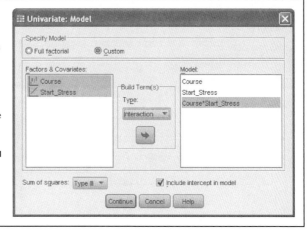

5 Click **OK** to output the analyses.

10.3.2.5. PASW Statistics Output (Part 2: Homogeneity of Regression Slopes)

Univarate Analysis of Variance

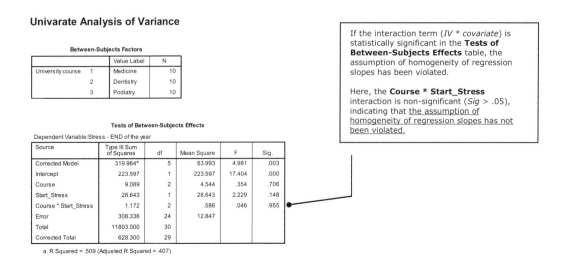

Between-Subjects Factors

		Value Label	N
University course	1	Medicine	10
	2	Dentistry	10
	3	Podiatry	10

Tests of Between-Subjects Effects

Dependent Variable:Stress - END of the year

Source	Type III Sum of Squares	df	Mean Square	F	Sig.
Corrected Model	319.964ᵃ	5	63.993	4.981	.003
Intercept	223.597	1	223.597	17.404	.000
Course	9.089	2	4.544	.354	.706
Start_Stress	28.643	1	28.643	2.229	.148
Course * Start_Stress	1.172	2	.586	.046	.955
Error	308.336	24	12.847		
Total	11803.000	30			
Corrected Total	628.300	29			

a. R Squared = .509 (Adjusted R Squared = .407)

> If the interaction term (*IV * covariate*) is statistically significant in the **Tests of Between-Subjects Effects** table, the assumption of homogeneity of regression slopes has been violated.
>
> Here, the **Course * Start_Stress** interaction is non-significant (*Sig* > .05), indicating that the assumption of homogeneity of regression slopes has not been violated.

10.3.2.6. PASW Statistics Procedure (Part 3: Linearity)

ANCOVA assumes that there is a linear relationship between the DV and the covariate. This assumption must be checked before proceeding with the ANCOVA.

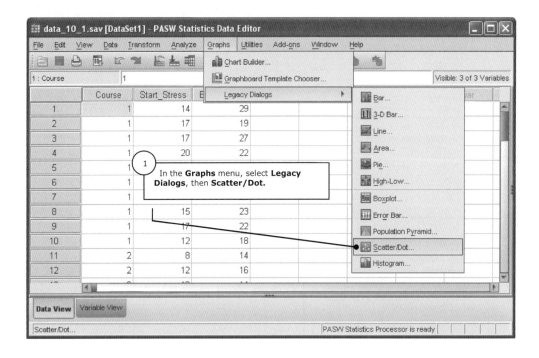

1. In the **Graphs** menu, select **Legacy Dialogs**, then **Scatter/Dot.**

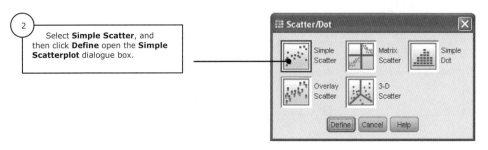

2. Select **Simple Scatter**, and then click **Define** open the **Simple Scatterplot** dialogue box.

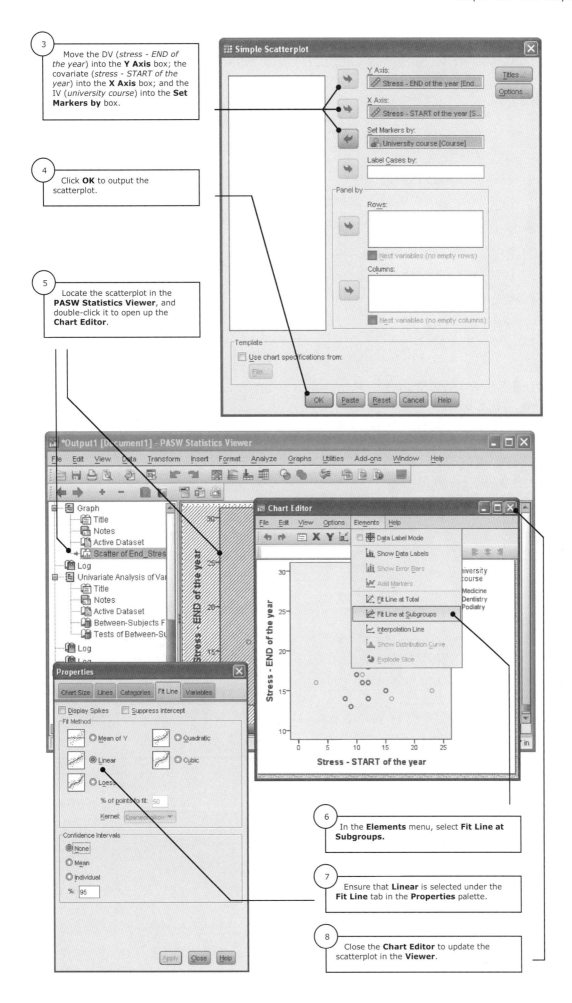

3 Move the DV (*stress - END of the year*) into the **Y Axis** box; the covariate (*stress - START of the year*) into the **X Axis** box; and the IV (*university course*) into the **Set Markers by** box.

4 Click **OK** to output the scatterplot.

5 Locate the scatterplot in the **PASW Statistics Viewer**, and double-click it to open up the **Chart Editor**.

6 In the **Elements** menu, select **Fit Line at Subgroups.**

7 Ensure that **Linear** is selected under the **Fit Line** tab in the **Properties** palette.

8 Close the **Chart Editor** to update the scatterplot in the **Viewer**.

10.3.2.7. PASW Statistics Output (Part 3: Linearity)

Graph

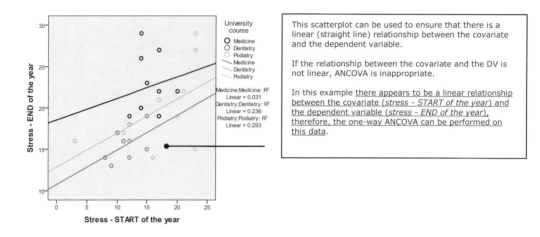

(i) Tip:
Your graph will be in colour, making it much easier to distinguish between the different groups of scores.

This scatterplot can be used to ensure that there is a linear (straight line) relationship between the covariate and the dependent variable.

If the relationship between the covariate and the DV is not linear, ANCOVA is inappropriate.

In this example there appears to be a linear relationship between the covariate (stress - START of the year) and the dependent variable (stress - END of the year), therefore, the one-way ANCOVA can be performed on this data.

10.3.2.8. PASW Statistics Procedure (Part 4: Homogeneity of Variance & the ANCOVA)

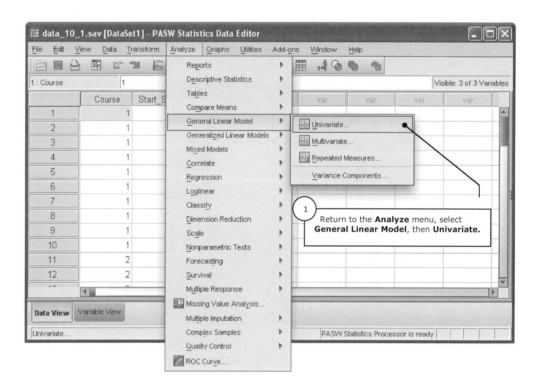

1 Return to the **Analyze** menu, select **General Linear Model**, then **Univariate.**

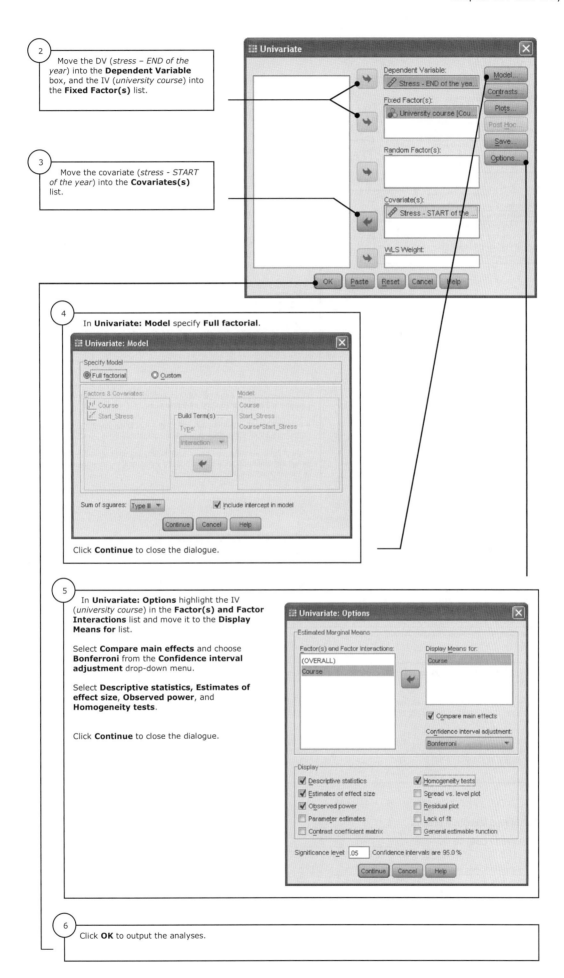

2 Move the DV (*stress – END of the year*) into the **Dependent Variable** box, and the IV (*university course*) into the **Fixed Factor(s)** list.

3 Move the covariate (*stress - START of the year*) into the **Covariates(s)** list.

4 In **Univariate: Model** specify **Full factorial**.

Click **Continue** to close the dialogue.

5 In **Univariate: Options** highlight the IV (*university course*) in the **Factor(s) and Factor Interactions** list and move it to the **Display Means for** list.

Select **Compare main effects** and choose **Bonferroni** from the **Confidence interval adjustment** drop-down menu.

Select **Descriptive statistics, Estimates of effect size, Observed power**, and **Homogeneity tests**.

Click **Continue** to close the dialogue.

6 Click **OK** to output the analyses.

10.3.2.9. PASW Statistics Output (Part 4: Homogeneity of Variance & the ANCOVA)

Univariate Analysis of Variance

Between-Subjects Factors

		Value Label	N
University course	1	Medicine	10
	2	Dentistry	10
	3	Podiatry	10

The **Between-Subjects Factors** table provides information about the number of participants in each of the groups.

In this example there were 10 students in each university course.

Descriptive Statistics

Dependent Variable:Stress - END of the year

University course	Mean	Std. Deviation	N
Medicine	22.50	3.749	10
Dentistry	15.60	1.955	10
Podiatry	19.80	5.007	10
Total	19.30	4.655	30

The **Descriptive Statistics** table provides basic descriptive information for each group of scores.

Here, participants in the medicine course (n = 10) had a mean stress at the end of the year score of 22.50.

Participants in the dentistry course had a mean stress at the end of the year score of 15.60, and the participants in the podiatry course had a mean stress at the end of the year score of 19.80.

It is clear that there is a difference between these means, with the medicine group having higher stress at the end of the year. The results from the ANCOVA will allow us to determine whether this difference is statistically significant.

The standard deviations indicate a larger spread of scores in the podiatry course.

Levene's Test of Equality of Error Variances[a]

Dependent Variable:Stress - END of the year

F	df1	df2	Sig.
2.246	2	27	.125

Tests the null hypothesis that the error variance of the dependent variable is equal across groups.

a. Design: Intercept + Start_Stress + Course

The assumption of homogeneity of variance can be assessed using **Levene's Test of Equality of Variances.**

As Levene's test is non-significant (Sig > .05), the assumption of homogeneity of variance has not been violated here.

Even if Levene's test had been statistically significant, the F ratio is generally quite robust with respect to violations of the homogeneity of variance assumption (Lindman, 1974).

Tests of Between-Subjects Effects

Dependent Variable:Stress - END of the year

Source	Type III Sum of Squares	df	Mean Square	F	Sig.	Partial Eta Squared	Noncent. Parameter	Observed Power[b]
Corrected Model	318.791[a]	3	106.264	8.927	.000	.507	26.780	.988
Intercept	397.605	1	397.605	33.400	.000	.562	33.400	1.000
Start_Stress	76.991	1	76.991	6.468	.017	.199	6.468	.687
Course	141.823	2	70.912	5.957	.007	.314	11.914	.838
Error	309.509	26	11.904					
Total	11803.000	30						
Corrected Total	628.300	29						

a. R Squared = .507 (Adjusted R Squared = .451)

b. Computed using alpha = .05

The **Tests of Between-Subjects Effects** table contains the main ANCOVA output.

The *Start_Stress* row indicates that the covariate (*stress - START of the year*) is significantly related to the DV (*stress – END of the year*), $F(1, 26) = 6.47$, $p = .017$.

The *Course* row indicates that, after controlling for stress at the start of the year, stress at the end of the year is significantly related to course of study, $F(2, 26) = 5.96$, $p = .007$.

Partial Eta Squared (η^2) is an index of effect size, and should be included in your results section.

However, see chapter 8 (section 8.4.2.2) for a discussion of some bias associated with partial eta-squared and a description of omega-squared, which may be used as an alternative measure of effect size.

Estimated Marginal Means

University course

Estimates

Dependent Variable:Stress - END of the year

University course	Mean	Std. Error	95% Confidence Interval	
			Lower Bound	Upper Bound
Medicine	22.250[a]	1.095	19.998	24.502
Dentistry	16.770[a]	1.184	14.336	19.204
Podiatry	18.880[a]	1.150	16.517	21.243

a. Covariates appearing in the model are evaluated at the following values: Stress - START of the year = 14.57.

The **Estimates** table reports group means adjusted for the effects of the covariate.

In other words, if everyone in the sample had the same level of stress at the beginning of the year, the end-of-year group means would be:

Medicine: 22.250
Dentistry: 16.770
Podiatry: 18.880

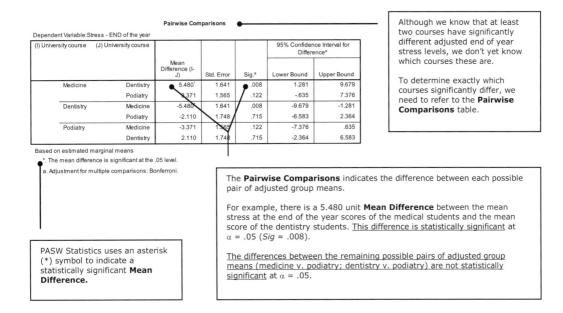

Pairwise Comparisons

Dependent Variable:Stress - END of the year

(I) University course	(J) University course	Mean Difference (I-J)	Std. Error	Sig.ᵃ	95% Confidence Interval for Differenceᵃ	
					Lower Bound	Upper Bound
Medicine	Dentistry	5.480*	1.641	.008	1.281	9.679
	Podiatry	3.371	1.565	.122	-.635	7.376
Dentistry	Medicine	-5.480	1.641	.008	-9.679	-1.281
	Podiatry	-2.110	1.748	.715	-6.583	2.364
Podiatry	Medicine	-3.371	1.565	.122	-7.376	.635
	Dentistry	2.110	1.748	.715	-2.364	6.583

Based on estimated marginal means

*. The mean difference is significant at the .05 level.

a. Adjustment for multiple comparisons: Bonferroni.

Although we know that at least two courses have significantly different adjusted end of year stress levels, we don't yet know which courses these are.

To determine exactly which courses significantly differ, we need to refer to the **Pairwise Comparisons** table.

The **Pairwise Comparisons** indicates the difference between each possible pair of adjusted group means.

For example, there is a 5.480 unit **Mean Difference** between the mean stress at the end of the year scores of the medical students and the mean score of the dentistry students. This difference is statistically significant at α = .05 (*Sig* = .008).

The differences between the remaining possible pairs of adjusted group means (medicine v. podiatry; dentistry v. podiatry) are not statistically significant at α = .05.

PASW Statistics uses an asterisk (*) symbol to indicate a statistically significant **Mean Difference.**

10.3.3. APA Style Results Write-Up

Results

A one-way analysis of covariance (ANCOVA) was used to compare the end-of-year stress levels of students undertaking three different university courses (medicine, dentistry and podiatry). A covariate was included to partial out the effects of participants' stress levels at the beginning of the year from the analysis.

Examination of the Shapiro-Wilk statistics and histograms for each group indicated that the ANCOVA assumption of normality was supported. Scatterplots indicated that the relationship between the covariate (stress at the start of the year) and the dependent variable (stress at the end of the year) was linear. Finally, the assumptions of homogeneity of regression slopes and homogeneity of variances were supported by the absence of a significant IV-by-covariate interaction, $F (2, 24) = 0.05$, $p = .955$, and a non-significant Levene's test, $F (2, 27) = 2.25$, $p = .125$, respectively.

The ANCOVA indicated that, that after accounting for the effects of stress at the beginning of the year, there was a statistically

Assumption testing is often not reported in journal articles (particularly when the assumptions are not violated). However, it is more commonly expected in student reports.

significant effect of university course on stress at the end of the year,

F (2,26) = 5.96, p = .007, partial η^2 = .314. Post-hoc testing revealed

Omega-squared may be included here, rather than partial eta-squared.

that the participants in the medicine course reported higher end-of-

year stress levels than the dentistry students, even after controlling for

start-of-year stress. The remaining pairwise comparisons were not

significant.

10.3.4. Summary

In this example, the one-way ANCOVA was statistically significant, and its assumptions were not violated. In the second illustrated example, reporting non-significant findings will be covered.

10.4. Illustrated Example Two

To investigate the hypothesis that heavy substance users have lower self-esteem than occasional substance users, a researcher recruited 40 adults on the basis of their self-reported substance use levels (either "high" or "low"), and asked each to complete a self-esteem questionnaire.

As the participants' ages varied quite substantially (from 19 through to 45 years of age), and the researcher was concerned that this variability could influence the outcome of her study, she decided to use age as a covariate variable in a one-way analysis of covariance (ANCOVA). ANCOVA will allow her to examine the effects of substance use on self-esteem while partialling out (or removing) the effect of age.

The data from this study are presented in Table 10.2.

Table 10.2

Self-Esteem Scores and Ages of Participants Self-Identifying as Either "Low" or "High" Substance Users (N=40)

📁 **Data:**
This is data file
data_10_2.sav
on the companion
website.

Low Substance Use			High Substance Use		
ID No.	Self Esteem	Age	ID No.	Self Esteem	Age
1	17	21	21	13	22
2	19	32	22	15	26
3	24	35	23	17	33
4	23	23	24	19	25
5	28	28	25	14	19
6	19	22	26	16	23
7	21	33	27	17	24
8	23	32	28	13	26
9	21	26	29	23	33
10	25	27	30	23	21
11	15	22	31	15	24
12	17	21	32	26	43
13	12	24	33	19	33
14	13	38	34	22	36
15	21	36	35	22	34
16	23	34	36	23	23
17	21	38	37	22	27
18	24	33	38	26	34
19	24	45	39	22	34
20	24	43	40	10	45

10.4.1. PASW Statistics Output (Part 1: Normality)

💻 *Syntax:*
Run these
analyses with
syntax_10_2.sps
on the companion
website.

Explore

Substance Use Level ●───────────────────────●

> The procedures for generating this output are the same as those used in *Illustrated Example One*.

Case Processing Summary

Substance Use Level		Cases					
		Valid		Missing		Total	
		N	Percent	N	Percent	N	Percent
Self-Esteem Score	Low substance use	20	100.0%	0	.0%	20	100.0%
	High substance use	20	100.0%	0	.0%	20	100.0%

Descriptives

Substance Use Level				Statistic	Std. Error
Self-Esteem Score	Low substance use	Mean		20.70	.935
		95% Confidence Interval for Mean	Lower Bound	18.74	
			Upper Bound	22.66	
		5% Trimmed Mean		20.78	
		Median		21.00	
		Variance		17.484	
		Std. Deviation		4.181	
		Minimum		12	
		Maximum		28	
		Range		16	
		Interquartile Range		7	
		Skewness		-.610	.512
		Kurtosis		-.129	.992
	High substance use	Mean		18.85	1.037
		95% Confidence Interval for Mean	Lower Bound	16.68	
			Upper Bound	21.02	
		5% Trimmed Mean		18.94	
		Median		19.00	
		Variance		21.503	
		Std. Deviation		4.637	
		Minimum		10	
		Maximum		26	
		Range		16	
		Interquartile Range		8	
		Skewness		-.168	.512
		Kurtosis		-1.029	.992

Tests of Normality

	Substance Use Level	Kolmogorov-Smirnov[a]			Shapiro-Wilk		
		Statistic	df	Sig.	Statistic	df	Sig.
Self-Esteem Score	Low substance use	.179	20	.094	.940	20	.245
	High substance use	.202	20	.033	.945	20	.298

a. Lilliefors Significance Correction

Due to the small size of the sample (*N* = 40), the **Shapiro-Wilk** test is considered a more appropriate test of normality than the Kolmogorov-Smirnov test.

If the Shapiro-Wilk statistic (*W*) is significant (i.e., *Sig* < .05), the group of scores is not normally distributed.

In this instance, *Sig* exceeded .05 for both *W* statistics. Thus, we can conclude <u>the assumption of normality is not violated</u>. This conclusion is further supported by **Skewness** and **Kurtosis** statistics that are relatively close to zero in the **Descriptives** table, and the reasonably bell-shaped **Histograms** below.

Self-Esteem Score

Histograms

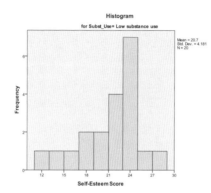

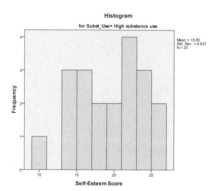

10.4.2. PASW Statistics Output (Part 2: Homogeneity of Regression Slopes)

Univariate Analysis of Variance

Between-Subjects Factors

		Value Label	N
Substance Use Level	1	Low substance use	20
	2	High substance use	20

If the interaction term (**Subst_Use * Age**) in the **Tests of Between-Subjects Effects** table is significant (*Sig* < .05) then the data have violated the homogeneity of regression slopes assumption.

In this example the interaction is not significant (*Sig* = .819) indicating that <u>this assumption has not been violated</u>.

Tests of Between-Subjects Effects

Dependent Variable:Self-Esteem Score

Source	Type III Sum of Squares	df	Mean Square	F	Sig.
Corrected Model	100.442[a]	3	33.481	1.787	.167
Intercept	440.085	1	440.085	23.487	.000
Subst_Use	.039	1	.039	.002	.964
Age	65.018	1	65.018	3.470	.071
Subst_Use * Age	.991	1	.991	.053	.819
Error	674.533	36	18.737		
Total	16417.000	40			
Corrected Total	774.975	39			

a. R Squared = .130 (Adjusted R Squared = .057)

10.4.3. PASW Statistics Output (Part 3: Linearity)

Graph

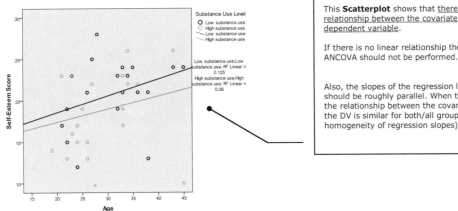

This **Scatterplot** shows that there is a linear relationship between the covariate and the dependent variable.

If there is no linear relationship then an ANCOVA should not be performed.

Also, the slopes of the regression lines should be roughly parallel. When they are, the relationship between the covariate and the DV is similar for both/all groups (i.e., homogeneity of regression slopes).

(i) *Tip:*
The regression lines were added to this **Scatterplot** in the manner described in section 10.3.2.6.

10.4.4. PASW Statistics Output (Part 4: Homogeneity of Variance & the ANCOVA)

Univariate Analysis of Variance

Between-Subjects Factors

		Value Label	N
Substance Use Level	1	Low substance use	20
	2	High substance use	20

The **Descriptive Statistics** table provides basic descriptive information for each group of scores.

Here, participants in the low substance use group (n = 20) had a mean self-esteem score of 20.70 (SD = 4.181), while participants in the high substance use group (n = 20) had a mean self-esteem score of 18.85 (SD = 4.637).

Descriptive Statistics

Dependent Variable:Self-Esteem Score

Substance Use Level	Mean	Std. Deviation	N
Low substance use	20.70	4.181	20
High substance use	18.85	4.637	20
Total	19.77	4.458	40

Levene's Test of Equality of Error Variances[a]

Dependent Variable:Self-Esteem Score

F	df1	df2	Sig.
.943	1	38	.338

Tests the null hypothesis that the error variance of the dependent variable is equal across groups.

a. Design: Intercept + Age + Subst_Use

Levene's test is not significant (Sig > .05), indicating that the assumption of homogeneity of variance has not been violated.

The **Tests of Between-Subjects Effects** table reports the ANCOVA results.

Tests of Between-Subjects Effects

Dependent Variable:Self-Esteem Score

Source	Type III Sum of Squares	df	Mean Square	F	Sig.	Partial Eta Squared	Noncent. Parameter	Observed Power[b]
Corrected Model	99.451[a]	2	49.725	2.724	.079	.128	5.447	.505
Intercept	440.587	1	440.587	24.132	.000	.395	24.132	.998
Age	65.226	1	65.226	3.573	.067	.088	3.573	.453
Subst_Use	25.296	1	25.296	1.386	.247	.036	1.386	.209
Error	675.524	37	18.257					
Total	16417.000	40						
Corrected Total	774.975	39						

a. R Squared = .128 (Adjusted R Squared = .081)

b. Computed using alpha = .05

The **Age** row indicates that the covariate is not significantly related to the DV.

The **Subst_Use** row indicates that, after partialling out the effects of the covariate (age), substance use is not significantly related to self-esteem.

In other words, the researcher's hypothesis was not supported by this data.

Estimated Marginal Means

Substance Use Level

Dependent Variable:Self-Esteem Score

Substance Use Level	Mean	Std. Error	95% Confidence Interval	
			Lower Bound	Upper Bound
Low substance use	20.574[a]	.958	18.633	22.515
High substance use	18.976[a]	.958	17.035	20.917

a. Covariates appearing in the model are evaluated at the following values: Age = 29.95.

10.4.5. APA Style Results Write-Up

Results

A one-way analysis of covariance variance (ANCOVA) was used to examine the impact of substance use levels on self-esteem. As age was expected to impact on this relationship, it was measured and included in the analysis as a covariate.

Before interpreting the outcome of the ANCOVA, its assumptions were tested. Shapiro-Wilk tests and histograms indicated support for the normality assumption. Scatterplots indicated that there was a linear relationship between the covariate (age) and the DV (self-esteem), and that the regression slopes were homogenous. Finally, Levene's test was statistically non-significant, indicating homogeneity of variances, $F(1, 38) = 0.94$, $p = .338$.

The ANCOVA indicated that age was not significantly related to self-esteem, $F(1, 37) = 3.57$, $p = .067$, partial $\eta^2 = .088$. Furthermore, the effect of substance use on self-esteem was statistically non-significant, $F(1, 37) = 1.39$, $p = .247$, partial $\eta^2 = .036$.

> The **Results** should not include any interpretations of your findings. These should be saved for the **Discussion** section of a research report.

10.5. One-Way ANCOVA Checklist

Have you:

- ✔ Checked that each group of data is approximately normally distributed?
- ✔ Checked for homogeneity of regression slopes?
- ✔ Checked for linearity?
- ✔ Checked for homogeneity of variance?
- ✔ Identified the main effect and taken note of the F value, degrees of freedom, and significance for your write-up?
- ✔ Identified the effect of the covariate variable?
- ✔ Written up your results in the APA style?

Chapter 11: Multivariate Analysis of Variance (MANOVA)

Chapter Overview

Included With:
MANOVA is a part of the *PASW Advanced Statistics* add-on module.

It is not included in *PASW Statistics*, and is consequently not available in *PASW Statistics Student Version*.

11.1. Purpose of the MANOVA

MANOVA is an extension of the analysis of variance (ANOVA) methods described in previous chapters. It simultaneously tests for a statistically significant difference between groups on multiple dependent variables.

The MANOVA is useful when an outcome of interest is complex, and cannot be defined using one single measure. For example, a researcher looking at "love" may find it difficult to adequately capture this construct using just one measure. Consequently, the researcher may decide to employ multiple measures of "love", and use a MANOVA to analyse them simultaneously.

11.2. Questions We Could Answer Using the MANOVA

1. Does a program designed to improve students' study habits actually work?

To address this question with a MANOVA, we could simultaneously compare two groups of students (those who have participated in this program, versus those who have not) on a set of measures (or dependent variables) believed to represent various aspects of study behaviour (e.g., attitudes towards study, time spent studying, grades, satisfaction with studying etc.).

2. Do everyday orientations towards people with mental illness differ between males and females?

Everyday orientations are unlikely to be adequately captured with a single measure. Consequently, a researcher answering a question like this may administer several measures to each participant including, for example, a measure of attitudes towards mental illness; a measure of stereotypes about mental illness; and a measure of feelings towards people with mental illness. Then, a MANOVA could be used to compare the males and females on all three DVs simultaneously.

11.3. Illustrated Example One

To evaluate the efficacy of a new program designed to reduce the impact of stress on health care workers, a researcher divided the employees of a local hospital into two groups. The first group were administered the program, while the second group were not. (They were put on a waiting list, and will have an opportunity to complete the program later.) All the employees were then asked to complete three questionnaires: (a) an 18-item general anxiety questionnaire; (b) a 30-item positive attitudes towards stress scale; and (c) a self-reported rating (between 1 and 10) of shoulder tension. This data is reproduced in Table 11.1.

A MANOVA will be used to examine the effects of participation in the program on the three DVs simultaneously.

Table 11.1

Data:
This is data file
data_11_1.sav
on the companion
website.

Impact of Stress on Health Care Workers Data (N = 40)

Experimental group: Participated in stress reduction program				**Control group:** Did not participate in stress reduction program			
ID	General Anxiety	Positive Attitudes	Tension	ID	General Anxiety	Positive Attitudes	Tension
1	3	14	2	21	7	8	5
2	8	29	9	22	12	15	7
3	1	15	6	23	2	13	4
4	2	17	3	24	15	13	6
5	3	17	7	25	18	23	8
6	12	13	3	26	9	5	6
7	11	16	6	27	12	22	7
8	13	14	6	28	12	12	4
9	15	18	7	29	14	21	8
10	12	25	7	30	13	16	8
11	11	17	2	31	17	15	5
12	11	23	8	32	9	14	5
13	1	15	7	33	10	19	7
14	5	14	8	34	11	17	3
15	19	27	10	35	18	20	8
16	8	21	6	36	14	16	5
17	6	23	5	37	9	12	3
18	3	18	5	38	7	16	6
19	6	19	6	39	6	12	5
20	7	19	4	40	3	8	2

11.3.1. Setting Up the PASW Statistics Data File

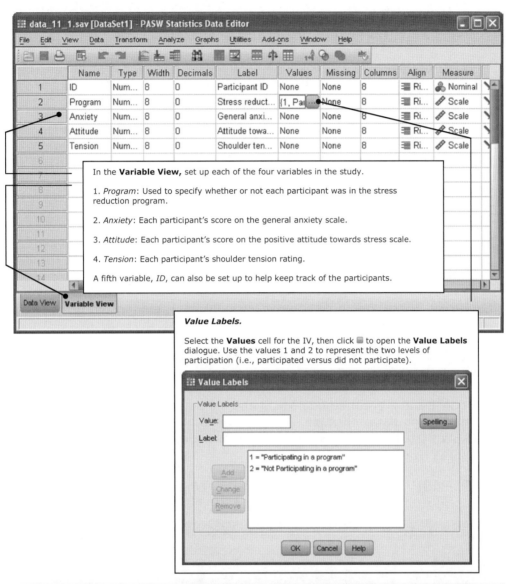

In the **Variable View,** set up each of the four variables in the study.

1. *Program*: Used to specify whether or not each participant was in the stress reduction program.

2. *Anxiety*: Each participant's score on the general anxiety scale.

3. *Attitude*: Each participant's score on the positive attitude towards stress scale.

4. *Tension*: Each participant's shoulder tension rating.

A fifth variable, *ID*, can also be set up to help keep track of the participants.

Value Labels.

Select the **Values** cell for the IV, then click ▦ to open the **Value Labels** dialogue. Use the values 1 and 2 to represent the two levels of participation (i.e., participated versus did not participate).

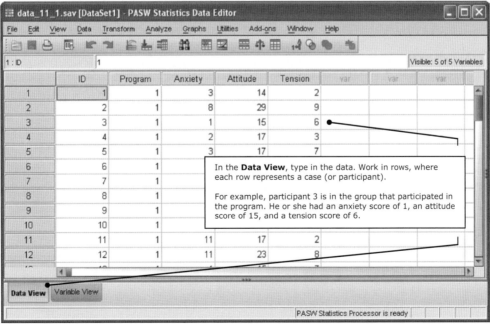

In the **Data View**, type in the data. Work in rows, where each row represents a case (or participant).

For example, participant 3 is in the group that participated in the program. He or she had an anxiety score of 1, an attitude score of 15, and a tension score of 6.

🖥 *Syntax:*
Run these
analyses with
syntax_11_1.sps
on the companion
website.

11.3.2. Analysing the Data

11.3.2.1. Assumptions

The following criteria should be met before conducting a MANOVA.

1. **Independence.** Each participant should participate only once in the research, and should not influence the participation of others.

2. **Cell sizes.** The number of cases in each cell must be greater than the number of dependent variables. If cell size is greater than 30 then the MANOVA is fairly robust against violations of normality and equality of variance.

3. **Normality.** There should be both univariate and multivariate normality of distributions.

4. **Multicollinearity.** Multicollinearity refers to high correlations between dependent variables, and can make interpretation of the MANOVA results extremely difficult.

5. **Linearity.** There should be a linear (straight line) relationship between the dependent variables.

6. **Homogeneity of Variance-Covariance Matrices.** This assumption is similar to the homogeneity of variance mentioned in the ANOVA chapters.

Assumption 1 and 2 are methodological and should have been addressed before and during data collection. Assumptions 3, 4, 5 and 6 can be tested using PASW Statistics and PASW Advanced Statistics.

11.3.2.2. PASW Statistics Procedure (Part 1: Univariate Normality)

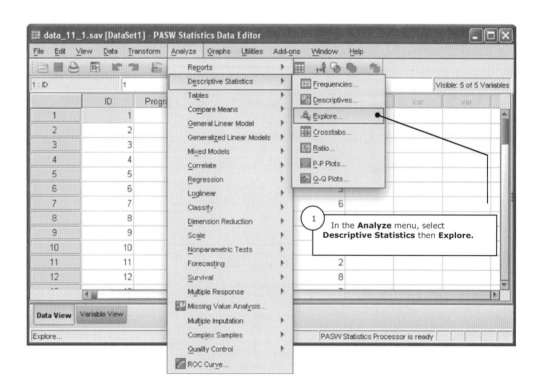

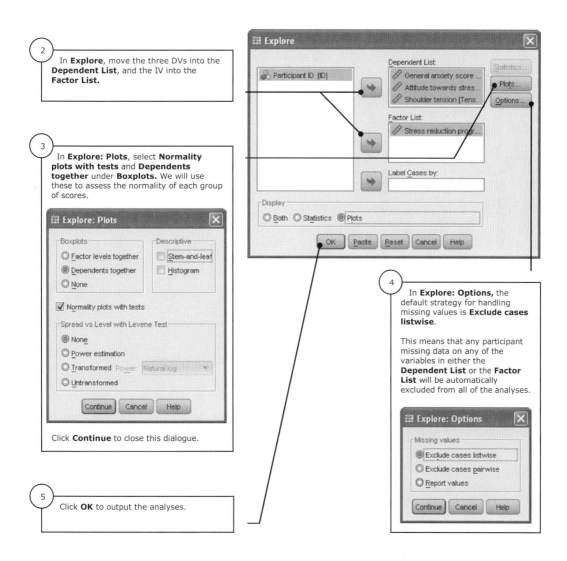

2 In **Explore**, move the three DVs into the **Dependent List**, and the IV into the **Factor List.**

3 In **Explore: Plots**, select **Normality plots with tests** and **Dependents together** under **Boxplots.** We will use these to assess the normality of each group of scores.

Click **Continue** to close this dialogue.

4 In **Explore: Options,** the default strategy for handling missing values is **Exclude cases listwise**.

This means that any participant missing data on any of the variables in either the **Dependent List** or the **Factor List** will be automatically excluded from all of the analyses.

5 Click **OK** to output the analyses.

11.3.2.3. PASW Statistics Output (Part 1: Univariate Normality)

Explore

Stress reduction program

Case Processing Summary

Stress reduction program		Cases					
		Valid		Missing		Total	
		N	Percent	N	Percent	N	Percent
General anxiety score	Participating in a program	20	100.0%	0	.0%	20	100.0%
	Not Participating in a program	20	100.0%	0	.0%	20	100.0%
Attitude towards stress	Participating in a program	20	100.0%	0	.0%	20	100.0%
	Not Participating in a program	20	100.0%	0	.0%	20	100.0%
Shoulder tension	Participating in a program	20	100.0%	0	.0%	20	100.0%
	Not Participating in a program	20	100.0%	0	.0%	20	100.0%

A series of non-significant **Shapiro-Wilk** tests indicates that the univariate normality assumption is not violated.

▶▶| *Link:*
The full PASW Statistics output also included six **Normal Q-Q Plots** and six **Detrended Normal Q-Q Plots**.

These have been described elsewhere in this text (for example, in chapter 4).

Tests of Normality

Stress reduction program		Kolmogorov-Smirnov[a]			Shapiro-Wilk		
		Statistic	df	Sig.	Statistic	df	Sig.
General anxiety score	Participating in a program	.134	20	.200*	.950	20	.362
	Not Participating in a program	.096	20	.200*	.970	20	.756
Attitude towards stress	Participating in a program	.174	20	.115	.911	20	.068
	Not Participating in a program	.125	20	.200*	.973	20	.814
Shoulder tension	Participating in a program	.177	20	.100	.956	20	.467
	Not Participating in a program	.129	20	.200*	.935	20	.195

a. Lilliefors Significance Correction

*. This is a lower bound of the true significance.

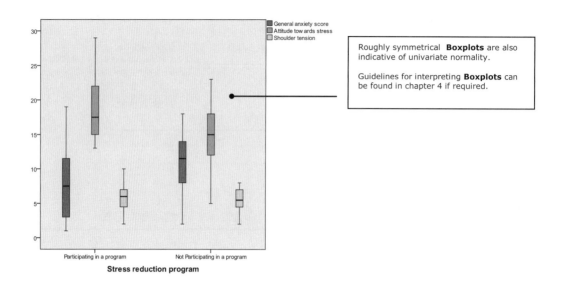

Roughly symmetrical **Boxplots** are also indicative of univariate normality.

Guidelines for interpreting **Boxplots** can be found in chapter 4 if required.

11.3.2.4. PASW Statistics Procedure (Part 2: Multicollinearity & Multivariate Outliers)

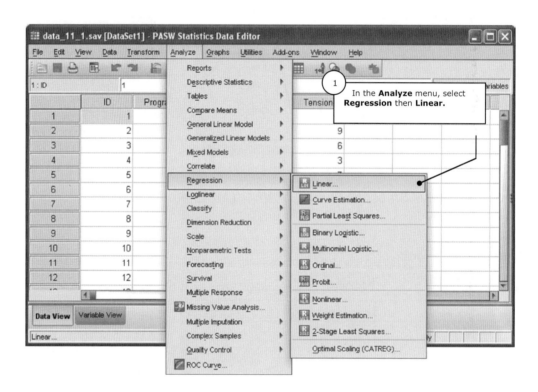

1 In the **Analyze** menu, select **Regression** then **Linear**.

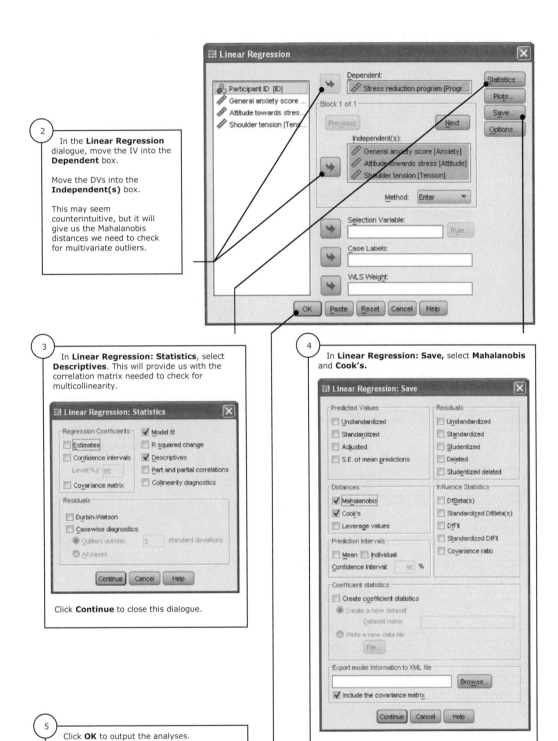

2 In the **Linear Regression** dialogue, move the IV into the **Dependent** box.

Move the DVs into the **Independent(s)** box.

This may seem counterintuitive, but it will give us the Mahalanobis distances we need to check for multivariate outliers.

3 In **Linear Regression: Statistics**, select **Descriptives**. This will provide us with the correlation matrix needed to check for multicollinearity.

Click **Continue** to close this dialogue.

4 In **Linear Regression: Save,** select **Mahalanobis** and **Cook's.**

Click **Continue** to close this dialogue.

5 Click **OK** to output the analyses.

11.3.2.5. PASW Statistics Output (Part 2: Multicollinearity & Multivariate Outliers)

Regression

Descriptive Statistics

	Mean	Std. Deviation	N
Stress reduction program	1.50	.506	40
General anxiety score	9.38	4.970	40
Attitude towards stress	16.77	5.010	40
Shoulder tension	5.73	2.000	40

Correlations

		Stress reduction program	General anxiety score	Attitude towards stress	Shoulder tension
Pearson Correlation	Stress reduction program	1.000	.311	-.389	-.063
	General anxiety score	.311	1.000	.305	.392
	Attitude towards stress	-.389	.305	1.000	.557
	Shoulder tension	-.063	.392	.557	1.000
Sig. (1-tailed)	Stress reduction program	.	.025	.007	.349
	General anxiety score	.025	.	.028	.006
	Attitude towards stress	.007	.028	.	.000
	Shoulder tension	.349	.006	.000	.
N	Stress reduction program	40	40	40	40
	General anxiety score	40	40	40	40
	Attitude towards stress	40	40	40	40
	Shoulder tension	40	40	40	40

> Multicollinearity is present when the DVs correlate strongly with each other (e.g., .90+).
>
> Here, the correlations between the DVs are:
>
> Anxiety and attitude:
> $r(38) = .305$, $p = .028$.
>
> Anxiety and tension:
> $r(38) = .392$, $p = .006$.
>
> Attitude and tension:
> $r(38) = .557$, $p < .001$.
>
> These correlations are all well shy of suggesting multicollinearity.
>
> **Dealing with Multicollinearity**
>
> When faced with multicollinearity, consider (a) deleting one of the offending variables, or (b) combining the collinear variables into a single composite variable.

Variables Entered/Removed[b]

Model	Variables Entered	Variables Removed	Method
1	Shoulder tension, General anxiety score, Attitude towards stress[a]	.	Enter

a. All requested variables entered.

b. Dependent Variable: Stress reduction program

> Ignore these tables.
>
> When looking for multivariate outliers, we are only interested in the **Residual Statistics** table.

Model Summary[b]

Model	R	R Square	Adjusted R Square	Std. Error of the Estimate
1	.599[a]	.358	.305	.422

a. Predictors: (Constant), Shoulder tension, General anxiety score, Attitude towards stress

b. Dependent Variable: Stress reduction program

ANOVA[b]

Model		Sum of Squares	df	Mean Square	F	Sig.
1	Regression	3.584	3	1.195	6.704	.001[a]
	Residual	6.416	36	.178		
	Total	10.000	39			

a. Predictors: (Constant), Shoulder tension, General anxiety score, Attitude towards stress

b. Dependent Variable: Stress reduction program

Residuals Statistics[a]

	Minimum	Maximum	Mean	Std. Deviation	N
Predicted Value	.79	2.17	1.50	.303	40
Std. Predicted Value	-2.327	2.197	.000	1.000	40
Standard Error of Predicted Value	.073	.209	.129	.033	40
Adjusted Predicted Value	.74	2.22	1.51	.308	40
Residual	-.833	.657	.000	.406	40
Std. Residual	-1.974	1.557	.000	.961	40
Stud. Residual	-2.033	1.628	-.006	1.006	40
Deleted Residual	-.884	.718	-.006	.445	40
Stud. Deleted Residual	-2.131	1.668	-.010	1.023	40
Mahal. Distance	.206	8.589	2.925	1.985	40
Cook's Distance	.000	.117	.024	.029	40
Centered Leverage Value	.005	.220	.075	.051	40

a. Dependent Variable: Stress reduction program

> A **Maximum Mahalanobis Distance** larger than the critical chi-square (χ^2) value for $df =$ the number of DVs in the MANOVA at $\alpha = .001$ indicates the presence of one or more multivariate outliers.
>
> The critical χ^2 value for $df = 3$ at $\alpha = .001$ is 16.266. (Tables of χ^2 critical tables are available in the appendices of most statistics textbooks.) As our **Maximum Mahalanobis Distance** is only 8.589, we need not be concerned about multivariate outliers.

(i) Tip:
The critical χ^2 values for 2-10 degrees of freedom at $\alpha = .001$ are as follows:

df	Critical χ^2
2	13.816
3	16.266
4	18.467
5	20.515
6	22.458
7	24.322
8	26.125
9	27.877
10	29.588

For $df > 10$, refer to the tables available in the appendices of most decent statistics texts (e.g., Howell, 2010b).

Dealing with Multivariate Outliers

First, return to the **PASW Statistics Data Editor**, where two additional variables (**MAH_1** and **COO_1**) will have been appended to your data file. Sort the data file by **MAH_1** (in the **Data** menu, select **Sort Cases**), so cases with the largest Mahalanobis distances are grouped together. Pay particular attention to those which also have **COO_1** values > 1. (**Cook's Distance** is a measure of influence. That is, it's a measure of the impact that each individual case has on the predictive efficacy of the regression model as a whole.)

Once all the multivariate outliers in the data file have been identified, you will need to decide on what to do with them. The most common strategies for dealing with multivariate outliers include (a) ignoring them; (b) deleting them; or (c) modifying them to reduce their impact on the regression model. Note that it is quite reasonable to use different strategies for different types of multivariate outliers (e.g., deleting those that are influential; ignoring those that are not). Be mindful that deleting one problem case can often reveal others, and that multivariate outliers have a reputation for hiding behind each other!

Refer to Tabachnick and Fidell (2007b) for a more comprehensive discussion of methods for dealing with multivariate outliers.

11.3.2.6. PASW Statistics Procedure (Part 3: Linearity)

MANOVA is based on the assumptions that any relationships between DVs are linear. This assumption should be tested before proceeding with the MANOVA.

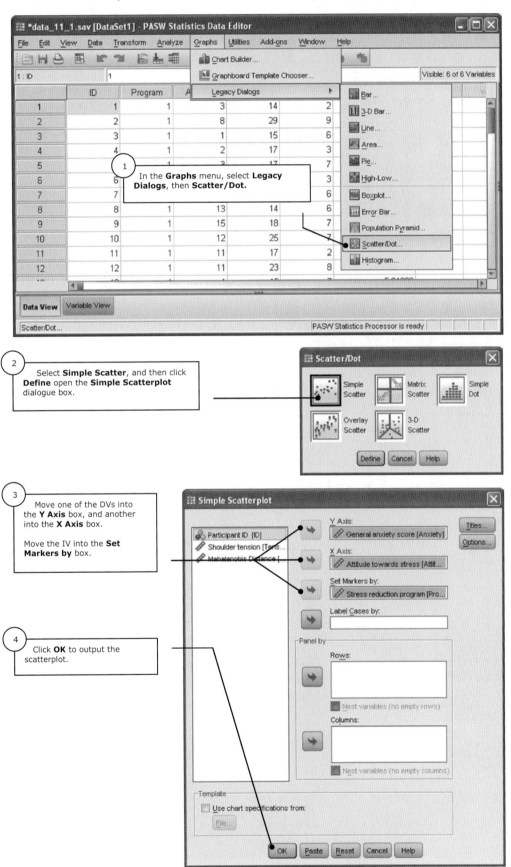

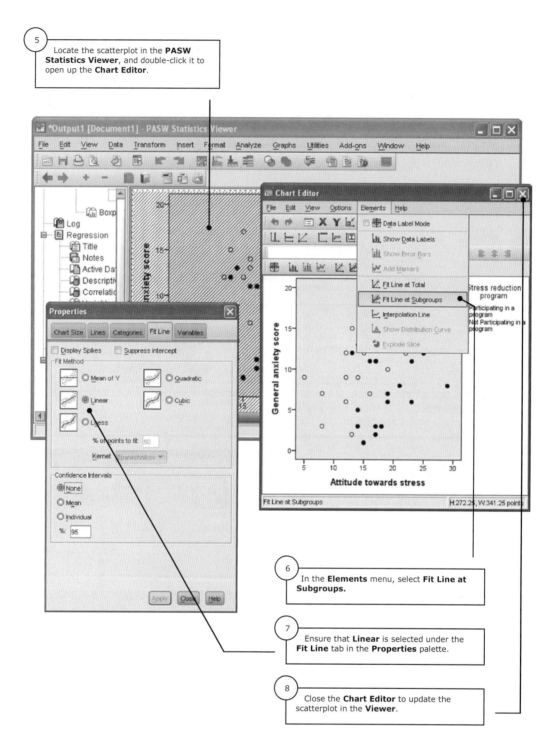

5 Locate the scatterplot in the **PASW Statistics Viewer**, and double-click it to open up the **Chart Editor**.

6 In the **Elements** menu, select **Fit Line at Subgroups.**

7 Ensure that **Linear** is selected under the **Fit Line** tab in the **Properties** palette.

8 Close the **Chart Editor** to update the scatterplot in the **Viewer**.

Repeat this process for each remaining pair of DVs (i.e., anxiety against tension and attitudes against tension).

11.3.2.7. PASW Statistics Output (Part 3: Linearity)

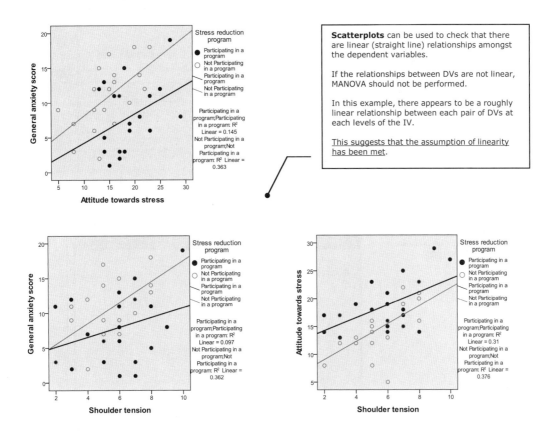

Scatterplots can be used to check that there are linear (straight line) relationships amongst the dependent variables.

If the relationships between DVs are not linear, MANOVA should not be performed.

In this example, there appears to be a roughly linear relationship between each pair of DVs at each levels of the IV.

<u>This suggests that the assumption of linearity has been met.</u>

11.3.2.8. PASW Advanced Statistics Procedure (Part 4: Homogeneity of Variance-Covariance & the MANOVA)

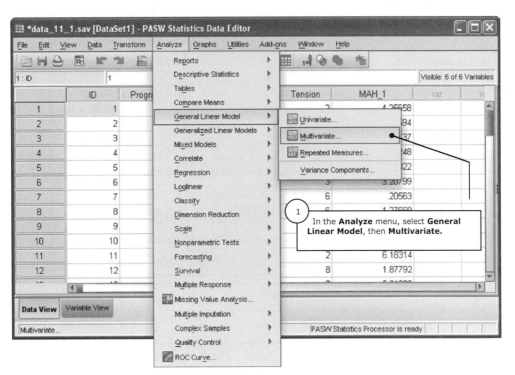

1. In the **Analyze** menu, select **General Linear Model**, then **Multivariate.**

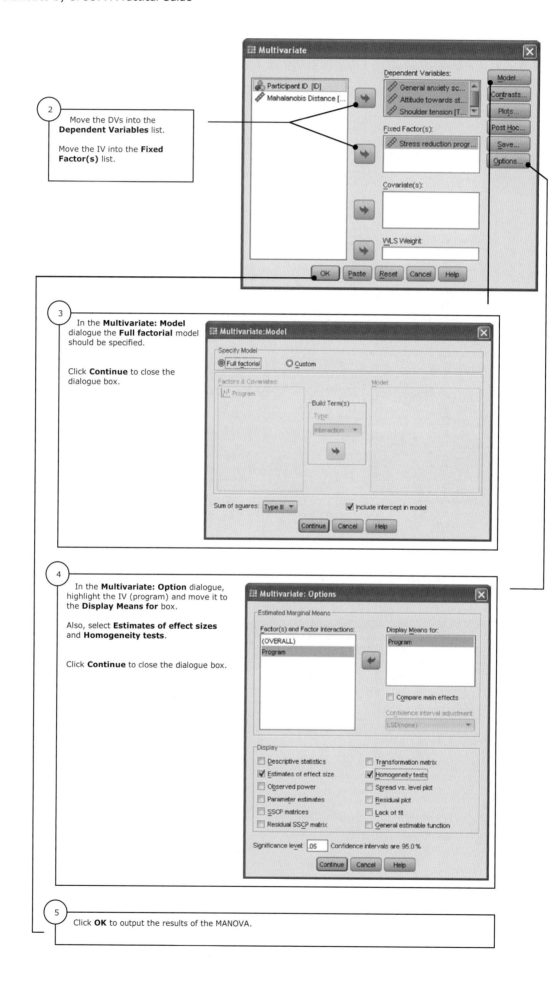

2
Move the DVs into the
Dependent Variables list.

Move the IV into the **Fixed
Factor(s)** list.

3
In the **Multivariate: Model**
dialogue the **Full factorial** model
should be specified.

Click **Continue** to close the
dialogue box.

4
In the **Multivariate: Option** dialogue,
highlight the IV (program) and move it to
the **Display Means for** box.

Also, select **Estimates of effect sizes**
and **Homogeneity tests**.

Click **Continue** to close the dialogue box.

5
Click **OK** to output the results of the MANOVA.

11.3.2.9. PASW Advanced Statistics Output (Part 4: Homogeneity of Variance-Covariance & the MANOVA)

General Linear Model

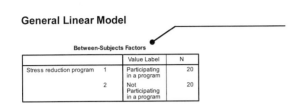

Between-Subjects Factors

		Value Label	N
Stress reduction program	1	Participating in a program	20
	2	Not Participating in a program	20

The **Between-Subjects Factors** table provides information about the number of participants in each of the groups. In this example there were 20 participants in each group.

Box's Test of Equality of Covariance Matrices[a]

Box's M	4.022
F	.613
df1	6
df2	10462.189
Sig.	.720

Tests the null hypothesis that the observed covariance matrices of the dependent variables are equal across groups.

a. Design: Intercept + Program

Box's Test of Equality of Covariance Matrices can be used to determine whether the data violate the assumption of homogeneity of variance-covariance matrices.

If **Box's M** is significant at α = .001, the assumption has been violated.

In this example the test is non-significant (*Sig* = .720) meaning that the assumption of homogeneity of variance-covariance matrices has not been violated.

Multivariate Tests[b]

Effect		Value	F	Hypothesis df	Error df	Sig.	Partial Eta Squared
Intercept	Pillai's Trace	.935	173.450[a]	3.000	36.000	.000	.935
	Wilks' Lambda	.065	173.450[a]	3.000	36.000	.000	.935
	Hotelling's Trace	14.454	173.450[a]	3.000	36.000	.000	.935
	Roy's Largest Root	14.454	173.450[a]	3.000	36.000	.000	.935
Program	Pillai's Trace	.358	6.704[a]	3.000	36.000	.001	.358
	Wilks' Lambda	.642	6.704[a]	3.000	36.000	.001	.358
	Hotelling's Trace	.559	6.704[a]	3.000	36.000	.001	.358
	Roy's Largest Root	.559	6.704[a]	3.000	36.000	.001	.358

a. Exact statistic

b. Design: Intercept + Program

Read the bottom half of the **Multivariate Tests** table.

There are four tests of significance provided for the MANOVA:

1. **Pillai's Trace**
2. **Wilks' Lambda**
3. **Hotelling's Trace**
4. **Roy's Largest Root**

As the IV in our example has only two levels, all four *F*-values are the same. The outcomes of these four tests would differ if the IV had more than two levels.

Pillai's Trace is the appropriate **Multivariate Test** for small samples.

The findings in this table indicate whether there is a significant difference between the groups on the combination of DVs. In this example there is a significant difference at α = .05.

For your write-up make note of:

F = 6.704
Sig = .001
Hypothesis df = 3
Error df = 36
Partial η^2 = .358

Levene's Test of Equality of Error Variances[a]

	F	df1	df2	Sig.
General anxiety score	.418	1	38	.522
Attitude towards stress	.001	1	38	.978
Shoulder tension	.273	1	38	.604

Tests the null hypothesis that the error variance of the dependent variable is equal across groups.

a. Design: Intercept + Program

If **Levene's Test of Equality of Error Variances** is non-significant (*Sig* > .05) then there is equality of variances.

In this example Levene's test is non-significant for all three dependent variables.

The significant **Pillai's Trace** in the **Multivariate Tests** table indicated that there was a difference, on a combination of the dependent variables, between people who did and did not participate in the stress reduction program.

The **Tests of Between-Subjects Effects** can now be used to interpret the results for each DV.

Tests of Between-Subjects Effects

Source	Dependent Variable	Type III Sum of Squares	df	Mean Square	F	Sig.	Partial Eta Squared
Corrected Model	General anxiety score	93.025ᵃ	1	93.025	4.062	.051	.097
	Attitude towards stress	148.225ᵇ	1	148.225	6.780	.013	.151
	Shoulder tension	.625ᶜ	1	.625	.153	.698	.004
Intercept	General anxiety score	3515.625	1	3515.625	153.494	.000	.802
	Attitude towards stress	11256.025	1	11256.025	514.871	.000	.931
	Shoulder tension	1311.025	1	1311.025	320.688	.000	.894
Program	General anxiety score	93.025	1	93.025	4.062	.051	.097
	Attitude towards stress	148.225	1	148.225	6.780	.013	.151
	Shoulder tension	.625	1	.625	.153	.698	.004
Error	General anxiety score	870.350	38	22.904			
	Attitude towards stress	830.750	38	21.862			
	Shoulder tension	155.350	38	4.088			
Total	General anxiety score	4479.000	40				
	Attitude towards stress	12235.000	40				
	Shoulder tension	1467.000	40				
Corrected Total	General anxiety score	963.375	39				
	Attitude towards stress	978.975	39				
	Shoulder tension	155.975	39				

a. R Squared = .097 (Adjusted R Squared = .073)

b. R Squared = .151 (Adjusted R Squared = .129)

c. R Squared = .004 (Adjusted R Squared = -.022)

The **Program** (i.e., the IV) section of the **Tests of Between Subjects Effects** table provides the results for the three univariate ANOVAs (one for each DV). To control for the inflated family-wise error rates that result from performing multiple tests on the same data, we need to evaluate the significance of these univariate ANOVAs at a Bonferroni adjusted alpha level.

To make this adjustment, we divide the family-wise alpha level ($\alpha = .05$) by the number of DVs in the MANOVA. In this example, there are three dependent variables:

$$p_{corrected} = \frac{.05}{3} = .017$$

At the Bonferroni adjusted alpha level of .017, only one dependent variable (positive attitudes towards stress) was significant (Sig = .013).

For your write-up make note of the details of the significant univariate ANOVA:

$F = 6.780$
$Sig = .013$
$df = (1, 38)$
$Partial\ \eta^2 = .151$

Estimated Marginal Means

The **Estimated Marginal Means** table reports a mean, standard error and confidence intervals for each group on each DV.

Stress reduction program

Dependent Variable	Stress reduction program	Mean	Std. Error	95% Confidence Interval Lower Bound	95% Confidence Interval Upper Bound
General anxiety score	Participating in a program	7.850	1.070	5.684	10.016
	Not Participating in a program	10.900	1.070	8.734	13.066
Attitude towards stress	Participating in a program	18.700	1.046	16.583	20.817
	Not Participating in a program	14.850	1.046	12.733	16.967
Shoulder tension	Participating in a program	5.850	.452	4.935	6.765
	Not Participating in a program	5.600	.452	4.685	6.515

11.3.3. APA Style Results Write-Up

Results

A multivariate analysis of variance (MANOVA) was used to examine the effectiveness of a program designed to reduce the stress levels of health care workers ($N = 40$).

Before conducting the MANOVA the data were examined using PASW Statistics to ensure all of its underlying assumptions were met.

Univariate normality was assessed with Shapiro-Wilk tests and boxplots, and could be assumed. Additionally, no multivariate outliers were found in the data, supporting the assumption of multivariate normality. Correlations between the dependent variables were not excessive, indicating that multicollinearity was not of concern. Furthermore, the relationships that did exist between the dependent variables were roughly linear. Finally, Box's M was non-significant at $\alpha = .001$, indicating that homogeneity of variance-covariance matrices could be assumed.

> Assumption testing is often not reported in journal articles (particularly when the assumptions are not violated). However, it is more commonly expected in student reports.

As all the underlying assumptions were supported by the data, a MANOVA was conducted. Findings showed that there was a significant effect of the program variable (participating in the program versus not participating in the program) on the combined dependent variables, $F (3, 36) = 6.70$, $p = .001$, partial $\eta^2 = .358$.

> Omega-squared may be preferred here over partial eta-squared (see chapter 8, section 8.4.2.2).

Analysis of the dependent variables individually showed no effects for the anxiety and shoulder tension variables. However, the attitudes variable was statistically significant at a Bonferroni adjusted alpha level of .017, $F (1, 38) = 6.78$, $p = .013$, partial $\eta^2 = .151$. The health care workers who participated in the stress reduction program reported significantly higher (i.e., more positive) attitudes towards stress ($M = 18.70$) than those who did not ($M = 14.85$).

> These findings indicate that individuals who participated in the stress reduction program have more positive attitudes towards stress than those who did not participate in the program.
>
> Explanation of the findings should be saved for the **Discussion** section of the research report.

11.3.4. Summary

In this example, a MANOVA was conducted on a small set of data. Assumptions underlying the test were examined and a statistically significant finding was reported for one of the dependent variables. In the second illustrated example, reporting non-significant findings will be covered.

11.4. Illustrated Example Two

To investigate the hypothesis that there is a difference between the overall performance of males and females at his school, Principal Johnson gathered up the math, reading and art results of 40 male and 40 female students. These results are reproduced in Table 11.2.

Table 11.2

📖 *Data:*
This is data file
data_11_2.sav
on the companion
website.

There are a couple of results missing from the data set.

Math, Reading and Art Results of 80 Students

	Females				Males		
ID	Math	Reading	Art	ID	Math	Reading	Art
1	78	78	88	41	56	45	23
2	94	94	75	42	94	99	78
3	92	88	88	43	92	89	67
4	93	66	76	44	93	66	56
5	98	88	87	45	67	77	65
6	78	78	67	46	78	67	98
7	76	55	65	47	76	68	77
8	98	88	88	48	67	87	88
9	74	38	68	49	63	54	55
10	73	98	56	50	61	76	68
11	79	86	74	51	64	74	67
12	78	67	56	52	58	.	50
13	58	79	86	53	55	57	67
14	56	46	33	54	66	89	80
15	98	58	88	55	45	58	34
16	74	56	78	56	44	56	55
17	73	78	67	57	77	78	64
18	.	89	45	58	76	76	76
19	78	66	87	59	67	56	66
20	58	35	65	60	77	81	76
21	56	67	56	61	66	87	89
22	78	74	77	62	45	66	54
23	76	87	88	63	78	78	99
24	47	35	55	64	76	77	89
25	69	98	98	65	45	47	51
26	76	88	74	66	74	76	67
27	76	73	56	67	76	83	78
28	83	69	76	68	76	87	54
29	87	78	57	69	98	76	67
30	76	58	68	70	87	67	56
31	67	56	56	71	55	34	43
32	78	98	66	72	56	53	66
33	100	74	88	73	78	87	76
34	73	66	75	74	67	89	87
35	56	54	55	75	55	67	66
36	98	78	89	76	98	78	87
37	87	58	56	77	87	67	77
38	45	43	46	78	45	65	67
39	56	78	76	79	56	56	69
40	77	76	76	80	77	67	67

11.4.1. PASW Statistics Output (Part 1: Univariate Normality)

Explore

Gender

Although there are 80 participants in the data file, the results of the **Explore** analyses are based on just 78 cases. Two cases have been excluded due to missing data.

Syntax:
Run these analyses with **syntax_11_2.sps** on the companion website.

Case Processing Summary

| | Gender | Cases | | | | | |
| | | Valid | | Missing | | Total | |
		N	Percent	N	Percent	N	Percent
Maths score	Female	39	97.5%	1	2.5%	40	100.0%
	Male	39	97.5%	1	2.5%	40	100.0%
Reading score	Female	39	97.5%	1	2.5%	40	100.0%
	Male	39	97.5%	1	2.5%	40	100.0%
Art score	Female	39	97.5%	1	2.5%	40	100.0%
	Male	39	97.5%	1	2.5%	40	100.0%

Tests of Normality

| | Gender | Kolmogorov-Smirnov[a] | | | Shapiro-Wilk | | |
		Statistic	df	Sig.	Statistic	df	Sig.
Maths score	Female	.159	39	.014	.940	39	.039
	Male	.124	39	.134	.953	39	.103
Reading score	Female	.103	39	.200*	.958	39	.150
	Male	.130	39	.096	.969	39	.360
Art score	Female	.126	39	.118	.945	39	.055
	Male	.134	39	.074	.959	39	.161

a. Lilliefors Significance Correction

*. This is a lower bound of the true significance.

A significant **Shapiro-Wilk** (*W*) statistic suggests non-normality.

Here, the **Shapiro-Wilk** test for the distribution of female math scores is significant, $W(39) = .940$, $p = .039$, which indicates a violation of the normality assumption.

The remaining five **Shapiro-Wilk** tests are not significant.

If the Assumption is Violated

Fortunately, MANOVA is fairly robust against violations of the normality assumption when group sizes exceed 30 or so.

Consequently, we'll continue with the analyses.

Link:
The full PASW Statistics output also included six **Normal Q-Q Plots** and six **Detrended Normal Q-Q Plots**.

These have been described elsewhere in this text (for example, in chapter 4).

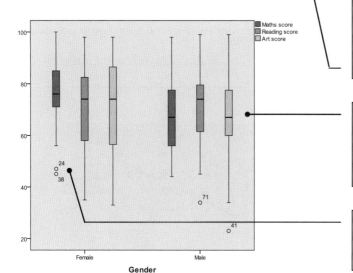

Boxplots can also be used to assess the assumption of univariate normality, and to check for the presence of univariate outliers and extreme scores (which are denoted by circles and asterisks respectively).

These two outliers are probably responsible for the significant **Shapiro-Wilk** test for the female math scores distribution.

11.4.2. PASW Statistics Output (Part 2: Multicollinearity & Multivariate Outliers)

Link:
Several other tables were produced along with this output. As they are not required for multicollinearity and multivariate outlier checking, they have been omitted from this worked example.

You can, however, find an example of each of these tables in section 11.3.2.5.

Correlations

		Gender	Maths score	Reading score	Art score
Pearson Correlation	Gender	1.000	-.216	.008	-.094
	Maths score	-.216	1.000	.497	.499
	Reading score	.008	.497	1.000	.555
	Art score	-.094	.499	.555	1.000
Sig. (1-tailed)	Gender	.	.029	.472	.207
	Maths score	.029	.	.000	.000
	Reading score	.472	.000	.	.000
	Art score	.207	.000	.000	.
N	Gender	78	78	78	78
	Maths score	78	78	78	78
	Reading score	78	78	78	78
	Art score	78	78	78	78

In the **Correlations** table the correlations between pairs of DVs are all statistically significant, but are not exceptionally high.

Consequently, we can assume that the DVs are not multicollinear.

Residuals Statisticsᵃ

	Minimum	Maximum	Mean	Std. Deviation	N
Predicted Value	1.17	1.76	1.50	.130	78
Std. Predicted Value	-2.574	1.988	.000	1.000	78
Standard Error of Predicted Value	.061	.185	.108	.030	78
Adjusted Predicted Value	1.19	1.77	1.50	.135	78
Residual	-.685	.720	.000	.486	78
Std. Residual	-1.381	1.452	.000	.980	78
Stud. Residual	-1.449	1.494	-.001	1.007	78
Deleted Residual	-.766	.762	-.001	.514	78
Stud. Deleted Residual	-1.460	1.507	-.001	1.009	78
Mahal. Distance	.183	9.687	2.962	2.149	78
Cook's Distance	.004	.071	.014	.013	78
Centered Leverage Value	.002	.126	.038	.028	78

a. Dependent Variable: Gender

A **Maximum Mahalanobis Distance** larger than the critical chi-square (χ^2) value for *df = the number of DVs in the MANOVA* at α = .001 indicates the presence of one or more multivariate outliers.

The critical χ^2 value for *df* = 3 at α = .001 is 16.266. (Tables of χ^2 critical tables are available in the appendices of most statistics textbooks.) As our **Maximum Mahalanobis Distance** is only 9.687, <u>we need not be concerned about multivariate outliers.</u>

11.4.3. PASW Statistics Output (Part 3: Linearity)

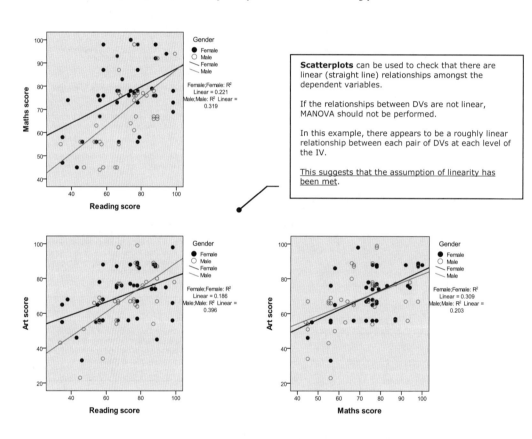

Scatterplots can be used to check that there are linear (straight line) relationships amongst the dependent variables.

If the relationships between DVs are not linear, MANOVA should not be performed.

In this example, there appears to be a roughly linear relationship between each pair of DVs at each level of the IV.

<u>This suggests that the assumption of linearity has been met.</u>

11.4.4. PASW Advanced Statistics Output (Part 4: Homogeneity of Variance-Covariance & the MANOVA)

General Linear Model

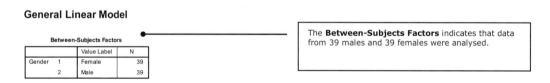

Between-Subjects Factors

		Value Label	N
Gender	1	Female	39
	2	Male	39

The **Between-Subjects Factors** indicates that data from 39 males and 39 females were analysed.

Box's Test of Equality of Covariance Matrices[a]

Box's M	6.195
F	.988
df1	6
df2	41848.755
Sig.	.431

Tests the null hypothesis that the observed covariance matrices of the dependent variables are equal across groups.

a. Design: Intercept + Gender

Box's Test of Equality of Covariance Matrices can be used to determine whether the data violate the homogeneity of variance-covariance matrices assumption.

If **Box's M** is significant at α = .001, the assumption has been violated.

In this example the test is non-significant (*Sig* = .431) meaning that the assumption of homogeneity of variance-covariance matrices has not been violated.

If the Assumption is Violated.

Box's M is very sensitive, and is particularly important when analysing small samples or working with very uneven group sizes. However, if group sizes are 30+ then the MANOVA is fairly robust against violations of the homogeneity of variance-covariance matrices assumption.

Multivariate Tests[b]

Effect		Value	F	Hypothesis df	Error df	Sig.	Partial Eta Squared
Intercept	Pillai's Trace	.970	808.392[a]	3.000	74.000	.000	.970
	Wilks' Lambda	.030	808.392[a]	3.000	74.000	.000	.970
	Hotelling's Trace	32.773	808.392[a]	3.000	74.000	.000	.970
	Roy's Largest Root	32.773	808.392[a]	3.000	74.000	.000	.970
Gender	Pillai's Trace	.066	1.750[a]	3.000	74.000	.164	.066
	Wilks' Lambda	.934	1.750[a]	3.000	74.000	.164	.066
	Hotelling's Trace	.071	1.750[a]	3.000	74.000	.164	.066
	Roy's Largest Root	.071	1.750[c]	3.000	74.000	.164	.066

a. Exact statistic

b. Design: Intercept + Gender

Pillai's Trace is considered the most appropriate **Multivariate Test** for small samples.

The findings in this table indicate whether there is a significant difference between the groups on the combination of DVs. In this example, there is not (*Sig* = .164).

For your write-up make note of:

F = 1.750
Sig = .164
Hypothesis df = 3
Error df = 74
Partial η^2 = .066

As **Levene's Test of Equality of Error Variances** is not significant for any of the DVs (*Sig* > .05), the homogeneity of variance assumption has not been violated.

If the Assumption is Violated

If homogeneity of variance cannot be assumed for one or more of the DVs, consider evaluating the outcome(s) of the corresponding univariate ANOVA(s) at a stricter alpha level (e.g., .001).

Levene's Test of Equality of Error Variances[a]

	F	df1	df2	Sig.
Maths score	.983	1	76	.325
Reading score	1.457	1	76	.231
Art score	.021	1	76	.884

Tests the null hypothesis that the error variance of the dependent variable is equal across groups.

a. Design: Intercept + Gender

Tests of Between-Subjects Effects

Source	Dependent Variable	Type III Sum of Squares	df	Mean Square	F	Sig.	Partial Eta Squared
Corrected Model	Maths score	827.128[a]	1	827.128	3.729	.057	.047
	Reading score	1.282[b]	1	1.282	.005	.943	.000
	Art score	157.962[c]	1	157.962	.677	.413	.009
Intercept	Maths score	413620.513	1	413620.513	1864.884	.000	.961
	Reading score	389232.051	1	389232.051	1536.655	.000	.953
	Art score	380661.551	1	380661.551	1630.473	.000	.955
Gender	Maths score	827.128	1	827.128	3.729	.057	.047
	Reading score	1.282	1	1.282	.005	.943	.000
	Art score	157.962	1	157.962	.677	.413	.009
Error	Maths score	16856.359	76	221.794			
	Reading score	19250.667	76	253.298			
	Art score	17743.487	76	233.467			
Total	Maths score	431304.000	78				
	Reading score	408484.000	78				
	Art score	398563.000	78				
Corrected Total	Maths score	17683.487	77				
	Reading score	19251.949	77				
	Art score	17901.449	77				

a. R Squared = .047 (Adjusted R Squared = .034)

b. R Squared = .000 (Adjusted R Squared = -.013)

c. R Squared = .009 (Adjusted R Squared = -.004)

As **Pillai's Trace** was not significant, you do not need to worry about the **Tests of Between-Subjects Effects** (univariate ANOVAs).

Estimated Marginal Means

Gender

Dependent Variable	Gender	Mean	Std. Error	95% Confidence Interval	
				Lower Bound	Upper Bound
Maths score	Female	76.077	2.385	71.327	80.827
	Male	69.564	2.385	64.814	74.314
Reading score	Female	70.513	2.548	65.437	75.589
	Male	70.769	2.548	65.693	75.845
Art score	Female	71.282	2.447	66.409	76.155
	Male	68.436	2.447	63.563	73.309

11.4.5. APA Style Results Write-Up

Results

Multivariate analysis of variance (MANOVA) was used to examine the effects of gender on overall school performance.

The Shapiro-Wilk test of univariate normality for the female math results data was statistically significant, $W(39) = .940, p = .039$. This is not considered problematic, as (a) MANOVA is considered robust with respect to univariate non-normality when group sizes exceed 30, and (b) a boxplot of this distribution suggested that the departure from normality was mild. The other five distributions were univariate normal. The remaining assumptions of no multivariate outliers, no multicollinearity, and homogeneity of variance-covariance matrices were satisfied.

> If an assumption is violated, describe how the violation was addressed.

The MANOVA was statistically non-significant, $F(3, 74) = 1.750, p = .164$, partial $\eta^2 = .066$, indicating the absence of any meaningful gender differences on overall school performance. Group means (and standard deviations) for each dependent variable are presented in Table 1.

Table 1

Descriptive Statistics for the Females (n = 39) and Males (n = 39) on Each Dependent Variable

Dependent Variable	Gender	M	SD
Math result	Female	76.08	14.44
	Male	69.56	15.33
Reading result	Female	70.51	17.36
	Male	70.77	14.33
Art result	Female	71.28	14.54
	Male	68.44	15.99

11.5. MANOVA Checklist

Have you:

- ✔ Checked that each group of data is approximately normally distributed?
- ✔ Checked for multicollinearity amongst the DVs?
- ✔ Checked for multivariate outliers?
- ✔ Ensured that any relationships between DVs are linear?
- ✔ Checked for homogeneity of variance-covariance?
- ✔ Identified the multivariate effect and (if applicable) any univariate effects.
- ✔ Taken note of the relevant F-values, degrees of freedom, significance levels and measures of effect size (i.e., partial η^2) for your write-up?
- ✔ Written up your results in the APA style?

Chapter 12: Correlation

Chapter Overview

12.1. Purpose of Correlation

Pearson's product-movement correlation coefficient (or Pearson's r) is used to assess the strength and direction of the linear association/relationship between two continuous variables.

In this chapter two common types of correlation will be discussed:

1. **Bivariate correlation**, used to measure the linear association between two continuous variables.

2. **Partial correlation**, used to measure the linear association between two continuous variables, after controlling for a third (and fourth, fifth etc.) continuous variable.

12.2. Questions We Could Answer Using Correlation

1. Is there a relationship between level of physical activity and weight?

A bivariate correlation can be used to assess the strength and direction of the association between these two variables which, when graphed with a scatterplot, might look something like this:

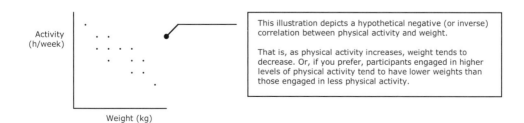

Activity (h/week)

Weight (kg)

This illustration depicts a hypothetical negative (or inverse) correlation between physical activity and weight.

That is, as physical activity increases, weight tends to decrease. Or, if you prefer, participants engaged in higher levels of physical activity tend to have lower weights than those engaged in less physical activity.

Beyond bivariate relationships, Pearson's *r* can also be used to address questions about the linear association between two variables, after controlling for a third.

2. How well does age predict income, after controlling for the effects of intelligence?

This is referred to as partial correlation.

12.3. Illustrated Example One: Bivariate Correlation

To investigate the relationship between mathematical ability and reading ability 35 students were asked to complete a math test and a reading ability scale.

The data from this study are presented in Table 12.1.

Table 12.1

Math Test and Reading Ability Scale Scores of 35 Students

□ **Data:**
This is data file
data_12_1.sav
on the companion
website.

ID	Math	Reading	ID	Math	Reading
1	49	12	19	55	16
2	50	12	20	55	17
3	50	11	21	55	17
4	51	11	22	55	18
5	51	12	23	56	22
6	51	13	24	58	18
7	52	13	25	56	18
8	52	13	26	57	20
9	51	16	27	60	19
10	52	16	28	57	18
11	53	15	29	56	24
12	53	16	30	56	24
13	54	17	31	60	19
14	54	15	32	57	18
15	54	16	33	58	20
16	54	16	34	60	19
17	55	17	35	58	21
18	55	17			

12.3.1. Setting Up the PASW Statistics Data File

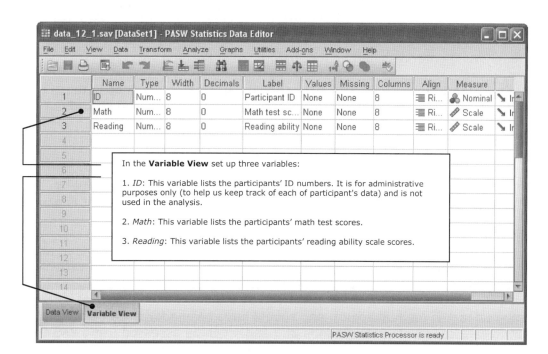

▶▶ *Link:*
Setting up a data file
is explained in greater
detail in chapter 1.

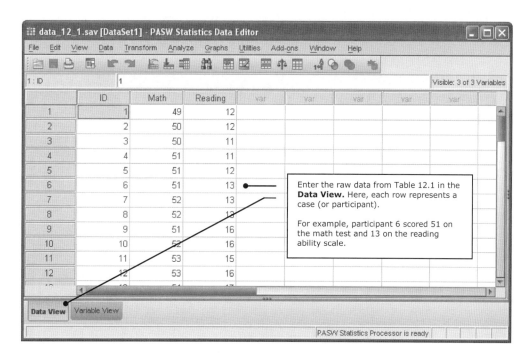

Syntax:
Run these
analyses with
syntax_12_1.sps
on the companion
website.

12.3.2. Analysing the Data

12.3.2.1. Assumptions

The following criteria should be met before using Pearson's r.

1. **Independence.** Each participant should participate only once in the research (i.e., they should provide just one set of data), and should not influence the participation of others.

2. **Normality.** Each variable should be normally distributed.

3. **Linearity.** There should be a linear (straight line) relationship between the variables. If the relationship between a pair of variables is not linear, it will not be adequately captured and summarised by Pearson's r.

4. **Homoscedasticity.** The error variance is assumed to be the same at all points along the linear relationship. That is, the variability in one variable should be similar across all values of the other variable.

Assumption 1 is methodological and should have been addressed before and during data collection. Assumptions 2, 3 and 4 can be assessed using PASW Statistics.

12.3.2.2. PASW Statistics Procedure (Part 1: Normality)

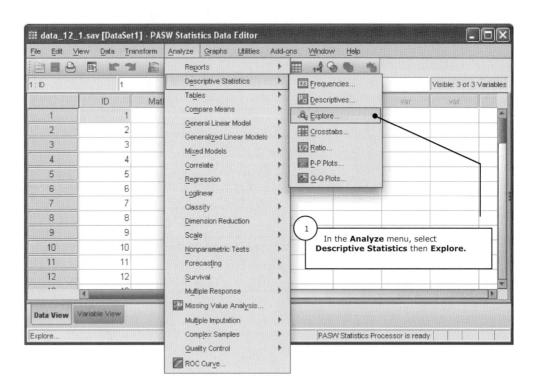

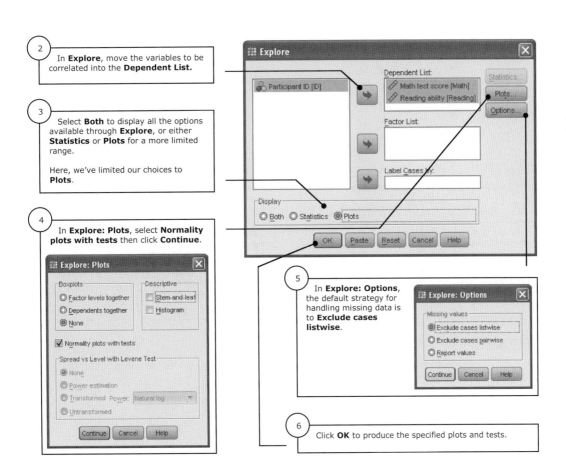

2

In **Explore**, move the variables to be correlated into the **Dependent List.**

3

Select **Both** to display all the options available through **Explore**, or either **Statistics** or **Plots** for a more limited range.

Here, we've limited our choices to **Plots**.

4

In **Explore: Plots**, select **Normality plots with tests** then click **Continue**.

5

In **Explore: Options**, the default strategy for handling missing data is to **Exclude cases listwise**.

6

Click **OK** to produce the specified plots and tests.

(i) *Tip:*
Move a variable by highlighting it, then clicking an arrow button:

12.3.2.3. PASW Statistics Output (Part 1: Normality)

Explore

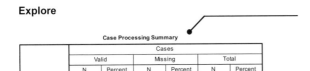

Case Processing Summary

	Cases					
	Valid		Missing		Total	
	N	Percent	N	Percent	N	Percent
Math test score	35	100.0%	0	.0%	35	100.0%
Reading ability	35	100.0%	0	.0%	35	100.0%

The **Case Processing Summary** shows how many cases were analysed, and how many were dropped due to missing data.

In this instance, there were 35 cases analysed, and none excluded because of missing data.

Tests of Normality

	Kolmogorov-Smirnov[a]			Shapiro-Wilk		
	Statistic	df	Sig.	Statistic	df	Sig.
Math test score	.100	35	.200*	.967	35	.361
Reading ability	.126	35	.171	.961	35	.251

a. Lilliefors Significance Correction

*. This is a lower bound of the true significance.

PASW Statistics computes two separate **Tests of Normality**.

The **Shapiro-Wilk** test is considered the more appropriate of the two when the sample is reasonably small.

A significant Shapiro-Wilk (*W*) statistic (i.e., *Sig* < .05) indicates that the data are not normally distributed.

Here, *W* is .967 (*Sig* = .361) for the math test variable and .961 (*Sig* = .251) for reading ability, suggesting that the normality assumption is not violated.

If the Assumption is Violated

If your data are not normally distributed, consider using Spearman's Rho or Kendall's Tau-B instead of Pearson's product-movement. Both are discussed in chapter 16.

Math test score

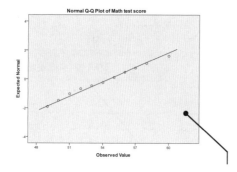

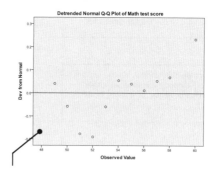

A **Normal Q-Q** (Quantile-Quantile) **Plot** graphs the observed sample data against the values we would expect if it were normally distributed (the diagonal line).

If the sample data are normally distributed the points on a **Normal Q-Q Plot** should cluster around the diagonal line, as they do in these plots.

A **Detrended Normal Q-Q** (Quantile-Quantile) **Plot** is a visual representation of the deviations from the diagonal line in the associated **Normal Q-Q Plot**.

If the data are normally distributed there should be a roughly even spread of points above and below the horizontal line, as there are here.

Reading ability

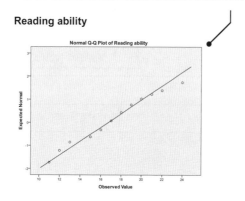

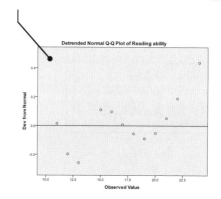

12.3.2.4. PASW Statistics Procedure (Part 2: Linearity & Homoscedasticity)

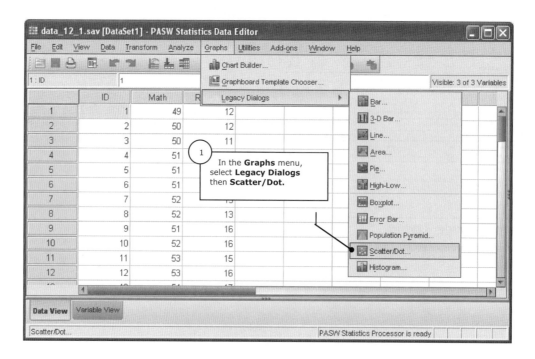

1 In the **Graphs** menu, select **Legacy Dialogs** then **Scatter/Dot.**

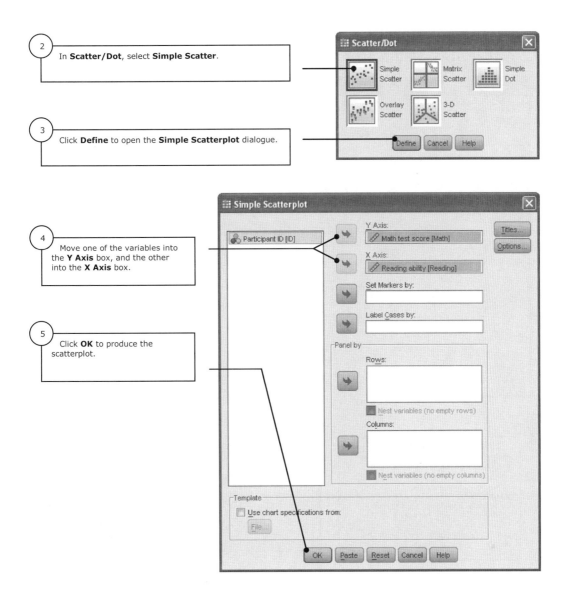

2

In **Scatter/Dot**, select **Simple Scatter**.

3

Click **Define** to open the **Simple Scatterplot** dialogue.

4

Move one of the variables into the **Y Axis** box, and the other into the **X Axis** box.

5

Click **OK** to produce the scatterplot.

12.3.2.5. PASW Statistics Output (Part 2: Linearity & Homoscedasticity)

Graph

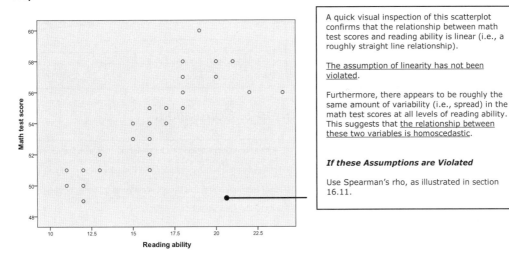

A quick visual inspection of this scatterplot confirms that the relationship between math test scores and reading ability is linear (i.e., a roughly straight line relationship).

The assumption of linearity has not been violated.

Furthermore, there appears to be roughly the same amount of variability (i.e., spread) in the math test scores at all levels of reading ability. This suggests that the relationship between these two variables is homoscedastic.

If these Assumptions are Violated

Use Spearman's rho, as illustrated in section 16.11.

ⓘ *Tip:*
A heteroscedastic relationship between two variables might look something like this:

At low levels of reading ability (at the left of the graph), there is very little variability in the math test data. However, as scores on the reading ability variable increase, so too does variability in the math test data.

12.3.2.6. PASW Statistics Procedure (Part 3: Correlation)

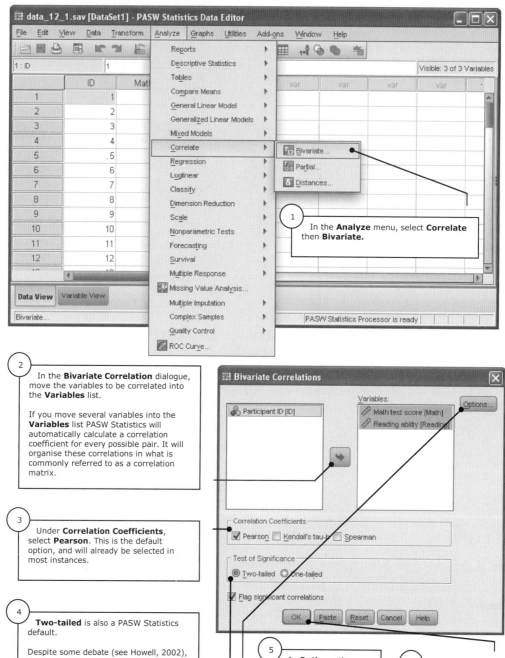

1 In the **Analyze** menu, select **Correlate** then **Bivariate.**

2 In the **Bivariate Correlation** dialogue, move the variables to be correlated into the **Variables** list.

If you move several variables into the **Variables** list PASW Statistics will automatically calculate a correlation coefficient for every possible pair. It will organise these correlations in what is commonly referred to as a correlation matrix.

3 Under **Correlation Coefficients**, select **Pearson**. This is the default option, and will already be selected in most instances.

4 **Two-tailed** is also a PASW Statistics default.

Despite some debate (see Howell, 2002), there appear to be relatively few circumstances in which a **one-tailed** test is more appropriate.

5 In **Options**, the defaults are suitable in most situations.

6 Click **OK** to run the analyses.

(i) **Tip:**
The **Correlations** table reports the significance level of Pearson's *r* as equal to zero.

In reality this value can never equal zero.

Double-click on the significance value in the **PASW Statistics Viewer** to see the actual significance level for this correlation (which, in this case, is 2.093 106159209e-008).

Instead of reporting that *p* = .000 in your results, report that *p* < .001. That is, the significance level is less than .001 (but is not quite zero).

12.3.2.7. PASW Statistics Output (Part 3: Correlation)

Correlations

Correlations

		Math test score	Reading ability
Math test score	Pearson Correlation	1	.787**
	Sig. (2-tailed)		.000
	N	35	35
Reading ability	Pearson Correlation	.787**	1
	Sig. (2-tailed)	.000	
	N	35	35

**. Correlation is significant at the 0.01 level (2-tailed).

A **Pearson Correlation** can be either positive or negative and range from -1 to +1, with zero indicating the complete absence of a linear relationship between the two variables.

In this example the **Pearson Correlation** (*r*) between the math test scores and reading ability is .787. This correlation is statistically significant (*Sig* < .05).

For your write-up, take note of the following:

Pearson Correlation = .787
N = 35
Sig (2-tailed) = .000

12.3.3. Follow Up Analyses

12.3.3.1. Effect Size

As Pearson's correlation coefficient (r) is itself an index of the strength of the relationship between two variables, a separate measure of effect size does not need to be calculated. Cohen (1988) has suggested that an r of around .1 can be considered small, an r of .3 can be considered medium, and an r of .5 can be considered large.

To calculate the proportion of variance in one variable that can be accounted for by variance in the second, simply square Pearson's r.

For example, $r = .787$ in the current example, so:

$$r^2 = .787 * .787$$
$$= .619$$

Multiply r^2 by 100 to calculate the *percentage* of variance in one variable that can be accounted for by the second.

AKA:
r^2 is sometimes referred to as the "coefficient of determination".

That is, 61.9% of the variability in our participants' math test scores can be predicted (or explained, or accounted for, depending on the terminology you prefer) by variability in their reading ability scores.

12.3.4. APA Style Results Write-Up

Results

To assess the size and direction of the linear relationship between math test scores and reading ability, a bivariate Pearson's product-movement correlation coefficient (r) was calculated. The bivariate correlation between these two variables was positive and strong, $r(33) = .787, p < .001$.

Degrees of freedom for a bivariate correlation are defined as $N - 2$ (where N is the total sample size).

$df = 35 - 2$
$= 33$

Prior to calculating r, the assumptions of normality, linearity and homoscedasticity were assessed, and found to be supported. Specifically, a visual inspection of the normal Q-Q and detrended Q-Q plots for each variable confirmed that both were normally distributed. Similarly, visually inspecting a scatterplot of math test scores against reading ability confirmed that the relationship between these variables was linear and heteroscedastic.

A **Results** section should not include any interpretation of your findings. Save these for the **Discussion** section of your research report.

12.3.5. Summary

In this example, bivariate correlation was used to investigate the relationship between math test scores and reading ability. On reflection, it is possible that a third variable (e.g., intelligence) may impact on both of these variables and that the bivariate correlation we've just observed is a spurious one (that is, an artefact of the relationship that each has with IQ).

Therefore, to further explore the relationship between math and reading abilities we can use partial correlation, which allows us to statistically control (or partial out) the effects of a third variable (in this case, intelligence).

12.4. Illustrated Example Two: Partial Correlation

Math test, reading ability and intelligence test scores for 35 students are presented in Table 12.2. These data will be used to assess the strength and direction of the partial correlation between math test scores and reading ability, while controlling for intelligence.

Table 12.2

Math Test, Reading Ability and Intelligence Test Scores For A Sample of 35 Students

□ Data:
This is data file
data_12_2.sav
on the companion
website.

ID	Math	Reading	Intelligence	ID	Math	Reading	Intelligence
1	49	12	99	19	55	16	102
2	50	12	99	20	55	17	102
3	50	11	100	21	55	17	102
4	51	11	100	22	55	18	102
5	51	12	100	23	56	22	102
6	51	13	100	24	58	18	102
7	52	13	100	25	56	18	103
8	52	13	100	26	57	20	103
9	51	16	101	27	60	19	103
10	52	16	101	28	57	18	104
11	53	15	101	29	56	24	104
12	53	16	101	30	56	24	104
13	54	17	101	31	60	19	104
14	54	15	101	32	57	18	105
15	54	16	101	33	58	20	105
16	54	16	102	34	60	19	105
17	55	17	102	35	58	21	106
18	55	17	102				

12.4.1. PASW Statistics Procedure

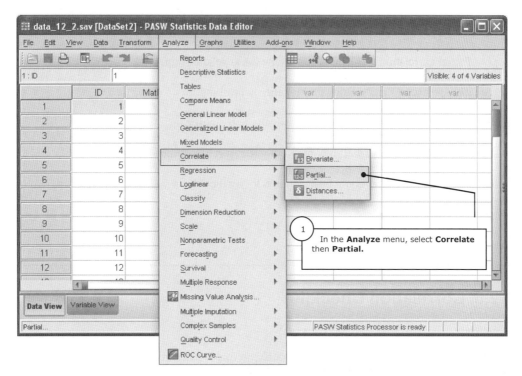

■ *Syntax:*
Run these
analyses with
syntax_12_2.sps
on the companion
website.

1 In the **Analyze** menu, select **Correlate** then **Partial.**

2 In the **Partial Correlations** dialogue move the variables to be correlated into the **Variables** list.

Move the control variable into the **Controlling for** list.

3 Click **OK** to output the analyses.

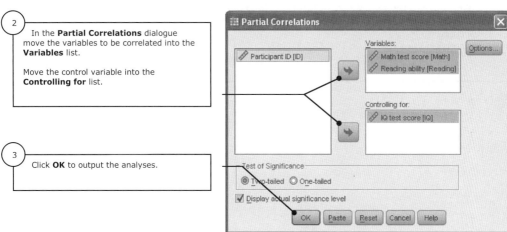

12.4.2. PASW Statistics Output

Partial Correlation

Correlations

Control Variables			Math test score	Reading ability
IQ test score	Math test score	Correlation	1.000	.226
		Significance (2-tailed)	.	.199
		df	0	32
	Reading ability	Correlation	.226	1.000
		Significance (2-tailed)	.199	.
		df	32	0

The **Correlations** table reports the partial correlation between math test scores and reading ability, controlling for intelligence.

In this example the correlation coefficient (Pearson's r) is .226, and is statistically non-significant (Sig = .199).

When intelligence was taken into account, the strong, significant relationship between math test scores and reading ability observed in the first illustrated example became statistically non-significant.

12.4.3. APA Style Results Write-Up

Results

Partial correlation was used to assess the linear relationship between math test scores and reading ability, after controlling for intelligence. The partial correlation was statistically non-significant, $r(32) = .23, p = .199$. After partialling out intelligence, just 5% of the variability in math test scores could be accounted for by the variability in reading abilities.

> Don't forget to report the results of your assumption testing, as we did in the previous example.

12.5. Correlation Checklist

Have you:

- ✔ Checked that each variable is approximately normally distributed?
- ✔ Checked that the relationships between variables are linear?
- ✔ Taken note of the r-value, N (and/or degrees of freedom) and significance value for your write-up?
- ✔ Calculated r^2?
- ✔ Written up your results in the APA style?

Chapter 13: Multiple Regression

Chapter Overview

13.1. Purpose of Multiple Regression

To examine the linear relationship between one continuous criterion (or dependent) variable, and two or more predictor (or independent) variables. Predictor variables can be either continuous or dichotomous.

There are three main types of multiple regression analysis:

1. Standard (or simultaneous) multiple regression.
2. Hierarchical (or sequential) multiple regression.
3. Stepwise (or statistical) multiple regression.

Standard and hierarchical multiple regression are demonstrated in illustrated examples one and two respectively. They differ with respect to: (a) how shared variance (due to correlated predictors) is handled; and (b) the order in which predictors are entered into the regression equation.

Stepwise multiple regression is a data, rather than theory driven procedure. Its use is quite controversial (Tabachnick & Fidell, 2007b), and it will not be demonstrated in this chapter. For further information about this technique, including detailed instructions for performing a stepwise multiple regression with PASW Statistics, see Field (2009).

> **(i) Tip:**
> If you have a categorical predictor with more than two levels, it can be converted into a series of dichotomies suitable for use in multiple regression with a procedure known as "dummy coding".
>
> This is illustrated in the margin tip beside section 13.3.2.4.

13.2. Questions We Could Answer Using Multiple Regression

Multiple regression is geared towards prediction, and researchers using multiple regression analyses typically ask questions in the form of, "How well can we predict criterion variable X, with a linear combination of predictor variables A, B and C?" or "How much of the variability in criterion variable X can be accounted for by predictor variables A, B and C?" For example:

1. Can job satisfaction be predicted by a linear combination of hours worked per week, annual income, job autonomy and perceived opportunities for promotion?

2. How much of the variance in fear of crime can be explained by age, gender, past victimisation experiences and neighbourhood incivilities?

Beyond these basic questions, researchers using multiple regression will also ask about the unique (in standard regression) or incremental (in hierarchical regression) importance of specific predictors.

Examples of questions focussed on incremental importance (and tested with hierarchical multiple regression) include:

3. Do perceived opportunities for promotion add predictive utility to a model of job satisfaction that already includes hours worked per week, annual income and job autonomy?

4. Can neighbourhood incivilities account for additional variance in fear of crime, beyond the variance already explained by age, gender and past victimisation experiences?

13.3. Illustrated Example One: Standard Multiple Regression

Deterrence theory suggests that everyday compliance with the law is shaped by perceived punishment certainty and severity. That is, when the likelihood (i.e., certainty) and personal costs (i.e., severity) of breaking the law are both high, law-breaking behaviour is low. Conversely, when perceived punishment certainty and severity decrease, non-compliance should increase.

To test this theory, 62 participants were asked to complete a three-part questionnaire. The first part asked about the frequency of participants' everyday law breaking behaviours (e.g., littering, speeding, driving while intoxicated, shop-lifting etc.). Part two addressed perceived certainty, and asked participants to consider the likelihood of being punished for each of behaviours identified in part one of the questionnaire. Finally, part three asked about the personal costs (i.e., severity) of punishment.

Responses to each part of this measure were averaged, and are presented in Table 13.1.

Table 13.1

Compliance, Perceived Certainty and Perceived Severity Data (N = 62)

📖 *Data:*
This is data file
data_13_1.sav
on the companion
website.

ID	Compliance	Certainty	Severity	ID	Compliance	Certainty	Severity
1	2	5	7	32	1	6	4
2	2	7	10	33	1	4	3
3	5	8	6	34	1	5	5
4	2	9	7	35	2	6	8
5	3	3	2	36	2	5	9
6	3	3	5	37	2	4	9
7	4	6	8	38	3	5	6
8	4	5	4	39	4	9	5
9	6	4	3	40	3	4	4
10	4	3	7	41	4	2	8
11	4	2	6	42	2	6	6
12	5	4	4	43	3	2	4
13	3	5	3	44	5	5	4
14	5	5	6	45	4	7	9
15	6	5	9	46	4	8	8
16	7	6	3	47	4	10	6
17	1	6	2	48	3	6	5
18	7	5	6	49	7	7	5
19	4	5	5	50	6	8	4
20	5	8	6	51	6	6	5
21	5	5	5	52	6	7	6
22	4	1	7	53	6	9	8
23	4	4	6	54	3	5	7
24	4	4	3	55	5	6	5
25	4	9	7	56	3	4	6
26	5	7	7	57	3	3	3
27	5	5	5	58	5	7	7
28	3	3	2	59	5	6	6
29	3	3	1	60	4	3	5
30	1	2	1	61	2	7	8
31	7	4	7	62	4	6	4

13.3.1. Setting Up the PASW Statistics Data File

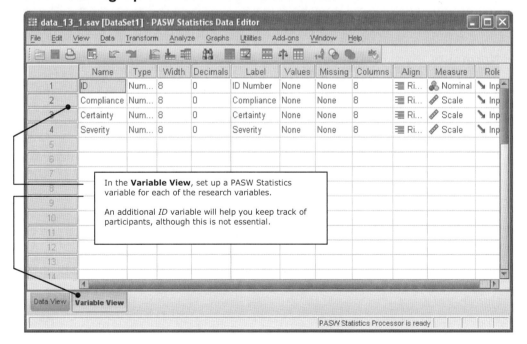

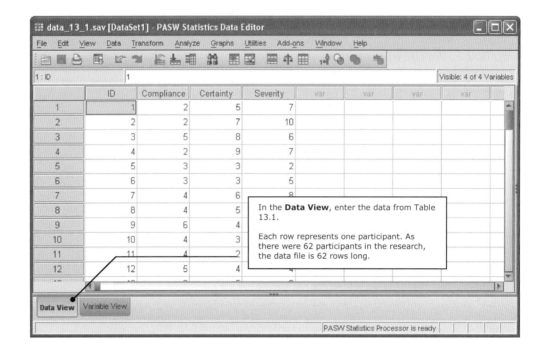

13.3.2. Analysing the Data

13.3.2.1. Assumptions

Syntax:
Run these
analyses with
syntax_13_1.sps
on the companion
website.

The following issues require consideration before running and interpreting a multiple regression analysis. The first is methodological, and addressed during the planning and data collection stages of research. The rest can be assessed through the multiple regression procedure.

1. **N (cases) : k (predictors) Ratio.** To produce a reliable regression model, a reasonable ratio of cases to predictors is required. Unfortunately, what is considered "reasonable" tends to vary throughout the literature! Consequently, we defer to Tabachnick and Fidell (2007b), who suggest that N should ideally be $50 + 8(k)$ for testing a full regression model, or $104 + k$ when testing individual predictors. (If you're doing both, which you probably are, calculate N both ways, then aim for the largest.) These "rules of thumb" assume medium sized effects.

① Tip:
Howell (2010b) cites
several more rules of
thumb for calculating
appropriate sample
sizes, including our
favourite, courtesy of
Darlington: "more is
better"!

2. **Normality.** Each continuous variable should be approximately normally distributed. PASW Statistics provides many graphical (e.g., histograms, stem-and-leaf plots, normal Q-Q plots, detrended normal Q-Q plots and boxplots) and statistical (e.g., skewness, kurtosis, Kolmogorov-Smirnov and Shapiro-Wilk statistics) means of assessing univariate normality.

▶▶ Link:
These methods for
assessing univariate
normality are all
illustrated in section
3.4 of chapter 3.

3. **Outliers.** Multiple regression is sensitive to outliers and other influential cases, which should be removed or reduced to lessen their impact on the final regression solution.

① Tip:
As an example of a
multivariate outlier,
imagine a 15-year-old
earning $70,000 per
annum.

There are two types of outliers: univariate and multivariate. Univariate outliers are cases with extreme values on single variables, and can be detected graphically with boxplots. Multivariate outliers are cases with unusual combinations of values across two or more predictor variables. They can be detected with **Mahalanobis distance**, as illustrated later in this chapter.

4. **Multicollinearity.** High correlations (e.g., $r \geq .85$) between predictors render a multiple regression model unstable, and very difficult to interpret. Multicollinearity can be detected with the **Tolerance** and **VIF** (variance inflation factor) statistics, as illustrated later in this chapter.

5. **Normality, Linearity and Homoscedasticity of Residuals.** It is assumed that the differences between the observed and predicted values on the criterion variable (referred to as "residuals") are normally distributed, and that their relationship with the predicted values on the criterion is linear. Finally, it is assumed that the variance in the residuals is homogenous across the full range of predicted values.

These three assumptions can be assessed using the graphical methods described later in this chapter.

13.3.2.2. PASW Procedure (Part 1: Normality & Univariate Outliers)

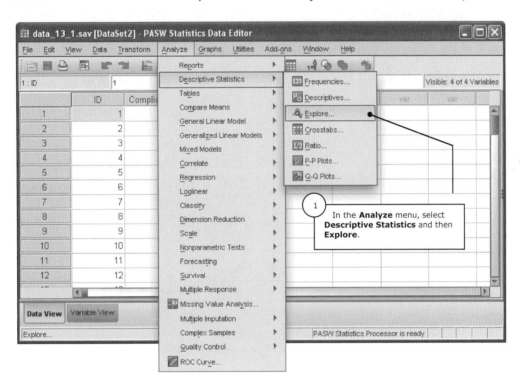

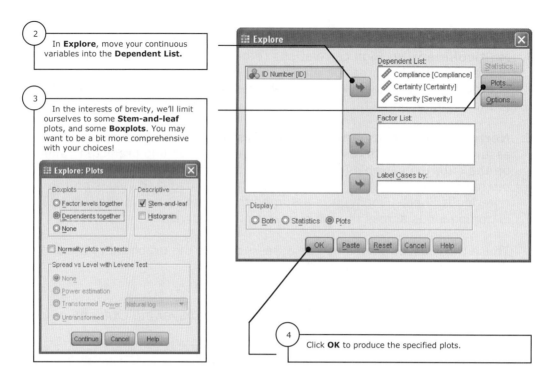

> **2**
> In **Explore**, move your continuous variables into the **Dependent List.**

> **3**
> In the interests of brevity, we'll limit ourselves to some **Stem-and-leaf** plots, and some **Boxplots.** You may want to be a bit more comprehensive with your choices!

> **4**
> Click **OK** to produce the specified plots.

13.3.2.3. PASW Output (Part 1: Normality & Univariate Outliers)

Stem-and-Leaf Plots

Compliance Stem-and-Leaf Plot

Frequency	Stem &	Leaf
5.00	1 .	00000
8.00	2 .	00000000
12.00	3 .	000000000000
16.00	4 .	0000000000000000
11.00	5 .	00000000000
6.00	6 .	000000
4.00	7 .	0000

Stem width: 1
Each leaf: 1 case(s)

Severity Stem-and-Leaf Plot

Frequency	Stem &	Leaf
2.00	1 .	00
3.00	2 .	000
6.00	3 .	000000
8.00	4 .	00000000
11.00	5 .	00000000000
12.00	6 .	000000000000
9.00	7 .	000000000
6.00	8 .	000000
4.00	9 .	0000
1.00	10 .	0

Stem width: 1
Each leaf: 1 case(s)

Severity Stem-and-Leaf Plot

Frequency	Stem &	Leaf
2.00	1 .	00
3.00	2 .	000
6.00	3 .	000000
8.00	4 .	00000000
11.00	5 .	00000000000
12.00	6 .	000000000000
9.00	7 .	000000000
6.00	8 .	000000
4.00	9 .	0000
1.00	10 .	0

Stem width: 1
Each leaf: 1 case(s)

These three **Stem-and-Leaf Plots** are all roughly symmetrical and bell shaped, indicating that <u>univariate non-normality is not a concern in this data set.</u> This conclusion is supported by the three roughly symmetrical boxplots to the left.

If the Assumption is Violated

Mild departures from normality are generally not of concern. However, researchers faced with more dramatic departures should consider variable transformation (see Field, 2009, or Tabachnick & Fidell, 2007b).

(i) Tip:
Before deleting cases or otherwise changing values in your data file, make a backup copy! You should then run your analyses on both the original and modified data, and examine the effects that your changes have on your results. If these effects are dramatic, report both sets of results.

Regardless of whether you ultimately report one set of results or two, *always* describe any changes you made to your data set (and why) in your write-up.

If there were any *outliers* on these distributions, each would be denoted with a circle and data file row number. (e.g., o[18]). On a boxplot, an outlier is defined as a score between 1.5 and 3 box lengths above or below the box boundaries.

If there were *extreme scores*, each would be denoted with an asterisk and a data file row number (e.g., *[22]). On a boxplot, an extreme score is defined as a score greater than 3 box lengths above or below the box boundaries.

If Outliers are Detected

Assuming that the outliers or extreme scores are not data entry errors, there are at least three courses of action available:

1. Delete the offending cases. You have much stronger justification for doing this if your outliers are not a part of the population that you intended to sample.
2. Change the offending data points to:
   - One unit higher than the largest non-outlier.
   - Three standard deviations above/below the mean of the distribution. (An outlier is typically defined as a score that is > 3.29 SD from the mean.)
3. Transform the variable(s) with outliers (see Field, 2009, or Tabachnick & Fidell, 2007b). Outliers tend to produce non-normality, and transformation can address both of these problems at once.

13.3.2.4. PASW Statistics Procedure (Part 2: The Remaining Assumptions & the Standard Multiple Regression)

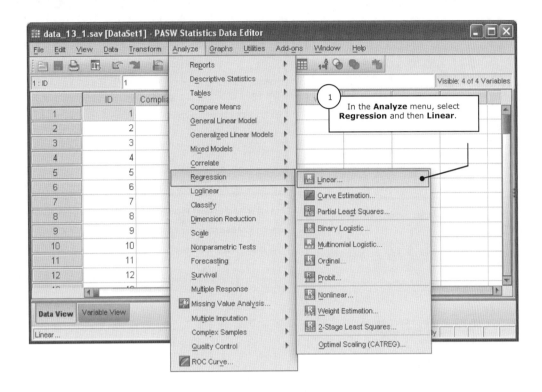

(i) Tip:
Although both of our predictor variables are continuous, you can also use categorical variables as predictors in multiple regression. However, if they have more than two levels, they will need to be dummy coded first.

Dummy coding involves converting a categorical variable with three or more levels into a series of dichotomous variables, each of which is then used to indicate the absence (usually coded as "0") or presence (coded as "1") of one level of the original categorical variable.

There will always be one fewer dichotomous "dummy" variables than there are levels of the original categorical variable.

For example, if we wanted to use employment status (EMP) – continuing (CO), fixed term (FT) or casual (CA) – as a predictor, we would first need to create two dummy variables: one to distinguish between CO (=1) and non-CO (=0) cases; and one to distinguish between FT (=1) and non-FT (=0) cases. These are variables D1 and D2, respectively:

EMP	D1	D2
CO	1	0
CO	1	0
FT	0	1
FT	0	1
CA	0	0
CA	0	0

These dichotomous dummy variables can then be entered into the regression model together.

(Note, do not include the original categorical variable in the multiple regression as well as the dummy variables.)

(1) In the **Analyze** menu, select **Regression** and then **Linear**.

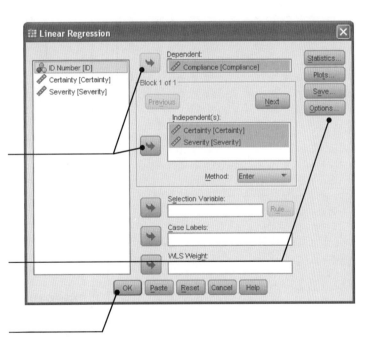

(2) Move the criterion variable into the **Dependent** box, and the predictor variable(s) into the **Independent(s)** list.

While the criterion must be measured on an interval or ratio scale, the predictor variables can be either continuous or categorical (refer to the tip in the page margin).

(3) Select the options you require in the **Statistics**, **Plots**, **Save** and **Options** dialogues, each of which is illustrated overleaf.

(4) Click **OK** to output the multiple regression results.

Linear Regression: Statistics

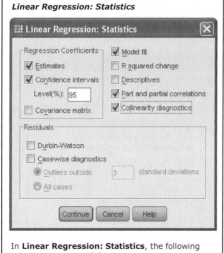

In **Linear Regression: Statistics**, the following selections are essential:

Estimates. Provides the information that is needed to assess the roles played by individual predictors in the regression model.

Model fit. Provides the information needed to assess the predictive utility of the entire regression model.

You may also want to select **Confidence Intervals**. **Part and partial correlations**, and **Collinearity diagnostics**.

Click **Continue** to close this dialogue.

Linear Regression: Plots

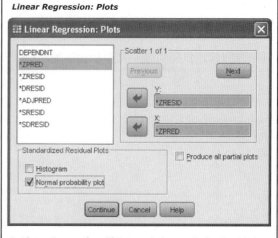

In **Linear Regression: Plots**, graph the standardised residuals against standardised predicted values on the criterion variable by moving ***ZRESID** into the **Y** box and ***ZPRED** into the **X** box.

Also, select **Normal probability plot**.

These will give us the information needed to assess the normality, linearity and homoscedasticity of residuals assumptions.

Click **Continue** to close this dialogue.

Linear Regression: Save

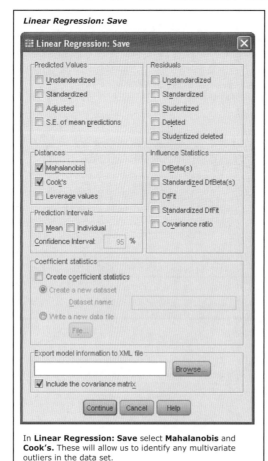

In **Linear Regression: Save** select **Mahalanobis** and **Cook's.** These will allow us to identify any multivariate outliers in the data set.

Click **Continue** to close this dialogue.

Linear Regression: Options

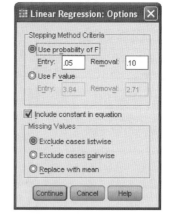

In **Linear Regression: Options**, the default selections are fine.

The **Stepping Method Criteria** are only applicable to stepwise procedures.

Exclude cases listwise means that cases missing data on any variables in the regression will be excluded from all the specified analyses (that is, PASW Statistics will simply ignore them).

Click **Continue** to close this dialogue.

13.3.2.5. PASW Statistics Output (Part 2: The Remaining Assumptions & the Standard Multiple Regression)

Regression

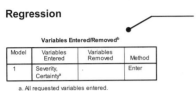

Variables Entered/Removed[b]

Model	Variables Entered	Variables Removed	Method
1	Severity, Certainty[a]	.	Enter

a. All requested variables entered.

b. Dependent Variable: Compliance

In a standard multiple regression, all predictor variables are entered into the regression simultaneously. Hence, we only have one **Model** to consider.

In *Illustrated Example Two*, the predictor variables will be entered in blocks, and this table will summarise the **Variables Entered** into multiple models.

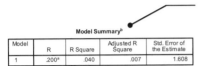

Model Summary[b]

Model	R	R Square	Adjusted R Square	Std. Error of the Estimate
1	.200[a]	.040	.007	1.608

a. Predictors: (Constant), Severity, Certainty

b. Dependent Variable: Compliance

The **Model Summary** table includes the following statistics:

R. The multiple correlation between the criterion variable and the set of predictor variables in the regression model. **R** is usually overlooked, in favour of the more interpretable R^2 and adjusted R^2 statistics.

R Square. R^2 represents the proportion of variance in the criterion that can be accounted for by the predictor variables in combination. Here, only 4% of the variance in *compliance* can be explained by *severity* and *certainty*.

Adjusted R Square. Compared to R^2, adjusted R^2 provides a more accurate estimate of the true extent of the relationship between the predictor variables and the criterion. In other words, it offers a better estimate of the population R^2. If we were to replicate this study many times with samples drawn from the same population we would, on average, account for around 0.7% of the variance in *compliance* with *severity* and *certainty*.

Some researchers report adjusted R^2 instead of R^2; some report it in addition to R^2; and many ignore it all together! We think that you should report both, and allow your readers to make the final decision regarding which to interpret.

The **ANOVA** table tells us whether or not the full regression model has predictive utility. That is, whether or not the predictors collectively account for a statistically significant proportion of the variance in the criterion variable. Or, to put it a third way, whether or not R^2 departs significantly from zero.

Here, the **ANOVA** is non-significant (*Sig* > .05), indicating that R^2 does not depart significantly from zero. In other words, punishment *severity* and *certainty* in combination could not account for any more of the variance in *compliance* than we would expect by chance.

When writing up the results of a multiple regression we usually report the following, along with R^2 and/or adjusted R^2:

F = 1.227
df (Regression) = 2
df (Residual) = 59
Sig = .300

① Tip:
The difference between R^2 and adjusted R^2 is sometimes referred to as "shrinkage".

If your *N:k* ratio is small (e.g., less than 10:1), expect to see a large amount of shrinkage. (In such circumstances, *always* include adjusted R^2 in your results.)

If your *N:k* ratio is greater than 30:1, the amount of shrinkage you see should be small (Weinberg & Abramowitz, 2008).

The first half of the **Coefficients** table details the role each individual predictor plays in the regression model. It is here that we find out which predictors say something unique about the criterion, and which are redundant.

The **Unstandardized Coefficients** or **B** weights indicate the predicted change in the criterion associated with a 1-unit change in the relevant predictor, after controlling for the effects of all the other predictors in the model.

For example, the unstandardised regression coefficient for *certainty* is .154. This means that, after controlling *severity*, a 1-unit increase in certainty will result in a predicted .154-unit increase in *compliance*.

The **Standardized Coefficients** or **Beta** (β) weights indicate the predicted change – in standard deviations (*SD*) – in the criterion associated with a 1 *SD* change in the relevant predictor, after controlling for the effects of the remaining predictors in the model.

For example, the standardised regression coefficient for *certainty* is .193. In other words, after controlling *severity*, a 1 *SD* increase in *certainty* will result in a .193 *SD* increase in *compliance*.

① Tip:
Beta (β) weights are *scale free*, and can be used to compare predictors within a regression model (e.g., to compare the impact of certainty and severity on compliance).

B weights are appropriate for comparing predictors between regression models (e.g., to compare the impact of certainty in the current sample with its impact in a prison sample).

ANOVA[b]

Model		Sum of Squares	df	Mean Square	F	Sig.
1	Regression	6.350	2	3.175	1.227	.300[a]
	Residual	152.618	59	2.587		
	Total	158.968	61			

a. Predictors: (Constant), Severity, Certainty

b. Dependent Variable: Compliance

Coefficients[a]

Model		Unstandardized Coefficients		Standardized Coefficients		
		B	Std. Error	Beta	t	Sig.
1	(Constant)	2.969	.697		4.262	.000
	Certainty	.154	.108	.193	1.424	.160
	Severity	.015	.105	.019	.143	.887

a. Dependent Variable: Compliance

The **t** statistics (and corresponding **Sig** levels) indicate whether or not each predictor accounts for a significant proportion of unique variance in the criterion. (Unique variance is variance that cannot be explained by other predictors in the regression model.)

The degrees of freedom for these *t* tests are $N - k - 1$ (where *N* is the sample size, and *k* is the number of predictors in the model).

For example, *certainty* cannot account for variance in *compliance* beyond that which can also be explained by *severity*. It is a non-significant predictor, $t(59)$ = 1.42, *p* = .160.

Severity is also non-significant.

The second half of the **Coefficients** table includes **95% Confidence Intervals for B**, **Correlations** and **Collinearity Statistics**.

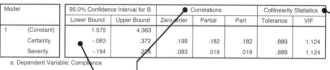

Coefficients[a] *cont.*

Model		95.0% Confidence Interval for B		Correlations			Collinearity Statistics	
		Lower Bound	Upper Bound	Zero-order	Partial	Part	Tolerance	VIF
1	(Constant)	1.575	4.363					
	Certainty	-.063	.372	.199	.182	.182	.889	1.124
	Severity	-.194	.224	.083	.019	.018	.889	1.124

a. Dependent Variable: Compliance

We can be 95% confident that the interval between the **Lower Bound** and **Upper Bound** contains the true population **B**.

When a confidence interval contains zero, the corresponding predictor is non-significant. This is the case for both predictors in this regression model.

The **Correlations** section of the **Coefficients** table provides three statistics for each predictor:

Zero-order. The bivariate Pearson's correlation between the predictor and criterion. For example, the Pearson's correlation between *certainty* and *compliance* is $r = .199$.

Partial. The partial correlation between the predictor and criterion.

Part. The semi-partial correlation (sr) between the predictor and criterion.

sr is a particularly useful statistic, as it can be squared to give the proportion of variance in the criterion that is uniquely explained by the predictor. In other words, sr^2 is the amount by which R^2 would decrease if the predictor were removed from the regression model.

Here, sr for *certainty* is .182, indicating that around 3.3% of the variance in *compliance* can be uniquely attributed to perceived punishment *certainty*.

Two measures of multicollinearity (high intercorrelations between predictor variables) are provided in the **Collinearity Statistics** section of the **Coefficients** table:

Tolerance. Predictor variables with tolerances < 0.1 are multicollinear with one or more other predictors, and definitely cause for concern. Menard (2002) argues that predictors with tolerances < 0.2 warrant closer inspection. Clearly, multicollinearity is not a problem in the current example!

VIF. VIF = 1/TOL. Therefore, predictor variables with VIFs > 10 are clearly multicollinear with one or more other predictors. You should probably also look twice at any predictors with VIFs > 5.

Addressing Multicollinearity

When faced with multicollinearity consider (a) deleting one of the offending variables, or (b) combining them into a single composite variable. The analyses can then be repeated with the modified data file.

(i) Tip:
The critical χ^2 values for 2-10 degrees of freedom at $\alpha = .001$ are as follows:

df	Critical χ^2
2	13.816
3	16.266
4	18.467
5	20.515
6	22.458
7	24.322
8	26.125
9	27.877
10	29.588

For *df* > 10, refer to the tables available in the appendices of most decent statistics texts (e.g., Howell, 2010b).

(i) Tip:
Run your analyses both with and without the problem cases. (Use **Select Cases** in the **Data** menu if you don't want to juggle multiple data files.) Hopefully, any differences will be minimal. If not, report both sets of results, along with a clear description of what you did, and why.

Residuals Statistics[a]

	Minimum	Maximum	Mean	Std. Deviation	N
Predicted Value	3.23	4.60	3.87	.323	62
Std. Predicted Value	-1.993	2.269	.000	1.000	62
Standard Error of Predicted Value	.211	.576	.340	.100	62
Adjusted Predicted Value	3.11	4.68	3.88	.331	62
Residual	-2.955	3.309	.000	.582	62
Std. Residual	-1.838	2.057	.000	.983	62
Stud. Residual	-1.894	2.098	-.003	1.007	62
Deleted Residual	-3.172	3.442	-.005	1.660	62
Stud. Deleted Residual	-1.938	2.163	-.003	1.020	62
Mahal. Distance	.067	6.853	1.968	1.730	62
Cook's Distance	.000	.101	.017	.024	62
Centered Leverage Value	.001	.112	.032	.028	62

a. Dependent Variable: Compliance

A **Maximum Mahalanobis Distance** larger than the critical chi-square (χ^2) value for *df = k* (i.e., the number of *predictor* variables in the model) at $\alpha = .001$ indicates the presence of one or more multivariate outliers.

The critical χ^2 value for *df = 2* at $\alpha = .001$ is 13.816. (Tables of χ^2 critical tables are available in the appendices of most statistics textbooks.) As our **Maximum Mahalanobis Distance** is only 6.853, we need not be concerned about multivariate outliers.

Dealing with Multivariate Outliers

First, return to the **PASW Statistics Data Editor**, where two additional variables (**MAH_1** and **COO_1**) will have been appended to your data file. Sort the data file by **MAH_1** (in the **Data** menu, select **Sort Cases**), so cases with the largest Mahalanobis distances are grouped together. Pay particular attention to those which also have **COO_1** values > 1. (**Cook's Distance** is a measure of influence. That is, it's a measure of the impact that each individual case has on the predictive efficacy of the regression model as a whole.)

Once all the multivariate outliers in the data file have been identified, you will need to decide on what to do with them. The most common strategies for dealing with multivariate outliers include (a) ignoring them; (b) deleting them; or (c) modifying them to reduce their impact on the regression model. Note that it is quite reasonable to use different strategies for different types of multivariate outliers (e.g., deleting those that are influential; ignoring those that are not). Be mindful that deleting one problem case can often reveal others, and that multivariate outliers have a reputation for hiding behind each other!

Refer to Tabachnick and Fidell (2007b) for a more comprehensive discussion of methods for dealing with multivariate outliers.

Charts

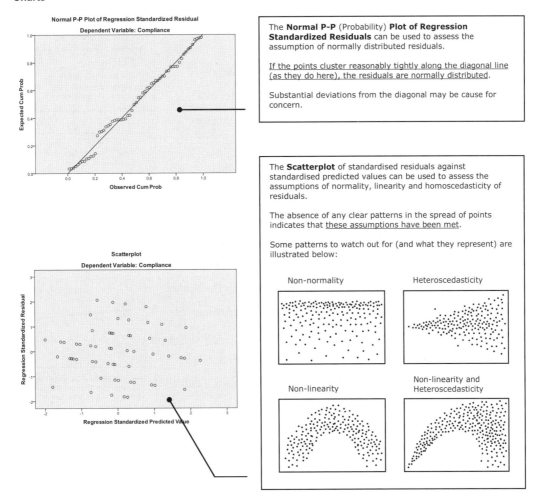

Normal P-P Plot of Regression Standardized Residual
Dependent Variable: Compliance

The **Normal P-P** (Probability) **Plot of Regression Standardized Residuals** can be used to assess the assumption of normally distributed residuals.

If the points cluster reasonably tightly along the diagonal line (as they do here), the residuals are normally distributed.

Substantial deviations from the diagonal may be cause for concern.

The **Scatterplot** of standardised residuals against standardised predicted values can be used to assess the assumptions of normality, linearity and homoscedasticity of residuals.

The absence of any clear patterns in the spread of points indicates that these assumptions have been met.

Some patterns to watch out for (and what they represent) are illustrated below:

Non-normality

Heteroscedasticity

Non-linearity

Non-linearity and Heteroscedasticity

Scatterplot
Dependent Variable: Compliance

13.3.3. Follow Up Analyses

13.3.3.1. Effect Size

Although R^2, defined as the proportion of variance in the criterion that can be accounted for by the predictors in combination, is an adequate index of effect size for multiple regression, you may have cause to also calculate Cohen's f^2:

$$f^2 = \frac{R^2}{1-R^2}$$

R^2 can be read directly off the **Model Summary** table, so:

$$f^2 = \frac{.040}{1-.040}$$

$$= .042$$

This is a small effect.

(i) *Tip:*
Jacob Cohen (1988) suggested that an f^2 of .02 (or R^2 of .0196) can be considered small, an f^2 of .15 (or R^2 of .13) can be considered medium, and an f^2 of .35 (or R^2 of .26) can be considered large.

13.3.4. APA Style Results Write Up

Before reporting the multiple regression, it can be useful to report some basic descriptive statistics (e.g., means and standard deviations) for each variable, as well as the bivariate correlations between them.

In the interests of brevity, we've omitted these statistics from the current example.

Results

To estimate the proportion of variance in everyday compliance with the law that can be accounted for by perceived punishment certainty and severity, a standard multiple regression analysis (MRA) was performed.

Prior to interpreting the results of the MRA, several assumptions were evaluated. First, stem-and-leaf plots and boxplots indicated that each variable in the regression was normally distributed, and free from univariate outliers. Second, inspection of the normal probability plot of standardised residuals as well as the scatterplot of standardised residuals against standardised predicted values indicated that the assumptions of normality, linearity and homoscedasticity of residuals were met. Third, Mahalanobis distance did not exceed the critical χ^2 for $df = 2$ (at $\alpha = .001$) of 13.82 for any cases in the data file, indicating that multivariate outliers were not of concern. Fourth, relatively high tolerances for both predictors in the regression model indicated that multicollinearity would not interfere with our ability to interpret the outcome of the MRA.

In combination, punishment certainty and severity accounted for a non-significant 4% of the variability in everyday compliance with the law, $R^2 = .04$, adjusted $R^2 = .007$, $F (2, 59) = 1.23$, $p = .300$. Unstandardised (B) and standardised (β) regression coefficients, and squared semi-partial (or 'part') correlations (sr^2) for each predictor in the regression model are reported in Table 1.

Open with a clear statement of how ("...a standard multiple regression...") and why ("To estimate the proportion of variance in everyday...") the data were analysed.

If an assumption is violated, describe how the violation was addressed, and whether or not you managed to fix the problem.

When writing up the actual standard multiple regression, you need to report two sets of findings:

1. The predictive utility of the entire model. Here, we did this with R^2, adjusted R^2 and the associated ANOVA test of statistical significance.

2. The predictive utility of each predictor in the model. Here, we did this by reporting B, β and sr^2 (and the results of the associated significance tests, denoted with asterisks) for each predictor in Table 1.

Table 1

Unstandardised (B) and Standardised (β) Regression Coefficients, and Squared Semi-Partial Correlations (sr^2) For Each Predictor in a Regression Model Predicting Everyday Compliance With the Law

Variable	B [95% CI]	β	sr^2
Certainty	.154 [-.063, .372]	.193	.033
Severity	.015 [-.194, .224]	.019	.001

Note. $N = 62$. CI = confidence interval.
* $p < .05$. ** $p < .01$.

In a table, indicate that a regression coefficient is statistically significant with one or more asterisks (*).

For example, if certainty was a significant (at the .05 level) predictor, we would report:

Variable	B [95%...
Certainty	.154 [-.0...*
Severity	.015 [-.1...

* $p < .05$.

Don't forget to include a note beneath the table indicating how the asterisks should be interpreted.

13.3.5. Summary

In the previous example of a standard multiple regression, all the predictors were simultaneously entered into the regression equation, and we were able to ask two basic questions: (a) how much variance in the criterion variable can be accounted for by the predictors in combination; and (b) to what extent does each predictor contribute uniquely to the regression model?

The next example demonstrates the use of hierarchical multiple regression. In hierarchical regression, the researcher adds predictors to a regression model in blocks, and asks about the incremental contribution they make (that is, the variance in the criterion they explain beyond the variance already accounted for by previous blocks of predictors).

13.4. Illustrated Example Two: Hierarchical Multiple Regression

A number of researchers have argued that deterrence theory (see section 13.3) provides an inadequate explanation for everyday compliance with the law. They note that perceived punishment certainty and severity typically account for a very small proportion of the variance in compliance, and that law-abidingness is primarily motivated by intrinsic factors, and not the possibility of punishment. One such intrinsic factor is legitimacy, which refers to an internalised sense of responsibility to support legal authorities and to abide by their rules and decisions (Tyler & Huo, 2002).

The 62 participants described in the previous illustrated example were asked to complete a measure of legitimacy in addition to the measures of compliance, certainty and severity discussed in section 13.3. Their legitimacy data are presented in Table 13.2, and can be appended to the data provided in Table 13.1.

We hypothesise that legitimacy will account for a significant proportion of variance in compliance, beyond that already accounted for by certainty and severity.

Table 13.2

⬜ **Data:**
This data (merged with the data in Table 13.1) is **data_13_2.sav** on the companion website.

Legitimacy Data To Be Added To the Data in Table 13.1 (N = 62)

ID	Legitimacy	ID	Legitimacy	ID	Legitimacy	ID	Legitimacy
1	2	17	2	33	3	49	5
2	2	18	5	34	1	50	6
3	6	19	7	35	4	51	3
4	4	20	5	36	3	52	4
5	1	21	5	37	3	53	4
6	4	22	4	38	3	54	4
7	5	23	6	39	5	55	6
8	6	24	3	40	4	56	5
9	5	25	4	41	3	57	3
10	5	26	4	42	3	58	7
11	2	27	5	43	2	59	6
12	1	28	2	44	5	60	3
13	3	29	1	45	4	61	5
14	5	30	1	46	3	62	5
15	2	31	6	47	6		
16	7	32	2	48	5		

13.4.1. Setting Up the PASW Statistics Data File

Add an extra variable to the data file created in section 13.3.1, and call it **Legitimacy**. Enter the data from Table 13.2.

13.4.2. Analysing the Data

🖥 **Syntax:**
Run these analyses with **syntax_13_2.sps** on the companion website.

13.4.2.1. Assumptions

The assumptions for hierarchical multiple regression are the same as those listed for the standard regression in section 13.3.2.1.

13.4.2.2. PASW Statistics Procedure

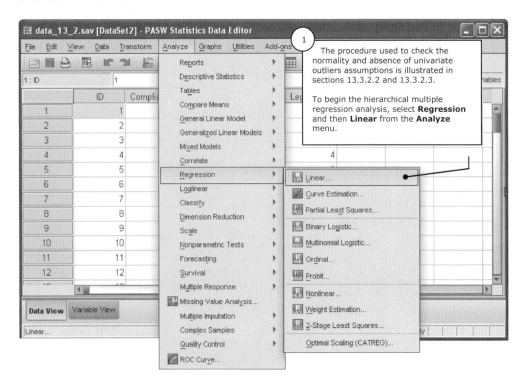

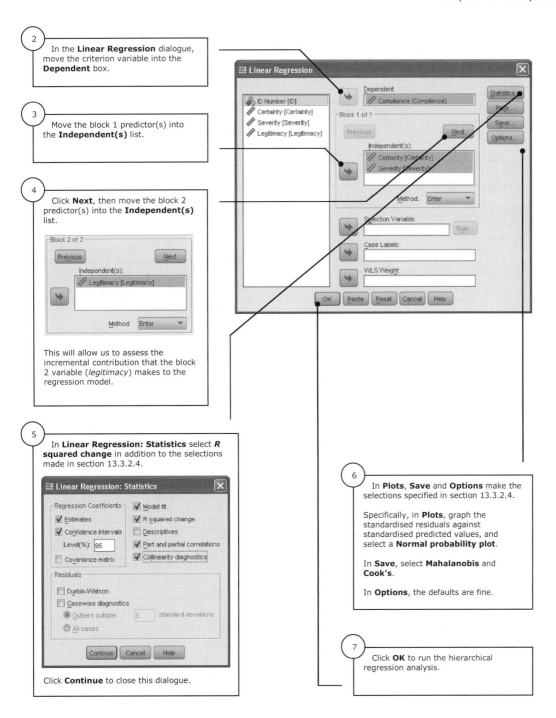

2 In the **Linear Regression** dialogue, move the criterion variable into the **Dependent** box.

3 Move the block 1 predictor(s) into the **Independent(s)** list.

4 Click **Next**, then move the block 2 predictor(s) into the **Independent(s)** list.

This will allow us to assess the incremental contribution that the block 2 variable (*legitimacy*) makes to the regression model.

5 In **Linear Regression: Statistics** select *R squared change* in addition to the selections made in section 13.3.2.4.

Click **Continue** to close this dialogue.

6 In **Plots**, **Save** and **Options** make the selections specified in section 13.3.2.4.

Specifically, in **Plots**, graph the standardised residuals against standardised predicted values, and select a **Normal probability plot**.

In **Save**, select **Mahalanobis** and **Cook's**.

In **Options**, the defaults are fine.

7 Click **OK** to run the hierarchical regression analysis.

13.4.2.3. PASW Statistics Output

Legitimacy

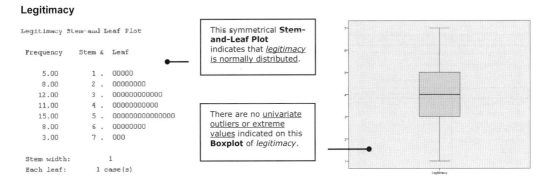

```
Legitimacy Stem-and-Leaf Plot

Frequency     Stem &  Leaf

    5.00        1 .  00000
    8.00        2 .  00000000
   12.00        3 .  000000000000
   11.00        4 .  00000000000
   15.00        5 .  000000000000000
    8.00        6 .  00000000
    3.00        7 .  000

Stem width:         1
Each leaf:      1 case(s)
```

This symmetrical **Stem-and-Leaf Plot** indicates that *legitimacy* is normally distributed.

There are no univariate outliers or extreme values indicated on this **Boxplot** of *legitimacy*.

▶▶| *Links*
Stem-and-Leaf Plots and **Boxplots** for the other variables in this regression are presented in section 13.3.2.3.

Regression

Variables Entered/Removed[b]

Model	Variables Entered	Variables Removed	Method
1	Severity, Certainty[a]	.	Enter
2	Legitimacy[a]	.	Enter

a. All requested variables entered.

b. Dependent Variable: Compliance

Because the variables were entered on two blocks, we have two **Models** to consider.

Model 1 is the same as the standard multiple regression in *Illustrated Example One*. The criterion variable is compliance, and the two predictor variables are *certainty* and *severity*.

Model 2 adds an additional predictor variable to the regression equation: *legitimacy*.

Model Summary[c]

Model	R	R Square	Adjusted R Square	Std. Error of the Estimate	Change Statistics R Square Change	F Change	df1	df2	Sig. F Change
1	.200[a]	.040	.007	1.608	.040	1.227	2	59	.300
2	.545[b]	.297	.261	1.388	.257	21.196	1	58	.000

a. Predictors: (Constant), Severity, Certainty

b. Predictors: (Constant), Severity, Certainty, Legitimacy

c. Dependent Variable: Compliance

The block 1 variables (*certainty* and *severity*) collectively accounted for 4% of the variability in compliance (R^2 = .040). This proportion of variance was statistically non-significant, $F(2, 59) = 1.23$, $p = .300$.

When *legitimacy* was entered into the regression model on block 2, R^2 increased from .040 to .297 (an **R Square Change** of .257). In other words, *legitimacy* accounted for an additional 25.7% of the variability in *compliance*, above and beyond the meagre 4% already accounted for by *certainty* and *severity*.

This incremental increase in R^2 on block 2 was statistically significant, $\Delta F(1,58) = 21.20$, $p < .001$.

(i) Tip:
R Square Change can be written as:

ΔR^2
R^2_{change}

Similarly, **F Change** can be written as:

ΔF
F_{change}

ANOVA[c]

Model		Sum of Squares	df	Mean Square	F	Sig.
1	Regression	6.350	2	3.175	1.227	.300[a]
	Residual	152.618	59	2.587		
	Total	158.968	61			
2	Regression	47.196	3	15.732	8.164	.000[b]
	Residual	111.772	58	1.927		
	Total	158.968	61			

a. Predictors: (Constant), Severity, Certainty

b. Predictors: (Constant), Severity, Certainty, Legitimacy

c. Dependent Variable: Compliance

Looking at the lower half of the **ANOVA** table (**Model 2**), the three predictors collectively accounted for a statistically significant proportion of the variance in *compliance*, $F(3, 58) = 8.16$, $p < .001$.

Coefficients[a]

Model		Unstandardized Coefficients B	Std. Error	Standardized Coefficients Beta	t	Sig.
1	(Constant)	2.969	.697		4.262	.000
	Certainty	.154	.108	.193	1.424	.160
	Severity	.015	.105	.019	.143	.887
2	(Constant)	1.825	.651		2.806	.007
	Certainty	-.012	.100	-.014	-.115	.909
	Severity	-.009	.090	-.012	-.099	.922
	Legitimacy	.546	.119	.552	4.604	.000

a. Dependent Variable: Compliance

When *certainty*, *severity* and *legitimacy* were combined in **Model 2**, *legitimacy* emerged as the only predictor capable of explaining a significant proportion of unique variance in *compliance*, $t(58) = 4.60$, $p < .001$.

Squaring the **Part** (i.e., semi-partial) correlation for *legitimacy* in **Model 2** indicates that it uniquely accounts for 25.7% of the variance in *compliance* ($sr^2 = .507 * .507 = .257$).

Coefficients[a] *cont.*

Model		95.0% Confidence Interval for B Lower Bound	Upper Bound	Correlations Zero-order	Partial	Part	Collinearity Statistics Tolerance	VIF
1	(Constant)	1.575	4.363					
	Certainty	-.063	.372	.199	.182	.182	.889	1.124
	Severity	-.194	.224	.083	.019	.018	.889	1.124
2	(Constant)	.523	3.128					
	Certainty	-.212	.189	.199	-.015	-.013	.775	1.291
	Severity	-.190	.172	.083	-.013	-.011	.886	1.128
	Legitimacy	.308	.783	.545	.517	.507	.842	1.187

a. Dependent Variable: Compliance

These **Collinearity Statistics** indicate that multicollinearity is not a problem in the current data set.

Excluded Variables[b]

Model		Beta In	t	Sig.	Partial Correlation	Collinearity Statistics Tolerance	VIF	Minimum Tolerance
1	Legitimacy	.552[a]	4.604	.000	.517	.842	1.187	.775

a. Predictors in the Model: (Constant), Severity, Certainty

b. Dependent Variable: Compliance

Residuals Statistics[a]

	Minimum	Maximum	Mean	Std. Deviation	N
Predicted Value	2.27	5.55	3.87	.880	62
Std. Predicted Value	-1.822	1.907	.000	1.000	62
Standard Error of Predicted Value	.213	.527	.342	.086	62
Adjusted Predicted Value	2.08	5.70	3.87	.885	62
Residual	-2.401	3.222	.000	1.354	62
Std. Residual	-1.730	2.321	.000	.975	62
Stud. Residual	-1.769	2.447	.000	1.008	62
Deleted Residual	-2.513	3.582	-.001	1.447	62
Stud. Deleted Residual	-1.803	2.562	.004	1.022	62
Mahal. Distance	.457	7.791	2.952	1.942	62
Cook's Distance	.000	.167	.017	.026	62
Centered Leverage Value	.007	.128	.048	.032	62

a. Dependent Variable: Compliance

As the largest **Mahalanobis Distance** is well below 16.266 (the critical χ^2 value for $df = 3$ at $\alpha = .001$), we need not be concerned about multivariate outliers.

Charts

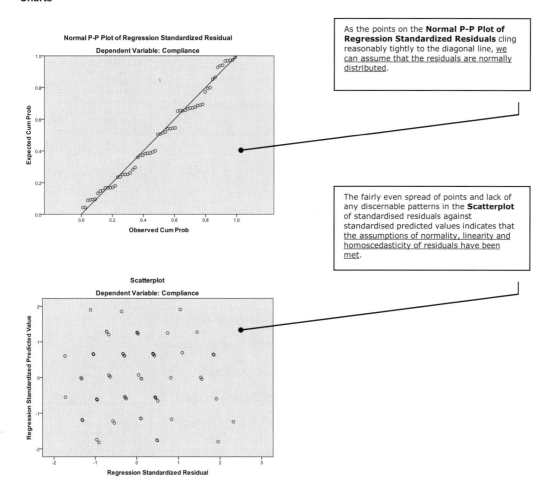

As the points on the **Normal P-P Plot of Regression Standardized Residuals** cling reasonably tightly to the diagonal line, <u>we can assume that the residuals are normally distributed</u>.

The fairly even spread of points and lack of any discernable patterns in the **Scatterplot** of standardised residuals against standardised predicted values indicates that <u>the assumptions of normality, linearity and homoscedasticity of residuals have been met</u>.

13.4.3. Follow Up Analyses

13.4.3.1. Effect Size

R^2 for the three-predictor model was .297. If required, this can be converted to Cohen's f^2 with:

$$f^2 = \frac{R^2}{1-R^2}$$

So:

$$f^2 = \frac{.297}{1-.297}$$

$$= .422$$

This is a large effect.

> **ⓘ Tip:**
> Jacob Cohen (1988) suggested that an f^2 of .02 (or R^2 of .0196) can be considered small, an f^2 of .15 (or R^2 of .13) can be considered medium, and an f^2 of .35 (or R^2 of .26) can be considered large.

13.4.4. APA Style Results Write Up

Results

As with the previous example, it is a good idea to report some basic descriptive statistics and bivariate correlations before describing the results of the multiple regression. These have been omitted from the current example in the interests of brevity.

To test the hypothesis that legitimacy can account for a significant proportion of the variance in everyday compliance with the law, beyond that already accounted for by perceived punishment certainty and severity, hierarchical multiple regression analysis (MRA) was employed.

Before interpreting the results of the MRA, a number of assumptions were tested, and checks were performed. First, stem-and-leaf plots and boxplots indicated that each variable in the regression was normally distributed and free from univariate outliers. Second, an inspection of the normal probability plot of standardised residuals and the scatterplot of standardised residuals against standardised predicted values indicated that the assumptions of normality, linearity and homoscedasticity of residuals were met. Third, Mahalanobis distance did not exceed the critical χ^2 for $df = 3$ (at $\alpha = .001$) of 16.27 for any cases in the data file, indicating that multivariate outliers were not of concern. Finally, relatively high tolerances for all three predictors in the final regression model indicated that multicollinearity would not interfere with our ability to interpret the outcome of the MRA.

The terms "step" and "block" are used interchangeably in reports of research using hierarchical regression.

On step 1 of the hierarchical MRA, certainty and severity accounted for a non-significant 4% of the variance in compliance, $R^2 = .04$, $F (2, 59) = 1.23$, $p = .300$. On step 2, legitimacy was added to the regression equation, and accounted for an additional 25.7% of the variance in compliance, $\Delta R^2 = .257$, $\Delta F (1, 58) = 21.20$, $p < .001$. In combination, the three predictor variables explained 29.7% of the variance in compliance, $R^2 = .297$, adjusted $R^2 = .261$, $F (3, 58) = 8.16$, $p < .001$. By Cohen's (1988) conventions, a combined effect of this magnitude can be considered "large" ($f^2 = .42$).

When writing up the results of a hierarchical MRA, you should do three things.

1. Report the incremental variance accounted for on each step of multiple regression, and whether or not these increases were statistically significant.

2. Report on the predictive utility of the entire model.

3. Report on the unique contribution that each predictor makes at each step of the multiple regression. Because these contributions are model specific (e.g., the unique contribution made by perceived certainty varies from model 1 to model 2), you must caveat them with reference to the specific model they refer to. We have done this in a table (overleaf).

Unstandardised (B) and standardised (β) regression coefficients, and squared semi-partial (or 'part') correlations (sr^2) for each predictor on each step of the hierarchical MRA are reported in Table 1.

Table 1

Unstandardised (B) and Standardised (β) Regression Coefficients,

and Squared Semi-Partial Correlations (sr²) For Each Predictor

Variable on Each Step of a Hierarchical Multiple Regression

Predicting Everyday Compliance With the Law (N = 62)

Variable	B [95% CI]	β	sr^2
Step 1			
Certainty	.154 [-.063, .372]	.193	.033
Severity	.015 [-.194, .224]	.019	.001
Step 2			
Certainty	-.012 [-.212, .189]	-.014	.001
Severity	-.009 [-.190, .172]	-.012	.001
Legitimacy	.546 [.308, .783]**	.552	.257

Note. CI = confidence interval.

** $p < .001$.

As can be seen in Table 1, the only significant predictor of compliance in the final regression model was legitimacy ($sr^2 = .257$).

13.5. Multiple Regression Checklist

If using standard multiple regression, have you:

✔ Checked for and dealt with any non-normality, outliers and multicollinearity concerns?
✔ Checked the assumptions of normality, linearity and homoscedasticity of residuals?
✔ Taken note of R^2 (and/or Adjusted R^2) for the full regression model?
✔ Used the ANOVA to determine whether or not R^2 is statistically significant, and taken note of the F-value, degrees of freedom and associated significance level?
✔ Considered the contribution each predictor makes to the regression model, taking note of B and β for each predictor, and the results of their associated significance tests (t tests)?
✔ Written up your results in the APA style, reporting on the predictive utility of both the full regression model and each specific predictor within the model?

If using hierarchical multiple regression, have you:

✔ Checked for and dealt with any non-normality, outliers and/or multicollinearity concerns?
✔ Checked the assumptions of normality, linearity and homoscedasticity of residuals?
✔ Taken note of ΔR^2 at each step of the regression, as well as R^2 (and/or Adjusted R^2) for the full model?
✔ Taken note of ΔF (and associated *df* and significance levels) at each step of the regression, as well the outcome of the full model significance test (in the ANOVA table).
✔ Considered the contribution each predictor makes at each step of the regression, taking note of B and β (and their associated significance tests) as applicable?
✔ Written up your results in the APA style, reporting on the incremental variance explained at each step of the model (and the roles played by particular predictors, as appropriate), as well as the predictive utility of the full model?

Chapter 14: Factor Analysis

💬 *AKA:*
Exploratory factor analysis.

Chapter Overview

14.1. Purpose of a Factor Analysis

Factor analysis is a data reduction technique whereby a large number of variables can be summarised into a more meaningful, smaller set of factors. Factor analysis can also be used to identify interrelationships between variables in a data set.

14.2. Questions We Could Answer Using a Factor Analysis

1. Is there a small set of factors underlying a questionnaire containing 150 items about attitudes towards school?

In this example, we're asking whether it is possible to reduce a large number of items/variables into a smaller, more meaningful set of factors.

Another question we could address with factor analysis is:

2. Is there as small set of common features (e.g., air conditioning, power steering etc.) underlying the car purchasing preferences of the general public?

14.3. Illustrated Example

To establish whether a small number of factors underlie (or can explain) community attitudes towards smoking, the following five-item *Attitudes Towards Smoking Questionnaire* was administered to 107 people.

> On a scale from 1 to 100, where 1 represents "strongly disagree" and 100 represents "strongly agree", please indicate the extent with which you agree with each of the following statements.
>
> 1. I think smoking is acceptable.
> 2. I don't care if people smoke around me.
> 3. I don't think people should smoke in restaurants.
> 4. I think people should have the right to smoke.
> 5. I don't think people should smoke around food.

Participants' responses to this questionnaire are reproduced in Table 14.1.

Table 14.1

□ *Data:*
This is data file
data_14_1.sav
on the companion
website.

Responses to a Five-Item "Attitudes Towards Smoking Questionnaire"
(N = 107)

ID	Qn1	Qn2	Qn3	Qn4	Qn5	ID	Qn1	Qn2	Qn3	Qn4	Qn5
1	69	69	23	69	23	55	62	62	65	62	65
2	65	66	34	65	23	56	37	37	56	37	65
3	61	61	45	61	33	57	77	77	30	77	65
4	63	63	47	62	33	58	57	22	60	57	65
5	23	23	45	23	34	59	74	74	61	74	66
6	76	76	23	76	34	60	55	35	73	55	66
7	76	76	23	76	34	61	61	61	66	61	66
8	65	57	47	65	34	62	64	64	62	64	66
9	74	74	45	74	34	63	55	34	67	55	67
10	68	68	23	68	35	64	67	67	54	67	67
11	69	69	23	69	36	65	35	42	66	35	67
12	68	68	45	68	37	66	23	25	78	23	67
13	65	65	43	65	43	67	56	24	66	56	67
14	45	23	65	35	43	68	39	39	78	39	67
15	76	76	44	76	44	69	84	84	51	84	67
16	69	69	25	69	45	70	63	63	60	62	67
17	24	24	56	24	45	71	62	62	77	62	67
18	65	65	54	65	45	72	59	59	78	59	68
19	87	87	45	87	45	73	78	78	68	68	68
20	68	68	53	68	46	74	87	87	77	87	68
21	67	67	51	67	46	75	65	78	63	65	68
22	76	76	62	68	46	76	56	27	83	56	68
23	34	34	62	34	47	77	33	22	87	33	69
24	33	33	60	33	47	78	34	34	60	34	70
25	65	65	60	65	48	79	44	23	70	44	72
26	23	23	72	23	49	80	34	34	56	34	74
27	46	25	49	37	49	81	62	62	65	61	74
28	61	61	57	61	50	82	58	34	74	58	74
29	46	24	48	35	53	83	56	33	91	56	76
30	77	77	67	77	53	84	62	62	63	62	76
31	76	76	50	76	54	85	43	23	77	43	76
32	76	76	76	68	54	86	33	33	62	33	77
33	53	27	65	39	54	87	62	62	67	62	77
34	67	67	54	67	54	88	79	79	67	68	77
35	35	35	56	35	55	89	67	67	83	67	77
36	87	87	39	87	55	90	25	25	62	25	78
37	65	65	77	65	55	91	54	33	74	54	78
38	63	63	61	63	55	92	67	67	76	67	78
39	45	25	62	34	55	93	78	78	89	78	78
40	25	27	69	25	56	94	75	75	99	75	78
41	84	84	63	84	56	95	78	78	79	68	78
42	68	68	39	68	57	96	54	22	98	54	79
43	79	79	61	79	57	97	56	25	78	56	80
44	55	34	61	55	58	98	57	34	72	57	81
45	53	33	57	43	58	99	73	73	72	73	81
46	76	76	52	76	58	100	21	23	76	21	84
47	79	79	67	79	58	101	56	23	60	56	84
48	82	82	70	82	60	102	67	67	87	67	84
49	55	55	84	55	62	103	56	56	87	56	86
50	88	88	69	88	62	104	75	75	87	68	87
51	76	76	87	76	62	105	83	83	98	83	88
52	67	67	48	67	64	106	60	60	99	60	91
53	70	70	61	70	65	107	56	33	91	56	98
54	55	34	65	55	65						

14.3.1. Setting Up The PASW Statistics Data File

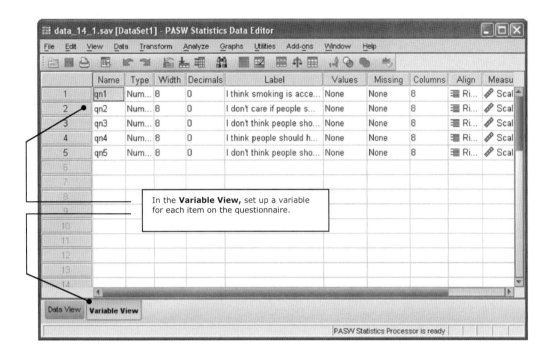

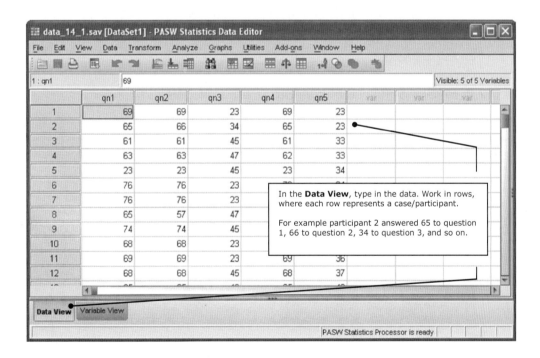

14.3.2. Analysing the Data

Syntax:
Run these
analyses with
syntax_14_1.sps
on the companion
website.

14.3.2.1. Assumptions

The following criteria should be met before conducting a Factor Analysis.

1. **Independence.** Each participant should participate only once in the research, and should not influence the participation of others.

2. **Sample size.** There should a minimum of five participants per variable in the study. Generally there should be at least 100 participants for a reliable factor analysis.

3. **Normality.** Each variable should be approximately normally distributed, although factor analysis is fairly robust against violations of this assumption.

4. **Linearity.** There should be roughly linear (straight-line) relationships between the variables, as factor analysis is based on the analysis of correlations.

5. **Multicollinearity.** Multicollinearity can exist when there is a high squared multiple correlations between the variables. Squared multiple correlations that are close to 1 can be problematic.

Assumption 1 and 2 are methodological, and should have been addressed before and during data collection. Assumptions 3, 4 and 5 can be tested using PASW Statistics.

14.3.2.2. PASW Statistics Procedure (Part 1: Normality)

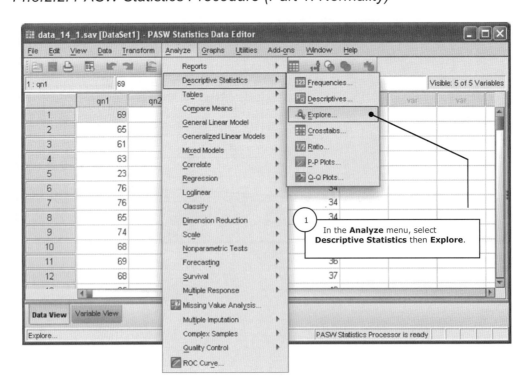

1   In the **Analyze** menu, select **Descriptive Statistics** then **Explore**.

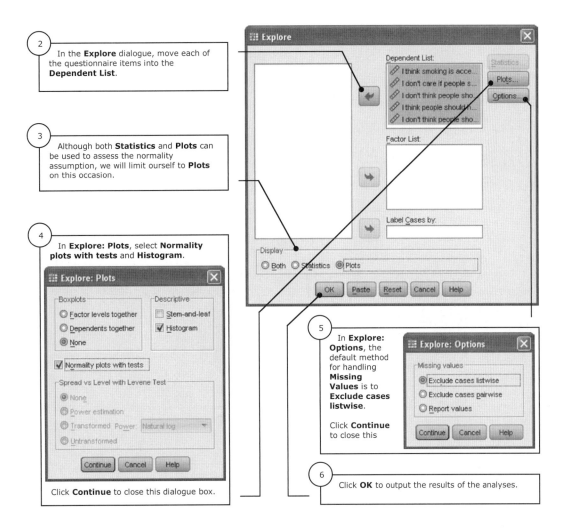

14.3.2.3. PASW Statistics Output (Part 1: Normality)

Explore

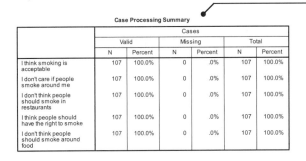

Case Processing Summary

	Cases					
	Valid		Missing		Total	
	N	Percent	N	Percent	N	Percent
I think smoking is acceptable	107	100.0%	0	.0%	107	100.0%
I don't care if people smoke around me	107	100.0%	0	.0%	107	100.0%
I don't think people should smoke in restaurants	107	100.0%	0	.0%	107	100.0%
I think people should have the right to smoke	107	100.0%	0	.0%	107	100.0%
I don't think people should smoke around food	107	100.0%	0	.0%	107	100.0%

The **Case Processing Summary** indicates how many participants answered each of the questionnaire items.

In this instance, all 107 participants completed the questionnaire. (That is, there was no **Missing** data.)

Tests of Normality

	Kolmogorov-Smirnov[a]			Shapiro-Wilk		
	Statistic	df	Sig.	Statistic	df	Sig.
I think smoking is acceptable	.122	107	.000	.941	107	.000
I don't care if people smoke around me	.186	107	.000	.877	107	.000
I don't think people should smoke in restaurants	.097	107	.016	.976	107	.048
I think people should have the right to smoke	.150	107	.000	.932	107	.000
I don't think people should smoke around food	.109	107	.003	.980	107	.111

a. Lilliefors Significance Correction

Looking at the results of the five **Shapiro-Wilk Tests of Normality**, it appears that responses to the first four questions may not be normally distributed.

Although not ideal, this is not too great a cause for concern because (a) these **Tests of Normality** can be quite sensitive to even trivial departures from normality; and (b) factor analysis is fairly robust against violations of the normality assumption anyway.

In situations like this, it can be useful to graph each distribution of scores with a histogram (or stem-and-leaf plot). We have done this overleaf.

▶▶ **Link:**
The full PASW Statistics output also includes a **Normal Q-Q Plot** and a **Detrended Normal Q-Q Plot** for each variable. These have been described elsewhere in this text.

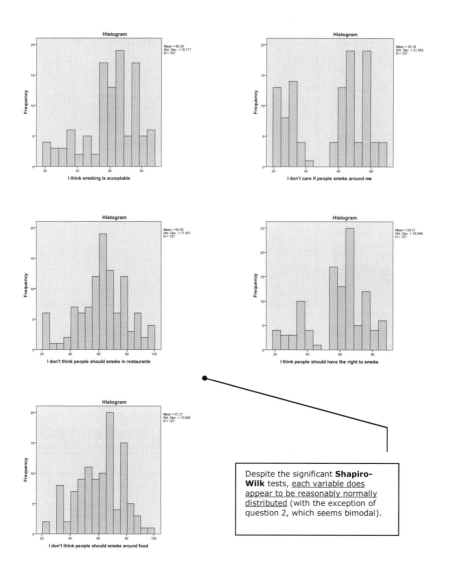

Despite the significant **Shapiro-Wilk** tests, <u>each variable does appear to be reasonably normally distributed</u> (with the exception of question 2, which seems bimodal).

14.3.2.4. PASW Statistics Procedure (Part 2: Factor Analysis)

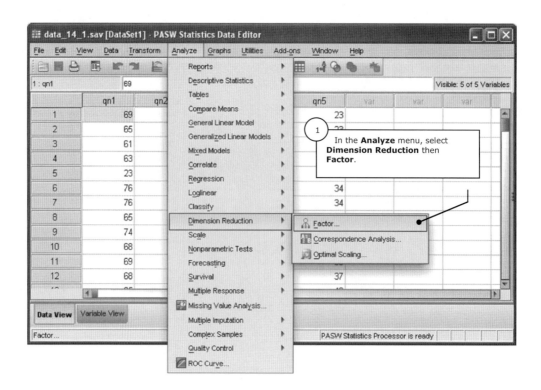

In the **Analyze** menu, select **Dimension Reduction** then **Factor**.

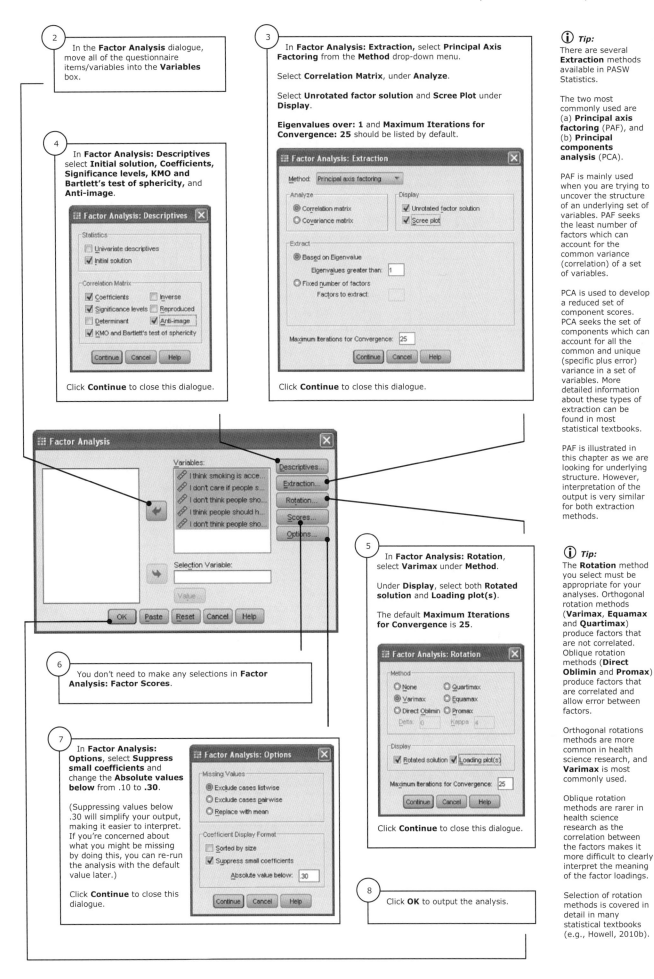

2 In the **Factor Analysis** dialogue, move all of the questionnaire items/variables into the **Variables** box.

3 In **Factor Analysis: Extraction,** select **Principal Axis Factoring** from the **Method** drop-down menu.

Select **Correlation Matrix**, under **Analyze**.

Select **Unrotated factor solution** and **Scree Plot** under **Display**.

Eigenvalues over: 1 and **Maximum Iterations for Convergence: 25** should be listed by default.

(i) Tip:
There are several **Extraction** methods available in PASW Statistics.

The two most commonly used are (a) **Principal axis factoring** (PAF), and (b) **Principal components analysis** (PCA).

PAF is mainly used when you are trying to uncover the structure of an underlying set of variables. PAF seeks the least number of factors which can account for the common variance (correlation) of a set of variables.

PCA is used to develop a reduced set of component scores. PCA seeks the set of components which can account for all the common and unique (specific plus error) variance in a set of variables. More detailed information about these types of extraction can be found in most statistical textbooks.

PAF is illustrated in this chapter as we are looking for underlying structure. However, interpretation of the output is very similar for both extraction methods.

4 In **Factor Analysis: Descriptives** select **Initial solution, Coefficients, Significance levels, KMO and Bartlett's test of sphericity,** and **Anti-image.**

Click **Continue** to close this dialogue.

Click **Continue** to close this dialogue.

6 You don't need to make any selections in **Factor Analysis: Factor Scores**.

5 In **Factor Analysis: Rotation,** select **Varimax** under **Method**.

Under **Display**, select both **Rotated solution** and **Loading plot(s)**.

The default **Maximum Iterations for Convergence** is **25**.

Click **Continue** to close this dialogue.

(i) Tip:
The **Rotation** method you select must be appropriate for your analyses. Orthogonal rotation methods (**Varimax**, **Equamax** and **Quartimax**) produce factors that are not correlated. Oblique rotation methods (**Direct Oblimin** and **Promax**) produce factors that are correlated and allow error between factors.

Orthogonal rotations methods are more common in health science research, and **Varimax** is most commonly used.

Oblique rotation methods are rarer in health science research as the correlation between the factors makes it more difficult to clearly interpret the meaning of the factor loadings.

Selection of rotation methods is covered in detail in many statistical textbooks (e.g., Howell, 2010b).

7 In **Factor Analysis: Options**, select **Suppress small coefficients** and change the **Absolute values below** from .10 to **.30.**

(Suppressing values below .30 will simplify your output, making it easier to interpret. If you're concerned about what you might be missing by doing this, you can re-run the analysis with the default value later.)

Click **Continue** to close this dialogue.

8 Click **OK** to output the analysis.

14.3.2.5. PASW Statistics Output (Part 2: Factor Analysis)

Factor Analysis

The **Correlation Matrix** table reports the bivariate correlation (Pearson's r) between each pair of variables in the analyses. (See chapter 12 for further details).

PASW Statistics uses this **Correlation Matrix** as the basis of the factor analysis.

Correlation Matrix

		I think smoking is acceptable	I don't care if people smoke around me	I don't think people should smoke in restaurants	I think people should have the right to smoke	I don't think people should smoke around food
Correlation	I think smoking is acceptable	1.000	.879	-.131	.982	-.081
	I don't care if people smoke around me	.879	1.000	-.207	.884	-.173
	I don't think people should smoke in restaurants	-.131	-.207	1.000	-.136	.762
	I think people should have the right to smoke	.982	.884	-.136	1.000	-.073
	I don't think people should smoke around food	-.081	-.173	.762	-.073	1.000
Sig. (1-tailed)	I think smoking is acceptable		.000	.089	.000	.204
	I don't care if people smoke around me	.000		.016	.000	.038
	I don't think people should smoke in restaurants	.089	.016		.081	.000
	I think people should have the right to smoke	.000	.000	.081		.226
	I don't think people should smoke around food	.204	.038	.000	.226	

As several of the correlations in the **Correlation Matrix** are above .3, the data are suitable for factor analysis.

If most of the correlations are small, it is unlikely that there is any underlying structure to them.

KMO and Bartlett's Test

Kaiser-Meyer-Olkin Measure of Sampling Adequacy.		.678
Bartlett's Test of Sphericity	Approx. Chi-Square	603.028
	df	10
	Sig.	.000

The **KMO and Bartlett's Test** table provides additional information about the factorability of the data.

The **Kaiser-Meyer-Olkin** (KMO) **Measure of Sampling Adequacy** reports the amount of variance in the data that can be explained by the factors. Higher values are better. Generally, values of .5 or lower are unacceptable, and values of .6 or above are acceptable.

In the current example the KMO value is .678, which suggests that the data are suitable for factor analysis. This value is the mean of the KMOs reported for each of the variables in the **Anti-Image Matrices** below.

Bartlett's Test of Sphericity also indicates how suitable the data are for factor analysis.

If **Bartlett's Test** is significant (*Sig* < .05) then the data are acceptable. A non-significant (*Sig* > .05) **Bartlett's Test** indicates the data are not suitable for factor analysis.

The **Anti-image Matrices** are used to further examine the suitability of the data for factor analysis.

Anti-image Matrices

		I think smoking is acceptable	I don't care if people smoke around me	I don't think people should smoke in restaurants	I think people should have the right to smoke	I don't think people should smoke around food
Anti-image Covariance	I think smoking is acceptable	.035	-.009	-.009	-.031	.008
	I don't care if people smoke around me	-.009	.204	.008	-.020	.037
	I don't think people should smoke in restaurants	-.009	.008	.411	.010	-.308
	I think people should have the right to smoke	-.031	-.020	.010	.033	-.014
	I don't think people should smoke around food	.008	.037	-.308	-.014	.409
Anti-image Correlation	I think smoking is acceptable	.670ᵃ	-.112	-.075	-.918	.070
	I don't care if people smoke around me	-.112	.947ᵃ	.028	-.246	.130
	I don't think people should smoke in restaurants	-.075	.028	.532ᵃ	.084	-.752
	I think people should have the right to smoke	-.918	-.246	.084	.657ᵃ	-.121
	I don't think people should smoke around food	.070	.130	-.752	-.121	.508ᵃ

a. Measures of Sampling Adequacy(MSA)

The values off the diagonal in the lower part of this table (the **Anti-image Correlation** part) are partial correlations. Small partial correlations suggest an underlying structure to the data.

The values marked with a superscript "a" (ᵃ) are the KMO values for each item. For example, KMO = .657 for "*I think people should have the right to smoke*".

A KMO value below .5 indicates that the variable does not have a strong relationship with the other variables in the matrix. You should consider dropping any variables with KMO values < .5 from the analyses.

As all the KMOs in the current example exceed .5, we do not need to consider dropping any variables from the analyses.

Communalities

	Initial	Extraction
I think smoking is acceptable	.965	.976
I don't care if people smoke around me	.796	.803
I don't think people should smoke in restaurants	.589	.757
I think people should have the right to smoke	.967	.989
I don't think people should smoke around food	.591	.769

Extraction Method: Principal Axis Factoring.

The **Communalities** table provides information about how much variance can be explained by each of the variables. The extracted communalities show how much variance is explained using the extracted factors. For example, the item "*I don't think people should smoke in restaurants*" has 75.7% of its variance explained by the factors in the analysis.

If a variable has a very low communality value it may be useful to re-run the factor analysis without it.

The **Total Variance Explained** table shows the amount of variance that can be explained by the factor analysis. This table indicates that <u>there were two strong underlying factors identified during the analysis.</u>

Total Variance Explained

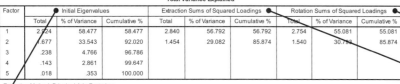

Factor	Initial Eigenvalues			Extraction Sums of Squared Loadings			Rotation Sums of Squared Loadings		
	Total	% of Variance	Cumulative %	Total	% of Variance	Cumulative %	Total	% of Variance	Cumulative %
1	2.924	58.477	58.477	2.840	56.792	56.792	2.754	55.081	55.081
2	1.677	33.543	92.020	1.454	29.082	85.874	1.540	30.793	85.874
3	.238	4.766	96.786						
4	.143	2.861	99.647						
5	.018	.353	100.000						

Extraction Method: Principal Axis Factoring.

The first section of the **Total Variance Explained** table displays the **Initial Eigenvalues** for all possible factors in the analysis.

These **Eigenvalues** (which show the amount of variance accounted for by each variable) are ordered from highest to lowest, so the first factor explains the greatest proportion of variance (58.477%) while the last explains the least (.353%).

Initially there will always be the same number of factors as there are variables in the study. However, only those factors that explain a sufficient amount of the variance will be retained in the final solution. These are the factors with **Eigenvalues** > 1.

In our example, <u>two factors will be retained.</u>

The **Extraction Sums of Squared Loadings** part of the **Total Variance Explained** table provides information about the two factors that will be retained in the final factor solution. Here, <u>85.874% of the variance can be explained by these two factors.</u>

The **Rotation Sums of Squared Loadings** part of the **Total Variance Explained** table provides information about the two extracted factors following rotation. Rotation does not impact on the total amount of variance explained by the factors in combination (85.874%).

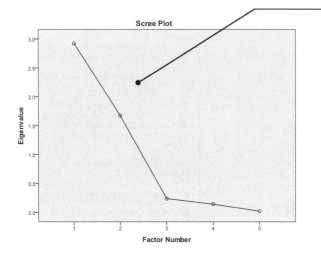

Scree Plot

The **Scree Plot** a graph of the **Initial Eigenvalues** reported in the **Total Variance Explained** table. It is another tool that you can use to decide how many factors should be interpreted.

Cattell (1966) first proposed the **Scree Plot** as a way of identifying the number of useful factors to retain. When interpreting this plot, the aim is to identify the point where the decrease of eigenvalues starts to abruptly level off.

This **Scree Plot** shows that <u>there are two factors with relatively high **Eigenvalues**, while the remaining three are quite small.</u>

Factor Matrix^a

	Factor	
	1	2
I think smoking is acceptable	.966	
I don't care if people smoke around me	.893	
I don't think people should smoke in restaurants	-.308	.813
I think people should have the right to smoke	.972	
I don't think people should smoke around food		.838

Extraction Method: Principal Axis Factoring.

a. 2 factors extracted. 10 iterations required.

The **Factor Matrix** shows the loadings between the factors and variables before rotation has occurred.

The higher the loading, the more likely the factor underlies the variable. For example, the item "*I think smoking is acceptable*" has a strong loading (.966) on **Factor 1**. Similarly, "*I don't think people should smoke around food*" loads highly onto **Factor 2** (.838). However, "*I don't think people should smoke in restaurants*" is harder to interpret because it loads onto both factors. Rotation will help clarify these loadings and make the factor solution easier to interpret.

ⓘ **Tip:**
As requested, loadings less than .3 have been suppressed, making these tables easier to interpret.

Rotated Factor Matrix^a

	Factor	
	1	2
I think smoking is acceptable	.987	
I don't care if people smoke around me	.884	
I don't think people should smoke in restaurants		.865
I think people should have the right to smoke	.994	
I don't think people should smoke around food		.876

Extraction Method: Principal Axis Factoring.
Rotation Method: Varimax with Kaiser Normalization.

a. Rotation converged in 3 iterations.

The **Rotated Factor Matrix** shows the loadings between the factors and variables after rotation has occurred.

It is now much easier to see how the variables load onto each of the factors.

It appears that three variables load onto **Factor 1** and the remaining two variables load onto **Factor 2**.

Factor Transformation Matrix

Factor	1	2
1	.969	-.248
2	.248	.969

Extraction Method: Principal Axis Factoring.
Rotation Method: Varimax with Kaiser Normalization.

The **Factor Transformation Matrix** illustrates the matrix used for the rotation. It does not need to be interpreted.

The **Factor Plot in Rotated Factor Space** is a graphical representation of the rotated factor solution. It shows that items 1, 2 and 4 load onto **Factor 1**, while items 3 and 5 load onto **Factor 2**.

This information (along with the **Rotated Factor Matrix**) can be used to begin the process of naming and interpreting each of the factors. When naming a factor, try to capture the aspect(s) that are common across every item that loads onto it.

In the current example three items load onto **Factor 1**:

1. *I think smoking is acceptable.*
2. *I don't care if people smoke around me.*
4. *I think people should have the right to smoke.*

Each of these items appears to relate to the respondents' approval of smoking by others. Therefore we could name the factor something like "*smoking by others*".

The remaining two items load onto **Factor 2**:

3. *I don't think people should smoke in restaurants.*
5. *I don't think people should smoke around food.*

These items appear related to smoking around food, so this factor could be called something like "*smoking near food*".

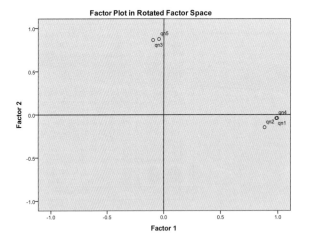

Factor Plot in Rotated Factor Space

14.3.3. APA Style Results Write-Up

Results

To investigate the underlying structure of a five-item questionnaire assessing attitudes toward smoking, data collected from 107 participants were subjected to principal axis factoring with varimax rotation.

Prior to running the principal axis factoring, examination of the data indicated that not every variable was perfectly normally distributed. Given the robust nature of factor analysis, these deviations were not considered problematic. Furthermore, a linear relationship was identified among the variables.

Two factors (with Eigenvalues exceeding 1) were identified as underlying the five questionnaire items (see Table 1). In total, these factors accounted for around 85% of the variance in the questionnaire data.

Table 1

Varimax Rotated Factor Structure of the Five Item Attitudes Towards Smoking Questionnaire

Item	Loadings	
	Factor 1[a]	Factor 2[b]
1. I think smoking is acceptable.	0.987	
2. I don't care if people smoke around me.	0.884	
3. I don't think people should smoke in restaurants.		0.865
4. I think people should have the right to smoke.	0.994	
5. I don't think people should smoke around food.		0.876
Percentage of Variance:	55.08%	30.79%

Note. [a] = "smoking by others"; [b] = "smoking near food".

14.3.4. Summary

In this example, principal axis factoring with varimax rotation was used to investigate the underlying structure of a questionnaire about attitudes towards smoking. Fortunately (as is typically the case in textbook examples), a clear two-factor solution emerged from our analyses.

However, factor analysis does not always produce such clean and interpretable results. For example, if we'd included an item like *"I smoke"* on the questionnaire, it may well have loaded substantially onto both factors, leaving you – the researcher and brains behind the whole operation – with some tough decisions to make. Those decisions are beyond the scope of this text, and we refer you to Tabachnick and Fidell (2007b) for extra guidance.

Always remember though, that a factor structure doesn't need to be perfect to be informative.

14.4. Factor Analysis Checklist

Have you:

- ✔ Checked that each variable is approximately normally distributed?
- ✔ Selected the appropriate extraction and rotation methods?
- ✔ Interpreted the rotated factor matrix and taken note of the number of factors and amount of variance explained for your write-up?
- ✔ Written up your results in the APA style?

Chapter 15: Reliability Analysis

Chapter Overview

15.1. Introduction

Reliability refers to the consistency or dependability of a measure over time, over questionnaire items, or over observers/raters. This chapter will focus on two of the most commonly cited indices of reliability:

1. **Cronbach's alpha.** Cronbach's alpha is a measure of *internal consistency*. It is used to assess the extent to which a set of questionnaire items tapping a single underlying construct covary.

2. **Cohen's kappa.** Cohen's kappa is a measure of *inter-rater reliability*. It is used to assess the extent to which two judges or raters agree over the classification of elements (behaviours, cases etc.) into mutually exclusive categories.

15.2. Cronbach's Alpha

Questionnaires are often used by researchers to measure personal characteristics (or "constructs") that are not directly observable. Examples of such constructs include anxiety, happiness, fear of crime, extraversion, and so on. The questionnaires we develop to measure these constructs are typically comprised of multiple items, each addressing slightly different (though clearly related) aspects of the construct.

For example, a questionnaire measuring anxiety might ask people to indicate how often they feel worried, short of breath, and unable to relax. Because these items are all assessing the same underlying construct (anxiety), people should respond to them consistently. That is, someone who is very anxious should report that they often feel worried, are often short of breath, and frequently feel unable to relax. Conversely, people who are not anxious should rarely feel worried, rarely experience shortness of breath, and only occasionally feel unable to relax.

Cronbach's alpha is an index of this consistency.

15.2.1. Illustrated Example One

The "Big Five" (or "Five Factor") model of personality asserts that personality can be accounted for, or understood in terms of five broad dimensions: openness to experience, conscientiousness, extraversion, agreeableness and neuroticism (Burger, 2008).

We administered the following 10-item measure of openness to experience (drawn from the International Personality Item Pool [IPIP], Goldberg et al., 2006) to 60 Australian university students over the age of 18:

Please use the rating scale below to indicate how accurately each of the following statements describes *you*. Describe yourself as you generally are now, not as you wish to be in the future. Please circle the number that corresponds to your most accurate response.

Rating Scale:	1.	Very inaccurate.
	2.	Moderately inaccurate.
	3.	Neither inaccurate nor accurate.
	4.	Moderately accurate.
	5.	Very accurate.

Statements:

1. Believe in the importance of art.	1	2	3	4	5
2. Tend to vote for conservative political candidates.	1	2	3	4	5
3. Have a vivid imagination.	1	2	3	4	5
4. Avoid philosophical discussions.	1	2	3	4	5
5. Tend to vote for liberal political candidates.	1	2	3	4	5
6. Do not like art.	1	2	3	4	5
7. Carry the conversation to a higher level.	1	2	3	4	5
8. Am not interested in abstract ideas.	1	2	3	4	5
9. Enjoy hearing new ideas.	1	2	3	4	5
10. Do not enjoy going to art museums.	1	2	3	4	5

The data we collected are reproduced in Table 15.1.

Table 15.1

Australian University Students' Responses to a 10-Item Measure of Openness to Experience

ID	Q1	Q2	Q3	Q4	Q5	Q6	Q7	Q8	Q9	Q10
1	5	3	4	1	3	1	5	1	5	5
2	5	1	3	1	1	1	4	1	5	2
3	5	1	4	1	1	1	4	1	5	2
4	3	3	4	3	3	2	4	1	5	3
5	5	2	5	1	2	1	4	3	4	1
6	4	1	4	1	1	1	4	1	5	1
7	5	2	5	1	2	1	3	1	5	1
8	5	3	4	1	3	1	5	2	5	1
9	5	3	4	1	3	1	4	4	4	1
10	5	1	5	1	3	1	5	1	5	1
11	5	3	4	1	1	1	4	1	5	1
12	5	1	5	2	4	1	3	2	4	1
13	1	4	5	2	1	2	4	3	4	3
14	2	2	4	2	2	2	4	2	5	3
15	5	2	5	1	4	1	3	1	5	1
16	3	3	5	1	3	2	3	2	5	2

Table 15.1 cont.

ID	Q1	Q2	Q3	Q4	Q5	Q6	Q7	Q8	Q9	Q10
17	4	3	4	3	3	2	3	2	4	2
18	4	2	4	2	1	1	3	2	5	1
19	4	1	5	2	4	1	3	1	5	1
20	4	3	4	3	3	1	4	3	4	2
21	1	3	3	3	3	4	3	3	4	3
22	2	4	2	2	4	3	3	2	4	3
23	5	2	4	1	1	1	4	1	5	2
24	3	2	2	1	3	3	4	2	3	4
25	3	4	4	2	4	3	4	2	4	3
26	4	1	4	1	1	1	4	3	5	1
27	1	4	4	2	3	5	5	3	4	5
28	1	4	4	3	2	3	4	3	2	3
29	2	3	3	4	3	1	4	4	5	2
30	5	2	5	1	3	1	1	5	5	3
31	1	3	3	4	4	3	2	4	4	3
32	4	4	4	2	2	2	4	2	4	2
33	2	5	3	5	5	2	2	3	4	2
34	4	3	4	4	3	2	4	4	5	1
35	4	4	5	1	3	1	4	2	5	1
36	3	3	4	2	1	1	3	1	4	1
37	5	3	3	1	3	1	4	1	5	2
38	4	3	3	2	3	2	2	2	2	2
39	3	2	5	1	4	2	4	1	5	3
40	5	1	4	3	1	1	5	1	5	1
41	5	2	5	1	1	1	4	1	5	1
42	4	4	3	2	2	3	4	4	4	3
43	5	1	4	1	3	1	5	1	5	3
44	2	4	3	2	3	3	4	2	5	4
45	5	1	5	1	2	1	5	1	5	4
46	4	3	4	2	3	2	4	2	4	2
47	4	3	3	3	3	1	3	5	4	1
48	5	3	5	1	3	1	4	2	4	2
49	2	1	3	2	1	2	3	4	4	3
50	5	1	5	1	5	1	4	1	5	1
51	2	4	4	3	4	2	2	2	4	4
52	4	3	4	2	3	2	4	2	4	2
53	5	1	5	1	1	1	5	1	5	1
54	5	3	5	1	3	1	4	1	5	2
55	3	3	1	3	3	1	3	4	4	5
56	4	3	4	2	2	2	4	2	4	3
57	4	5	5	1	3	2	4	1	5	5
58	2	2	4	2	3	2	3	2	5	5
59	1	4	3	2	4	4	3	4	4	5
60	2	4	3	2	2	3	3	3	4	4

15.2.1.1. Setting Up The PASW Statistics Data File

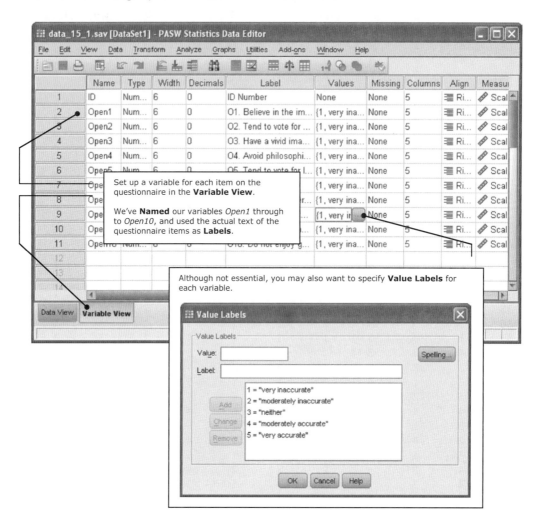

Set up a variable for each item on the questionnaire in the **Variable View**.

We've **Named** our variables *Open1* through to *Open10*, and used the actual text of the questionnaire items as **Labels**.

Although not essential, you may also want to specify **Value Labels** for each variable.

Value Labels

Value Labels

Value:

Label:

Spelling...

1 = "very inaccurate"
2 = "moderately inaccurate"
3 = "neither"
4 = "moderately accurate"
5 = "very accurate"

Add

Change

Remove

OK   Cancel   Help

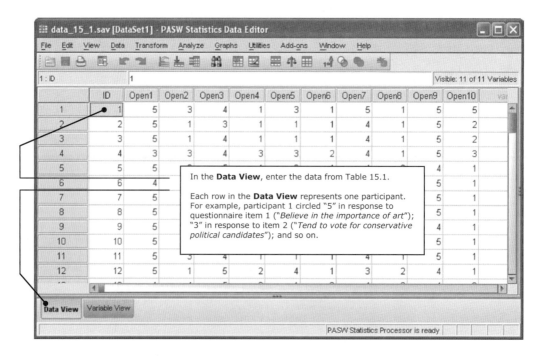

In the **Data View**, enter the data from Table 15.1.

Each row in the **Data View** represents one participant. For example, participant 1 circled "5" in response to questionnaire item 1 ("*Believe in the importance of art*"); "3" in response to item 2 ("*Tend to vote for conservative political candidates*"); and so on.

15.2.1.2. PASW Statistics Procedure (Reversing Negatively Scaled Items)

Five of the items on our questionnaire (items 2, 4, 6, 8 and 10) are negatively scaled, and will need to be reversed prior to any analyses.

Syntax:
Run these analyses with **syntax_15_1.sps** on the companion website.

▶▶| **Link:**
Recoding is discussed in chapter 2 (section 2.3).

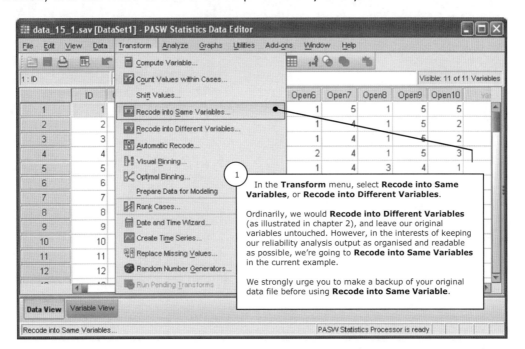

1 — In the **Transform** menu, select **Recode into Same Variables**, or **Recode into Different Variables**.

Ordinarily, we would **Recode into Different Variables** (as illustrated in chapter 2), and leave our original variables untouched. However, in the interests of keeping our reliability analysis output as organised and readable as possible, we're going to **Recode into Same Variables** in the current example.

We strongly urge you to make a backup of your original data file before using **Recode into Same Variable**.

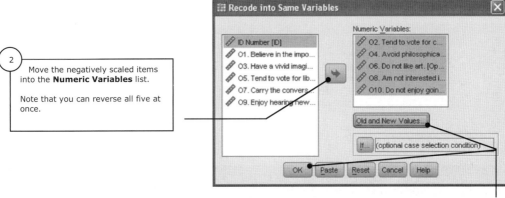

2 — Move the negatively scaled items into the **Numeric Variables** list.

Note that you can reverse all five at once.

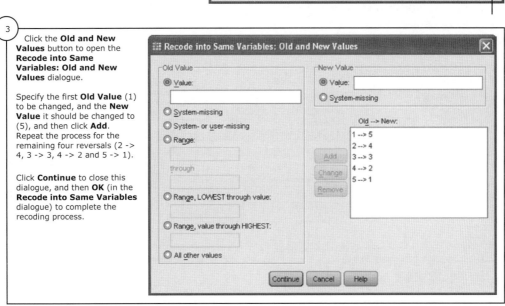

3 — Click the **Old and New Values** button to open the **Recode into Same Variables: Old and New Values** dialogue.

Specify the first **Old Value** (1) to be changed, and the **New Value** it should be changed to (5), and then click **Add**. Repeat the process for the remaining four reversals (2 -> 4, 3 -> 3, 4 -> 2 and 5 -> 1).

Click **Continue** to close this dialogue, and then **OK** (in the **Recode into Same Variables** dialogue) to complete the recoding process.

We've re-saved our data file with a new name, to prevent any future confusion.

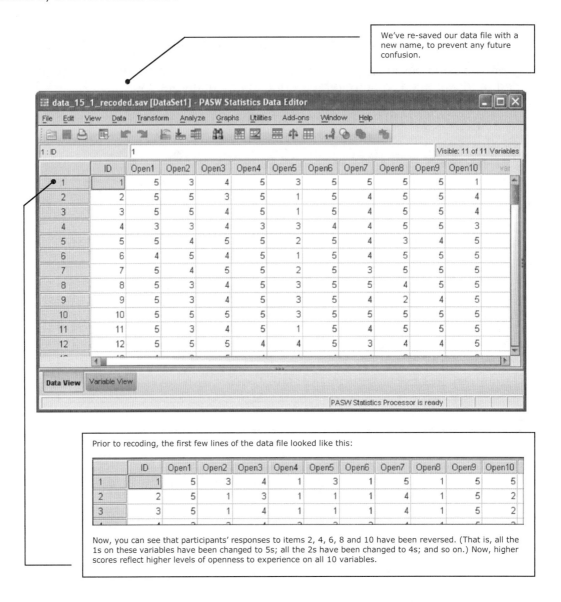

Prior to recoding, the first few lines of the data file looked like this:

	ID	Open1	Open2	Open3	Open4	Open5	Open6	Open7	Open8	Open9	Open10
1	1	5	3	4	1	3	1	5	1	5	5
2	2	5	1	3	1	1	1	4	1	5	2
3	3	5	1	4	1	1	1	4	1	5	2

Now, you can see that participants' responses to items 2, 4, 6, 8 and 10 have been reversed. (That is, all the 1s on these variables have been changed to 5s; all the 2s have been changed to 4s; and so on.) Now, higher scores reflect higher levels of openness to experience on all 10 variables.

15.2.1.3. PASW Statistics Procedure (Cronbach's Alpha)

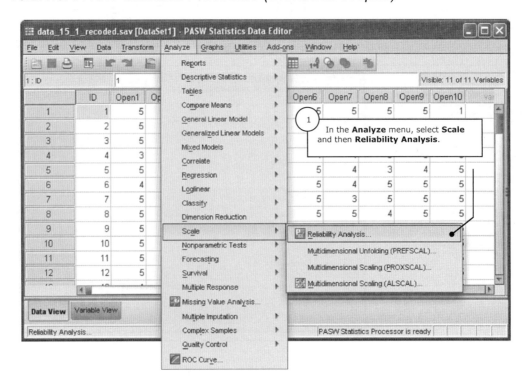

1 In the **Analyze** menu, select **Scale** and then **Reliability Analysis**.

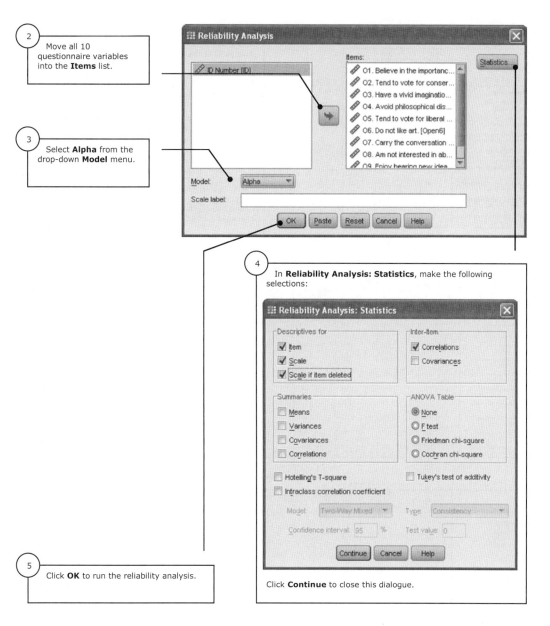

2 Move all 10 questionnaire variables into the **Items** list.

3 Select **Alpha** from the drop-down **Model** menu.

(i) Tip:
If you elected to **Recode into Different Variables** in section 15.2.1.2., ensure that you now move the new (recoded) variables into the **Items** list, and not the original (negatively scaled) variables.

4 In **Reliability Analysis: Statistics**, make the following selections:

Click **Continue** to close this dialogue.

5 Click **OK** to run the reliability analysis.

15.2.1.4. PASW Statistics Output

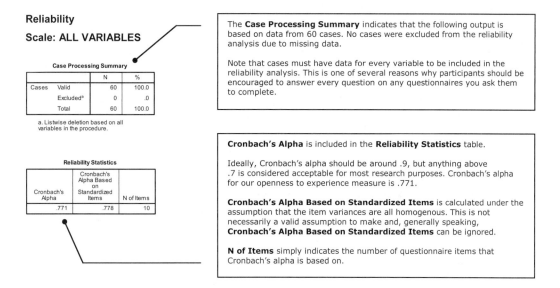

Reliability

Scale: ALL VARIABLES

Case Processing Summary

		N	%
Cases	Valid	60	100.0
	Excluded^a	0	.0
	Total	60	100.0

a. Listwise deletion based on all variables in the procedure.

Reliability Statistics

Cronbach's Alpha	Cronbach's Alpha Based on Standardized Items	N of Items
.771	.778	10

The **Case Processing Summary** indicates that the following output is based on data from 60 cases. No cases were excluded from the reliability analysis due to missing data.

Note that cases must have data for every variable to be included in the reliability analysis. This is one of several reasons why participants should be encouraged to answer every question on any questionnaires you ask them to complete.

Cronbach's Alpha is included in the **Reliability Statistics** table.

Ideally, Cronbach's alpha should be around .9, but anything above .7 is considered acceptable for most research purposes. Cronbach's alpha for our openness to experience measure is .771.

Cronbach's Alpha Based on Standardized Items is calculated under the assumption that the item variances are all homogenous. This is not necessarily a valid assumption to make and, generally speaking, **Cronbach's Alpha Based on Standardized Items** can be ignored.

N of Items simply indicates the number of questionnaire items that Cronbach's alpha is based on.

Item Statistics

	Mean	Std. Deviation	N
O1. Believe in the importance of art.	3.65	1.376	60
O2. Tend to vote for conservative political candidates.	3.35	1.132	60
O3. Have a vivid imagination.	3.97	.901	60
O4. Avoid philosophical discussions.	4.15	.971	60
O5. Tend to vote for liberal political candidates.	2.63	1.089	60
O6. Do not like art.	4.28	.940	60
O7. Carry the conversation to a higher level.	3.68	.854	60
O8. Am not interested in abstract ideas.	3.85	1.162	60
O9. Enjoy hearing new ideas.	4.45	.699	60
O10. Do not enjoy going to art museums.	3.63	1.301	60

The **Item Statistics** table was produced because we ticked **Item** under **Descriptives for** in the **Reliability Analysis: Statistics** dialogue.

It provides means and standard deviations for each of the items in the reliability analysis.

The **Inter-Item Correlation Matrix** reports the strength and direction of the relationship between all possible pairs of questionnaire items.

If the measure is internally consistent, we would expect moderate to strong positive correlations between all the items. Quickly scanning through this table indicates that item 5 (*"Tend to vote for liberal political candidates"*) may be problematic, as it is negatively correlated with the other nine questionnaire items!

Inter-Item Correlation Matrix

	O1. Believe in the importance of art.	O2. Tend to vote for conservative political candidates.	O3. Have a vivid imagination.	O4. Avoid philosophical discussions.	O5. Tend to vote for liberal political candidates.	O6. Do not like art.	O7. Carry the conversation to a higher level.	O8. Am not interested in abstract ideas.	O9. Enjoy hearing new ideas.	O10. Do not enjoy going to art museums.
O1. Believe in the importance of art.	1.000	.504	.442	.573	-.200	.772	.323	.444	.413	.552
O2. Tend to vote for conservative political candidates.	.504	1.000	.277	.398	-.320	.542	.204	.337	.397	.399
O3. Have a vivid imagination.	.442	.277	1.000	.451	-.064	.391	.228	.416	.401	.380
O4. Avoid philosophical discussions.	.573	.398	.451	1.000	-.284	.342	.365	.501	.348	.125
O5. Tend to vote for liberal political candidates.	-.200	-.320	-.064	-.284	1.000	-.261	-.309	-.151	-.136	-.216
O6. Do not like art.	.772	.542	.391	.342	-.261	1.000	.135	.334	.473	.599
O7. Carry the conversation to a higher level.	.323	.204	.228	.365	-.309	.135	1.000	.413	.300	.062
O8. Am not interested in abstract ideas.	.444	.337	.416	.501	-.151	.334	.413	1.000	.397	.187
O9. Enjoy hearing new ideas.	.413	.397	.401	.348	-.136	.473	.300	.397	1.000	.203
O10. Do not enjoy going to art museums.	.552	.399	.380	.125	-.216	.599	.062	.187	.203	1.000

Item-Total Statistics

	Scale Mean if Item Deleted	Scale Variance if Item Deleted	Corrected Item-Total Correlation	Squared Multiple Correlation	Cronbach's Alpha if Item Deleted
O1. Believe in the importance of art.	34.00	24.373	.776	.756	.693
O2. Tend to vote for conservative political candidates.	34.30	29.061	.529	.398	.739
O3. Have a vivid imagination.	33.68	30.390	.564	.392	.738
O4. Avoid philosophical discussions.	33.50	30.119	.539	.557	.740
O5. Tend to vote for liberal political candidates.	35.02	40.051	-.321	.266	.842
O6. Do not like art.	33.37	28.982	.685	.719	.722
O7. Carry the conversation to a higher level.	33.97	33.016	.312	.328	.766
O8. Am not interested in abstract ideas.	33.80	28.473	.536	.383	.737
O9. Enjoy hearing new ideas.	33.20	32.129	.529	.370	.748
O10. Do not enjoy going to art museums.	34.02	28.830	.450	.495	.752

▶▶ *Link:*
If you're unfamiliar with correlation or multiple regression, refer to chapters 12 and 13 respectively.

Scale Statistics

Mean	Variance	Std. Deviation	N of Items
37.65	36.808	6.067	10

The **Item-Total Statistics** table includes lots of useful information, including:

Corrected Item-Total Correlation. This column shows the Pearson's correlation between each item and the sum of the remaining nine items.

For example, item 1 correlates .776 with the sum of items 2-9.

Squared Multiple Correlation. This column shows R^2 for each item regressed on the remaining nine items.

For example, when item 1 is regressed on items 2-9 (i.e., item 1 is the criterion variable in a multiple regression with nine predictors – items 2-9) the resultant R^2 is .756.

Cronbach's Alpha if Item Deleted. This column shows Cronbach's alpha calculated with the remaining nine items.

For example, if item 1 were ignored, Cronbach's alpha for the remaining nine items would be .693.

This table was produced because we selected **Scale if Item Deleted** under **Descriptives for** in the **Reliability Analysis: Statistics** dialogue.

Looking at the **Item-Total Statistics** (and keeping in mind the pattern of correlations in the **Inter-Item Correlation Matrix**), item 5 does not seem to fit well with the other items on our questionnaire. It has a relatively weak negative correlation with the sum of the other nine items ($r = -.321$), and Cronbach's alpha for the scale improves markedly (from .771 to .842) when item 5 is ignored.

These factors indicate that we should look closely at item 5, with a view to either re-writing it, or deleting it from our questionnaire.

On reflection, our problems with item 5 probably relate to the ambiguity of "liberal". In Australia (where the questionnaire was administered) the Liberal Party is the dominant conservative political party, and voting for a Liberal candidate means voting for someone on the political right. In the US (where the questionnaire was developed) liberal means politically progressive, and voting for a liberal candidate means voting for someone on the political left.

Item 5 will need to be re-written or removed before this questionnaire is suitable for use in Australian research.

15.2.1.5. APA Style Results Write Up

Results ●————————————

Cronbach's alpha for the 10-item Openness to Experience

questionnaire was .77. Although this can be considered adequate for

research purposes, a closer examination of the questionnaire item-total

statistics indicated that alpha would increase to .84 if item 5 were

removed. This item asked whether participants "tend to vote for

liberal political candidates" and – due to multiple and contradictory

interpretations for the word 'liberal'– was probably ambiguous for a

substantial portion of our Australian sample. Consequently, this item

was dropped from the questionnaire, and all subsequent analyses are

based on participants' responses to the remaining nine items.

> Depending on the focus of your research report, it may be more appropriate to report psychometric data like this in your **Method** section, alongside descriptions of each of the measures used in the study.

15.3. Cohen's Kappa

Cohen's kappa is a measure of inter-rater reliability. It was designed specifically to index the extent to which two judges or raters agree over the classification of elements (behaviours, cases etc.) into mutually exclusive categories. It is commonly used to assess the reliability of behavioural observations and the coding of interview transcripts.

> (i) *Tip:*
> To assess the level of agreement between three or more raters, you will need to use Fleiss' (e.g., Fleiss, Nee, & Landis, 1979) generalised kappa.
>
> Like Cohen's kappa, Fleiss' generalised kappa was designed for use with nominal (categorical) data. Unlike Cohen's kappa, it is not currently available through the PASW Statistics drop-down menus. Refer to King (2004) for further information.

15.3.1. Illustrated Example Two

As part of her PhD research, Kate is using the "Strange Situation" (see Ainsworth, Blehar, Waters, & Wall, 1978) to classify infants into one of three attachment styles/categories (secure, insecure-avoidant, and insecure-resistant). As the Strange Situation scoring system is quite complex, Kate has sensibly asked her PhD supervisor to also classify a portion of her sample.

If Kate and her supervisor agree about the attachment classifications of the infants they both (independently) observe, Kate can be reasonably confident about the reliability (or consistency) of her use of the Strange Situation scoring system.

Each Strange Situation session was videotaped, ensuring that both Kate and her supervisor had access to exactly the same information when making their classifications. Kate's supervisor randomly selected 18 videotapes to classify. His classifications, along with Kate's are presented in Table 15.2.

☐ *Data:*
This is data file
data_15_2.sav
on the companion
website.

Table 15.2

Attachment Style Classifications of 18 Infants Made by Two
Independent Raters

Infant ID	Rater 1 (Kate)	Rater2 (Supervisor)
1	1	1
2	1	2
3	1	1
4	1	1
5	2	2
6	3	3
7	3	3
8	2	1
9	1	1
10	1	1
11	1	1
12	1	1
13	1	1
14	2	2
15	2	2
16	3	3
17	1	1
18	1	1

Note. For both raters, 1 = secure attachment; 2 = insecure-avoidant attachment; 3 = insecure-resistant attachment.

15.3.1.1. Setting Up The PASW Statistics Data File

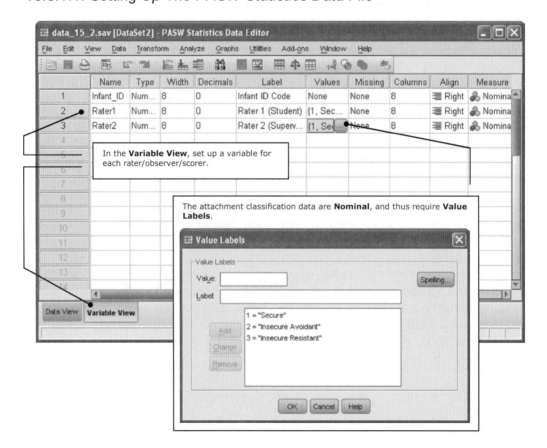

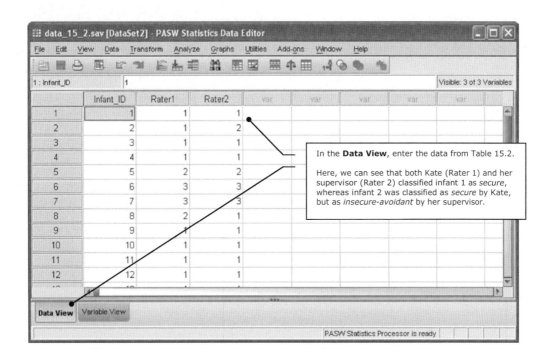

In the **Data View**, enter the data from Table 15.2.

Here, we can see that both Kate (Rater 1) and her supervisor (Rater 2) classified infant 1 as *secure*, whereas infant 2 was classified as *secure* by Kate, but as *insecure-avoidant* by her supervisor.

15.3.1.2. PASW Statistics Procedure

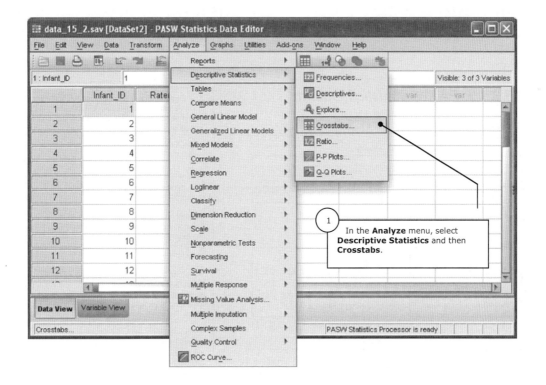

1   In the **Analyze** menu, select **Descriptive Statistics** and then **Crosstabs**.

Syntax:
Run these analyses with **syntax_15_2.sps** on the companion website.

2 In the **Crosstabs** dialogue, use the arrow buttons to move one set of ratings/classifications into the **Row(s)** box, and the other into the **Column(s)** box.

3 **Clustered bar charts** may help you to visually identify any particular sources or patterns of disagreement between raters, although this kind of information can also be gleaned from the **Crosstabulation** table included in the output by default.

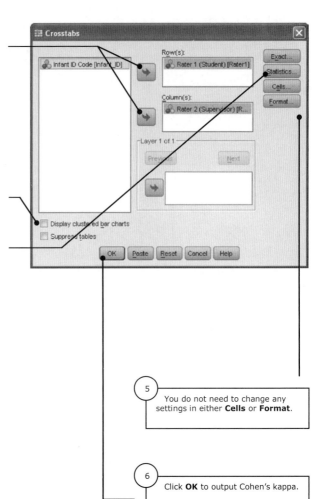

4 Select **Kappa** in the **Crosstabs: Statistics** dialogue.

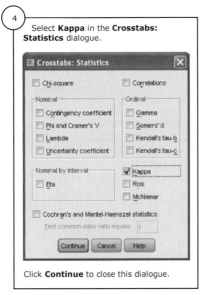

Click **Continue** to close this dialogue.

5 You do not need to change any settings in either **Cells** or **Format**.

6 Click **OK** to output Cohen's kappa.

15.3.1.3. PASW Statistics Output

Crosstabs

The **Case Processing Summary** indicates that 18 pairs of ratings were analysed. No cases were excluded from the calculation of kappa due to missing data.

Case Processing Summary

	Cases					
	Valid		Missing		Total	
	N	Percent	N	Percent	N	Percent
Rater 1 (Student) * Rater 2 (Supervisor)	18	100.0%	0	.0%	18	100.0%

The **Crosstabulation** table simply summarises the data set.

Values on the diagonal (starting in the upper left corner of the matrix) represent agreement frequencies, while values off the diagonal represent disagreement frequencies.

Here, we can see that 10 infants were classified as *secure* by both Kate and her supervisor; 3 infants were classified as *insecure-avoidant* by both; and 3 were classified as *insecure-resistant* by both.

Kate and her supervisor disagreed over the classification of the remaining 2 infants. In each of these cases, the disagreement centred on whether the infant was *secure* or *insecure-avoidant*.

Rater 1 (Student) * Rater 2 (Supervisor) Crosstabulation

Count

		Rater 2 (Supervisor)			Total
		Secure	Insecure Avoidant	Insecure Resistant	
Rater 1 (Student)	Secure	10	1	0	11
	Insecure Avoidant	1	3	0	4
	Insecure Resistant	0	0	3	3
Total		11	4	3	18

Symmetric Measures

		Value	Asymp. Std. Error[a]	Approx. T[b]	Approx. Sig.
Measure of Agreement	Kappa	.798	.137	4.564	.000
N of Valid Cases		18			

a. Not assuming the null hypothesis.

b. Using the asymptotic standard error assuming the null hypothesis.

Kappa is reported in the **Symmetric Measures** table. In the current example, it is .798, which Fleiss (1981, as cited in Robson, 2002) would regard as "excellent".

PASW Statistics also provides a test of significance for kappa, although this can be ignored for our purposes.

15.3.1.4. APA Style Results Write Up

Results ●────────────

> Depending on the focus of your research report, it may be more appropriate to discuss inter-rater reliability in your **Method** section (e.g., wherever you first describe the classification process used to sort the infants into the three attachment groups).

To assess inter-rater reliability, 18 infants (randomly selected from the full sample) were classified as secure, insecure-avoidant or insecure-resistant by both the researcher and a second, independent rater (the researcher's PhD supervisor). The two raters were in agreement regarding attachment classifications for 16 of the 18 infants. For the remaining two cases, disagreement was centred on whether the infants were best classified as secure or insecure-avoidant. Cohen's kappa was calculated with this data, and found to reflect a high level of inter-rater agreement (K = .80).

15.4. Reliability Analysis Checklist

If using Cronbach's alpha to assess the internal consistency of a measure, have you:

- ✔ Reversed participants' responses to any negatively scaled questionnaire items (if applicable)?
- ✔ Selected **Inter-Item Correlations** and **Descriptives for Scale if item deleted** in the **Reliability Analysis: Statistics** dialogue? With these statistics you will be able to determine if any items on your questionnaire are having an unduly negative influence on Cronbach's alpha, and what alpha would look like if these items were removed.
- ✔ Taken note of Cronbach's alpha (both before and after items are removed from the questionnaire, if applicable)?
- ✔ Considered the most appropriate place in your report for discussing the reliability of your measures?
- ✔ Written up your reliability analysis in the APA style?

If using Cohen's kappa to assess inter-rater reliability, have you:

- ✔ Studied the **Crosstabulation** table to identify an specific sources of disagreement between raters?
- ✔ Taken note of Cohen's kappa?
- ✔ Considered the most appropriate place in your report for discussing inter-rater reliability?
- ✔ Written up your reliability analysis in the APA style?

Chapter 16: Non-Parametric Procedures

Chapter Overview

16.1. Introduction

Unlike the parametric procedures described in previous chapters, non-parametric tests do not assume that the sampled population(s) is/are normally distributed, and do not require interval or ratio (i.e., **Scale**) data.

The non-parametric tests covered in this chapter (and their parametric equivalents) are listed in the following decision tree.

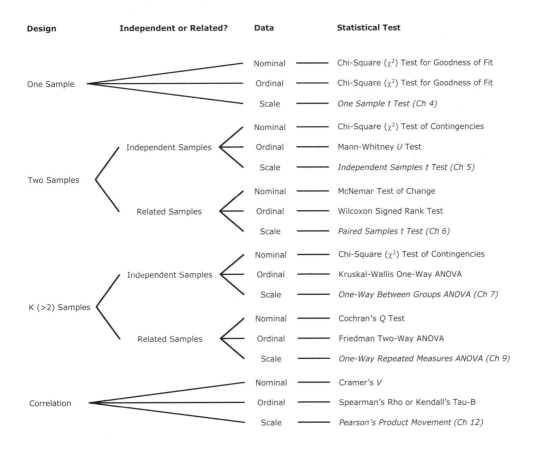

(i) Tip:

Reading the Decision Tree:

Work from left to right, by first selecting your basic design and then, if your design involves multiple samples, whether they are independent or not. At the third decision point select the type of data you are working with, which will lead to the appropriate statistical test. The tests for scale (i.e., interval or ratio) data are parametric, and are covered elsewhere in this book. The non-parametric tests for nominal and ordinal data are featured in this chapter.

Non-parametric tests are extremely useful in situations where severe violations of the normality assumption would make interpreting a parametric test problematic, and in situations where you wish to analyse categorical or ranked data. However, be aware that they are generally less powerful than their parametric counterparts, and that parametric tests are preferred when your data meet their requirements.

 AKA:
One-sample chi-square test.

16.2. Chi-Square (χ^2) Test for Goodness of Fit

The chi-square test for goodness of fit is used to assess whether observed group/category membership frequencies differ from hypothesised or expected membership frequencies. The hypothesised frequencies can be either equal (i.e., all categories will have the same number of members) or unequal. Both are illustrated below.

16.2.1. Questions We Could Answer Using The Chi-Square (χ^2) Test for Goodness of Fit

1. When given the choice, do students prefer some brands of cola to others?

If students had no cola preferences they would select each brand with equal frequency. To collect the data needed to test this proposition, we could count up how many bottles of each brand were sold at a university cafeteria over a specified period of time. The chi-square test could then be used to assess whether the purchasing pattern we've observed differs from the pattern that we would expect to see if the students didn't care which cola they drank (i.e., an approximately equal number of each brand purchased).

In the next question, the expected group frequencies are unequal.

2. Is left-handedness more prevalent amongst individuals with autism than in the general population? (10% of the general population are left-handed.)

16.2.2. Illustrated Example

Five hundred teenagers from a local high school were asked about their plans after graduation. Their responses were grouped into five categories, which are listed in Table 16.1 along with the number of students nominating each.

Table 16.1

The Post-Graduation Plans of 500 High School Students

Post-Graduation Plan (Value Code)	Frequency
Attend University (1)	240
Vocational Training (e.g., TAFE) (2)	110
Start an Apprenticeship (3)	40
Find Full-Time Employment (4)	80
Other (5)	30

Note. The values in parentheses represent the codes that will be used to represent each category in PASW Statistics.

□ Data:
This is data file **data_16_1.sav** on the companion website.

16.2.2.1. Setting Up The PASW Statistics Data File

The labour intensive version of this data file would be 500 rows long, and only 1 column wide! The first 240 rows would each contain a "1" (for plans to attend university), the next 110 rows would each contain a "2" (for vocational training), and so on.

Fortunately, there is a much quicker way of setting up a data file like this. It uses the PASW Statistics **Weight Cases** feature.

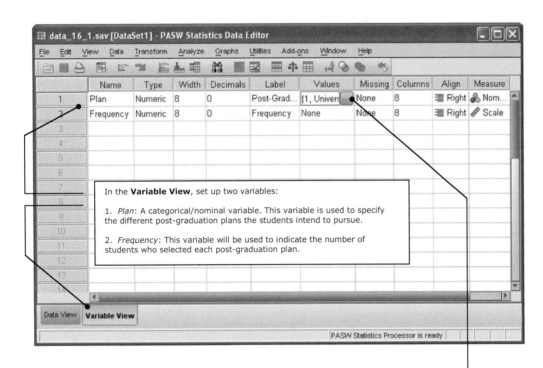

In the **Variable View**, set up two variables:

1. *Plan*: A categorical/nominal variable. This variable is used to specify the different post-graduation plans the students intend to pursue.

2. *Frequency*: This variable will be used to indicate the number of students who selected each post-graduation plan.

Post-graduation plan is a nominal variable, which requires **Value Labels**.

Select the **Values** cell for the plan variable, and then click ▣ to open the **Value Labels** dialogue. Use the values (or codes) "1" through "5" to represent the five options.

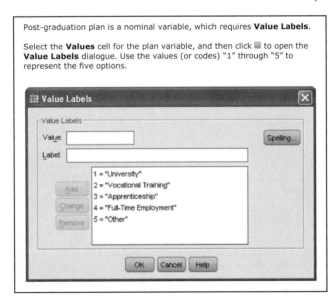

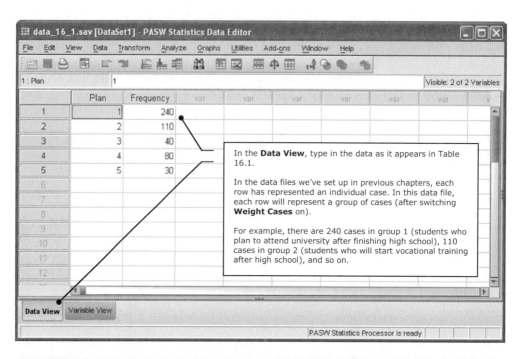

In the **Data View**, type in the data as it appears in Table 16.1.

In the data files we've set up in previous chapters, each row has represented an individual case. In this data file, each row will represent a group of cases (after switching **Weight Cases** on).

For example, there are 240 cases in group 1 (students who plan to attend university after finishing high school), 110 cases in group 2 (students who will start vocational training after high school), and so on.

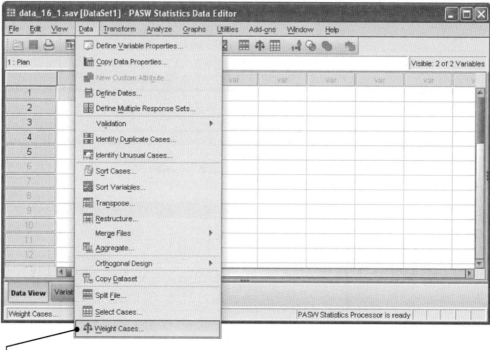

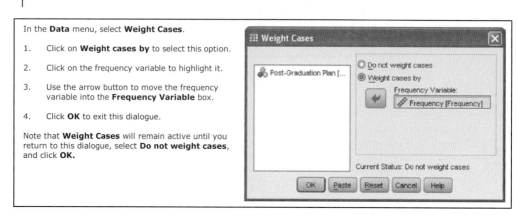

In the **Data** menu, select **Weight Cases**.

1. Click on **Weight cases by** to select this option.

2. Click on the frequency variable to highlight it.

3. Use the arrow button to move the frequency variable into the **Frequency Variable** box.

4. Click **OK** to exit this dialogue.

Note that **Weight Cases** will remain active until you return to this dialogue, select **Do not weight cases**, and click **OK.**

16.2.2.2. Analysing the Data

16.2.2.2.1. Assumptions

Generally speaking, non-parametric tests make fewer assumptions than their parametric cousins. The chi-square test is no exception here, with only two assumptions requiring your attention. The first is methodological and addressed during data collection; the second can be checked within the chi-square PASW Statistics output.

1. **Independence.** Each participant should participate only once in the research, and should not influence the participation of others.

2. **Expected Frequencies.** No more than 20% of the expected cell frequencies should be lower than five.

16.2.2.2.2. PASW Statistics Procedure

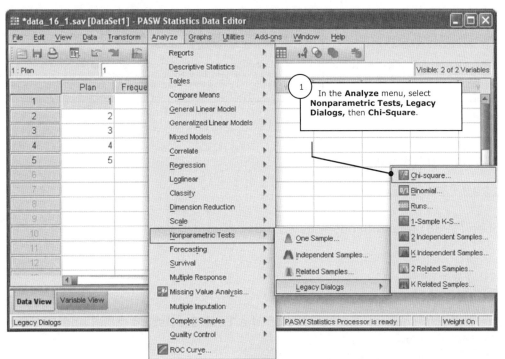

2 In **Chi-Square Test**, move the *post-graduation plan* variable into the **Test Variable List**.

3 Select **All categories equal**. This simply means that our hypothesised frequencies are equal (that is, that students choose all five options with equal frequency).

4 For our purposes, the default **Options** are fine.

5 Click **OK** to output the test results.

Syntax:
Run these analyses with **syntax_16_1.sps** on the companion website.

(i) Tip:
In the lower right corner of the **Data Editor**,

Weight On

indicates that **Weight Cases** is active.

See section 16.2.2.1 for further detail.

(i) Tip:
If your hypothesised/expected frequencies are unequal, select **Values** and then **Add** each to the **Expected Values** list.

This needs to be done in **Value Label** order (lowest to highest).

For example, we may want to compare our sample's plans to national statistics on the actual proportions of students taking up each option:

University: 30%
Vocational train: 40%
Apprenticeship: 16%
FT employment: 12%
Other: 2%

So, if we took 500 randomly selected graduates from across the country, we would expect 150 (30% of 500) to be attending university; 200 to be in vocational training (40% of 500); and so on.

These **Expected Values** (150, 200, …) then need to be added to the list like so:

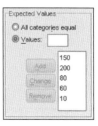

16.2.2.2.3. PASW Statistics Output

NPar Tests

Chi-Square Tests

Frequencies

Post-Graduation Plan

	Observed N	Expected N	Residual
University	240	100.0	140.0
Vocational Training	110	100.0	10.0
Apprenticeship	40	100.0	-60.0
Full-Time Employment	80	100.0	-20.0
Other	30	100.0	-70.0
Total	500		

The first output table has three columns of data:

Observed N: The number of cases we observed in each category. This column basically just replicates/summarises the data file. 240 students reported plans to attend university, 110 reported plans to begin vocational training, and so on.

Expected N: Our hypothesised frequencies. The pattern of responses we would expect if all five post-graduation plans were equally popular.

Residual: The difference between each category's **Observed N** and **Expected N**.

Test Statistics

	Post-Graduation Plan
Chi-square	286.000ᵃ
df	4
Asymp. Sig.	.000

a. 0 cells (.0%) have expected frequencies less than 5. The minimum expected cell frequency is 100.0.

The **Test Statistics** table contains the results of the chi-square test, and answers the basic question, "do the observed frequencies differ from the expected frequencies?"

Because the **Asymp. Sig** is less than .05 (our nominated alpha level), the answer to this question is yes!

Make a note of these figures for the write up:

Chi-Square = 286.000
df = 4
Asymp. Sig = .000

The note below the **Test Statistics** table informs us that 0% of categories had expected frequencies below five.

As this is less than 20%, the expected frequencies assumption has not been violated.

If the Assumption is Violated.

If more than 20% of your expected frequencies are below five, our best recommendations are to (a) increase the size of your sample; and/or (b) merge some categories.

16.2.2.3. Follow Up Analyses

16.2.2.3.1. Effect Size

Cohen (1988) recommends w as an index of effect size:

$$w = \sqrt{\frac{\chi^2}{N}}$$

Where χ^2 (chi-square) can be read off the **Test Statistics** table, and N is the total sample size.

So,

$$w = \sqrt{\frac{286.000}{500}}$$

$$= \sqrt{0.572}$$

$$= 0.756$$

Cohen (1988) suggested that a w of 0.1 be considered a small effect, 0.3 be considered medium, and 0.5 be considered large.

16.2.2.4. APA Style Results Write-Up

Results

A chi-square test for goodness of fit (with $\alpha = .05$) was used to assess whether some post-graduation plans were more popular than others in a sample of 500 high school students. Table 1 lists the percentages of the sample nominating each option.

Table 1

Summary of the Post-Graduation Plans of a Sample of High School Students (N = 500)

Post-Graduation Plan	Percentage (%) of Sample
Attend University	48
Vocational Training (e.g., TAFE)	22
Find Full-Time Employment	16
Start an Apprenticeship	8
Other	6
Total	**100**

> A table can be useful when reporting descriptive or summary statistics. However, don't repeat the content of a table in the body of your text. Report your data in one or the other!

> When tabulating the proportions, frequencies, or percentages of categories (which have no inherent order), organise them from largest-to-smallest (or smallest-to-largest).
>
> This will give readers a much more intuitive sense of the relative importance or value of each category than randomly or alphabetically organised data will.

The chi-square test was statistically significant, χ^2 (4, $N = 500$) = 286.00, $p < .001$, indicating that some post-graduation plans were reported with significantly greater frequency than others. As an index of effect size, Cohen's w was 0.76, which can be considered large.

> If Table 1 was not included in this write-up, it would be useful to indicate which plans were reported with "greater frequency than others" somewhere around here.

> A large effect indicates that the pattern of observed frequencies deviated substantially from the pattern of frequencies one would expect to see if the students just chose randomly from the list of plans.

AKA:
Chi-square test for independence or relatedness; Multi-dimensional chi-square.

16.3. Chi-Square (χ^2) Test of Contingencies

The chi-square test of contingencies is most commonly used to assess whether two categorical (nominal) variables are related. That is, whether group or category membership on one variable is influenced by (or contingent on) group membership on a second variable.

16.3.1. Questions We Could Answer Using The Chi-Square (χ^2) Test of Contingencies

1. Is there a relationship between support for increasing foreign aid to developing countries and voting intentions (e.g., Labor, Liberal, Democrat, or other) in the next federal election?

This question is essentially asking whether support for increasing foreign aid is contingent on one's political preference. Or, alternatively, whether political preference is related to whether or not one supports increasing foreign aid.

A statistically significant chi-square would indicate that the voting profile of the "supporters" group differs from the profile of the "non-supporters" group. (It may be that supporters are more likely to be Labor voters, whereas non-supporters are most likely to vote Liberal, or vice versa.)

2. Are smokers more or less likely than non-smokers to fall ill at least once during winter?

16.3.2. Illustrated Example

Last year, the doctors and nurses at St. John's Hospital provided care and treatment to 430 stroke survivors. One hundred and ten of these patients were also diagnosed with aphasia (a language disorder caused by brain injury) during their stay at the hospital.

A student nurse at St. John's is examining factors that influence the likelihood of developing aphasia following a stroke. Overseas research indicates that one of these factors may be handedness (i.e., whether the patient is left or right handed).

The student would like to find out whether handedness is related to the likelihood of developing aphasia in this sample of Australian stroke survivors. Her data are reproduced in Table 16.2.

Table 16.2

The Frequency of Left and Right Handed Stroke Survivors Who Were and Were Not Also Diagnosed with Aphasia

⬜ **Data:**
This is data file **data_16_2.sav** on the companion website.

Aphasia Diagnosis	Handedness	Frequency
1	1	36
1	2	74
2	1	44
2	2	276

Note. Value labels for the aphasia diagnosis variable: 1 = diagnosed with aphasia; 2 = not diagnosed with aphasia. Value labels for the handedness variable: 1= left handed; 2 = right handed.

16.3.2.1. Setting Up The PASW Statistics Data File

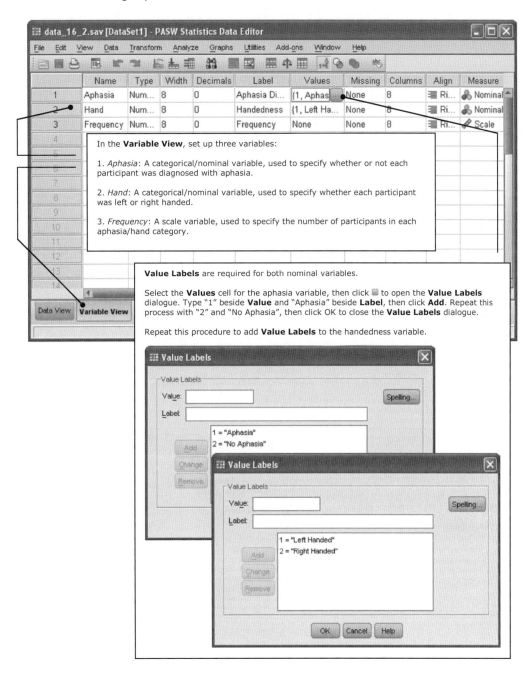

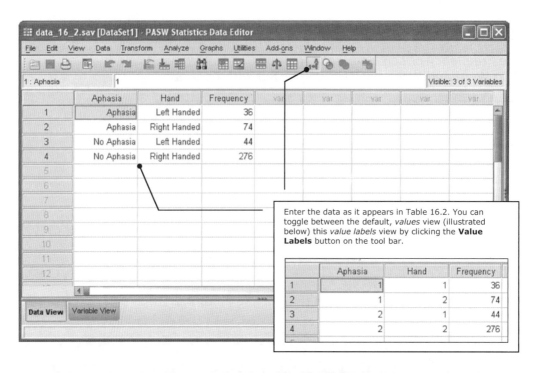

Enter the data as it appears in Table 16.2. You can toggle between the default, *values* view (illustrated below) this *value labels* view by clicking the **Value Labels** button on the tool bar.

	Aphasia	Hand	Frequency
1	1	1	36
2	1	2	74
3	2	1	44
4	2	2	276

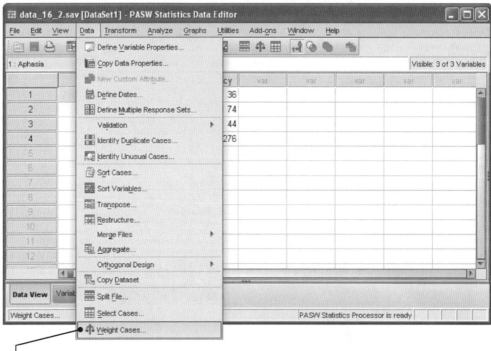

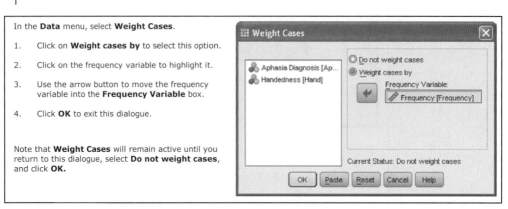

In the **Data** menu, select **Weight Cases**.

1. Click on **Weight cases by** to select this option.

2. Click on the frequency variable to highlight it.

3. Use the arrow button to move the frequency variable into the **Frequency Variable** box.

4. Click **OK** to exit this dialogue.

Note that **Weight Cases** will remain active until you return to this dialogue, select **Do not weight cases**, and click **OK.**

16.3.2.2. Analysing the Data

16.3.2.2.1. Assumptions

Only two assumptions require your attention. The first should be addressed during data collection; the second can be checked within the chi-square PASW Statistics output.

1. **Independence.** Each participant should participate only once in the research, and should not influence the participation of others.

2. **Expected Frequencies.** In a 2 x 2 design, all expected frequencies should be at least five. In larger designs, no more than 20% of the expected cell frequencies should be lower than five.

16.3.2.2.2. PASW Statistics Procedure

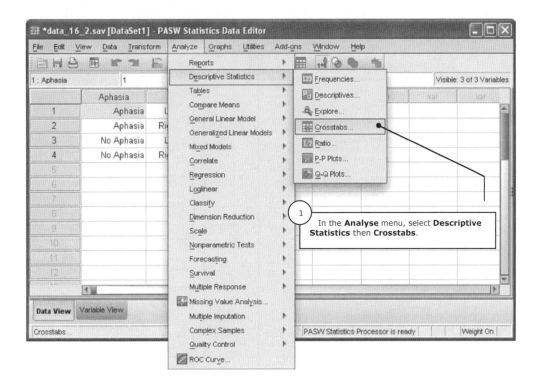

In the **Analyse** menu, select **Descriptive Statistics** then **Crosstabs**.

Syntax:
Run these analyses with **syntax_16_2.sps** on the companion website.

(i) *Tip:*
In the lower right corner of the **Data Editor,**

Weight On

indicates that **Weight Cases** is active.

2 In undertaking this research, we're asking whether two categorical variables are related, or independent. Move one of these variables into the **Row(s)** list, and the other into the **Column(s)** list.

3 Select **Display clustered bar charts**. This will provide a useful visual summary of our data.

4 Click the **Statistics**, **Cells** and/or **Format** buttons for additional options.

ⓘ *Tip:*

If you specify a (categorical) **Layer** variable, PASW Statistics will produce a set of statistics and graphs for each of its levels.

For example, if gender is added as a **Layer** variable, PASW Statistics will produce two separate sets of output. One set will focus on the handedness-aphasia relationship in the male data, while the other will focus on that same relationship in the female data.

5 In **Crosstabs: Statistics** select **Chi-square** and **Phi and Cramer's V**.

Chi-square will provide several chi-square tests. The most commonly reported of these is Pearson's chi-square.

Phi and Cramer's V are two measures of association (or effect size) suitable for use with nominal data.

After making your selections, click **Continue** to close this dialogue.

6 The options selected in **Crosstabs: Cell Display** determine what is included in the **Crosstabulation** table overleaf.

The selections illustrated below will provide you with the detail needed to properly interpret the outcome of a chi-square test.

Click **Continue** to close this dialogue.

7 In **Crosstabs: Table Format** you can specify whether the rows in the **Crosstabulation** table should be presented in **Ascending** or **Descending** order.

Generally, the default **Row Order** will be fine.

Click **Continue** to close this dialogue.

8 Click **OK** to output the test results.

16.3.2.2.3. PASW Statistics Output

Crosstabs

The **Case Processing Summary** indicates that data from 430 cases were included in these analyses.

Case Processing Summary

	Cases					
	Valid		Missing		Total	
	N	Percent	N	Percent	N	Percent
Aphasia Diagnosis * Handedness	430	100.0%	0	.0%	430	100.0%

Aphasia Diagnosis was selected as the **Row(s)** variable. It has two levels: (1) **Aphasia**; and (2) **No Aphasia**.

Handedness was selected as the **Column(s)** variable. It also has two levels: (1) **Left Handed**; and (2) **Right Handed**.

Within this 2 x 2 **Crosstabulation** matrix there are four cells. Each cell describes a particular group of cases:

(A) Left handed individuals diagnosed with aphasia.

(B) Right handed individuals diagnosed with aphasia.

(C) Left handed individuals not diagnosed with aphasia.

(D) Right handed individuals not diagnosed with aphasia.

Aphasia Diagnosis * Handedness Crosstabulation

			Handedness		Total
			Left Handed	Right Handed	
Aphasia Diagnosis	Aphasia	Count	(A) 36	(B) 74	110
		Expected Count	20.5	89.5	110.0
		% within Aphasia Diagnosis	32.7%	67.3%	100.0%
		% within Handedness	45.0%	21.1%	25.6%
		% of Total	8.4%	17.2%	25.6%
	No Aphasia	Count	(C) 44	(D) 276	320
		Expected Count	59.5	260.5	320.0
		% within Aphasia Diagnosis	13.8%	86.3%	100.0%
		% within Handedness	55.0%	78.9%	74.4%
		% of Total	10.2%	64.2%	74.4%
Total		Count	80	350	430
		Expected Count	80.0	350.0	430.0
		% within Aphasia Diagnosis	18.6%	81.4%	100.0%
		% within Handedness	100.0%	100.0%	100.0%
		% of Total	18.6%	81.4%	100.0%

Let's focus on cell (A), which describes the left handed cases with aphasia:

Count: There are 36 cases in this cell (i.e., 36 members of our sample were both left handed and diagnosed with aphasia).

Expected Count: If the null hypothesis (that handedness is not related to the likelihood of developing aphasia) were true, we would expect to see just 20.5 left handed cases with aphasia in this sample of 430 patients.

An expected cell count is calculated by multiplying together the total counts for the row and column in which the cell is located, and then dividing by the total sample size. Here,

(110 x 80) / 430 = 20.5

By default, PASW Statistics rounds the expected counts to 1 decimal place.

% within Aphasia Diagnosis: In total, 110 cases were diagnosed with aphasia. Of these, 32.7% (i.e., 36 cases) were also left handed.

% within Handedness: In total, 80 cases were left handed. Of these, 45% (i.e., 36 cases) were also diagnosed with aphasia.

% of Total: 8.4% of the full sample were both left handed, and diagnosed with aphasia.

Using the **Crosstabulation** table to make comparisons between cells, we can begin to develop a sense of whether or not our two variables are related.

For example, around one quarter (25.6%) of total sample developed post-stroke aphasia. However, the proportion of "lefties" with aphasia (45%) is more than double the proportion of "righties" diagnosed with this disorder (21.1%).

Move down to the **Chi-Square Tests** table to gauge whether this pattern is real and replicable, or likely due to chance factors.

The **Pearson Chi-Square** is statistically significant. It is unlikely that we would observe data like ours if handedness is not actually related to the likelihood of developing aphasia.

Consequently, we can conclude that <u>handedness probably is related to the likelihood of developing aphasia after a stroke.</u>

The following figures will be required for the write-up:

Pearson Chi-Square Value = 19.468
Pearson Chi-Square df = 1
Pearson Chi-Square Asymp. Sig (2-sided) = .000
N of Valid Cases = 430

Note "b" below the **Chi-Square Tests** table indicates that no (0%) cells had expected frequencies below five. <u>The expected frequencies assumption has not been violated.</u>

If the Assumption is Violated

If you have low expected frequencies (and a 2 x 2 design), consider using **Fisher's Exact Test.** Otherwise, increase your sample size and/or reduce the number of cells in your crosstabulation matrix.

▶▶ *Link:*
Below the **Pearson Chi-Square** several alternative statistics are listed. In most situations they will all lead you to the same basic conclusion (i.e., statistically significant or non-significant).

Interested readers can learn about these alternatives in Howell (2010b), or one of many other modern statistics texts.

Chi-Square Tests

	Value	df	Asymp. Sig. (2-sided)	Exact Sig. (2-sided)	Exact Sig. (1-sided)
Pearson Chi-Square	19.468[a]	1	.000		
Continuity Correction[b]	18.235	1	.000		
Likelihood Ratio	17.832	1	.000		
Fisher's Exact Test				.000	.000
Linear-by-Linear Association	19.423	1	.000		
N of Valid Cases	430				

a. 0 cells (.0%) have expected count less than 5. The minimum expected count is 20.47.

b. Computed only for a 2x2 table

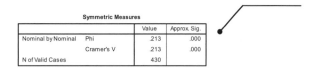

Phi and **Cramer's V** are both used as measures of effect size.

Phi (ϕ) is only appropriate for 2 x 2 designs, and is interpreted like a correlation coefficient (see chapter 12). Squaring phi will provide an estimate of the proportion of variance that is common to the two variables. Here, $\phi^2 = .045$, indicating that just under 5% of the variability in the aphasia diagnosis data can be accounted for by handedness.

Cramer's V is an extension of phi that is suitable for larger designs.

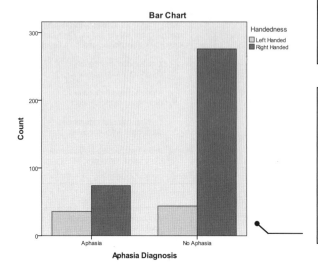

A clustered **Bar Chart** like this may help illustrate the patterns first observed in the **Crosstabulation** table.

It indicates that the left handed patients were roughly equally split between the aphasia and non-aphasia diagnoses, whereas a substantial majority (over three quarters) of right handed patients were not diagnosed with aphasia.

Or alternatively, that there were approximately twice as many righties as lefties diagnosed with aphasia, yet righties outnumbered lefties by a ratio of over 6 to 1 in the non-aphasia category.

16.3.2.3. Follow Up Analyses

16.3.2.3.1. Effect Size

Phi and Cramer's *V* (in the **Symmetric Measures** table) can be used as effect size indices for the chi-square test of contingencies. These statistics are both closely related to Cohen's *w*, which can be calculated for a contingency table of any size using the following formula:

$$w = \sqrt{\frac{\chi^2}{N}}$$

Where χ^2 (chi-square) can be read off the **Chi-Square Test** table, and *N* is the total sample size.

So,

$$w = \sqrt{\frac{19.468}{430}}$$

$$= \sqrt{0.045}$$

$$= 0.212$$

Cohen (1988) suggested that a *w* of 0.1 be considered a small effect, 0.3 be considered medium, and 0.5 be considered large.

> **Tip:**
> For the basic 2 x 2 design, Phi, Cramer's *V* and Cohen's *w* will be identical.
>
> Furthermore, *V* and *w* will be identical for any 2 x k design.

> **Tip:**
> As an alternative, you may want to calculate an odds ratio. In the current example, the odds of developing aphasia after a stroke are just over three times higher for lefties than they are for righties. Refer to Howell (2010b) for an expanded discussion of odds ratios.

16.3.2.4. APA Style Results Write-Up

Results

A Pearson's chi-square test of contingencies (with $\alpha = .05$)
was used to evaluate whether handedness is related to whether or not
patients develop aphasia following a stoke. The chi-square test was
statistically significant, χ^2 (1, $N = 430$) = 19.47, $p < .001$, although the
association between handedness and post-stroke aphasia was actually
quite small, $\phi = .21$. As illustrated in Figure 1, the left handed patients
were significantly more likely to develop post-stroke aphasia than the
right handed patients.

> In Microsoft Word, you can locate the chi (χ) and phi (ϕ) characters by selecting **Symbol** in the **Insert** menu.

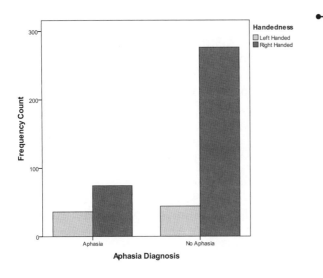

> A graph or table can often illustrate patterns in a data set with much greater clarity than a paragraph of text.
>
> However, when using them, there are a few basic rules that should be followed:
>
> 1. Don't repeat content of a graph/table in the body of your report. Use one or the other.
>
> 2. Refer to the graph or table in the body of your report (preferably in the paragraph immediately prior). Don't just leave it hanging in space.
>
> 3. Ensure that you've selected an appropriate type of graph/table for the information being conveyed. Ideally, a graph or table (along with its caption or title) should make sense without reference to the text surrounding it.

Figure 1. Clustered bar chart illustrating the number of left and right
handed patients who did and did not develop aphasia after a stroke.

Tip:
The Mann-Whitney
test is equivalent to
the Wilcoxon rank sum
test and the Kruskal-
Wallis ANOVA for two
groups.

16.4. Mann-Whitney *U* Test

The Mann-Whitney *U* test is typically used to compare two independent samples of ordinal (ranked) data. It is also useful to researchers who are unable to calculate an independent samples *t* statistic (chapter 4) due to severe violations of the normality assumption.

16.4.1. Questions We Could Answer Using The Mann-Whitney *U* Test

1. Do males and females differ in the emphasis they place on the importance of the physical attractiveness of prospective romantic partners?

To answer this question, we could first identify a set of characteristics that people commonly look for in prospective romantic partners (e.g., physical attractiveness, financial security, sense-of-humour and so on), then ask each participant to sort these characteristics from least to most important. Some participants would probably rank physical attractiveness very highly (e.g., first or second), whereas others would consider it much less important.

We could then use the Mann-Whitney *U* test to assess whether these rankings are related to gender. That is, whether one gender tends to rate physical attractiveness as more or less important than the other.

2. Every year, the ACORN Foundation publishes a list of the "Top 100" schools across the country. Are schools' rankings on this list related to whether they're private or public?

16.4.2. Illustrated Example

To assess whether gender is related to academic seniority, 25 male and 25 female academics were randomly sampled from a university staff directory. Their current academic levels are reported in Table 16.3.

Table 16.3

Data:
This is data file
data_16_3.sav
on the companion
website.

Academic Levels of Male and Female University Employees (N = 50)

Male				Female			
ID	Level	ID	Level	ID	Level	ID	Level
1	5	14	4	26	2	39	3
2	4	15	4	27	4	40	3
3	4	16	2	28	3	41	2
4	3	17	3	29	4	42	2
5	2	18	4	30	3	43	1
6	1	19	2	31	2	44	1
7	5	20	3	32	1	45	1
8	3	21	4	33	5	46	2
9	3	22	3	34	3	47	4
10	3	23	3	35	1	48	2
11	3	24	3	36	2	49	3
12	5	25	4	37	2	50	3
13	1			38	2		

Note. Value labels for academic level variable: 1 = Associate Lecturer; 2 = Lecturer; 3 = Senior Lecturer; 4 = Associate Professor; 5 = Professor.

16.4.2.1. Setting Up The PASW Statistics Data File

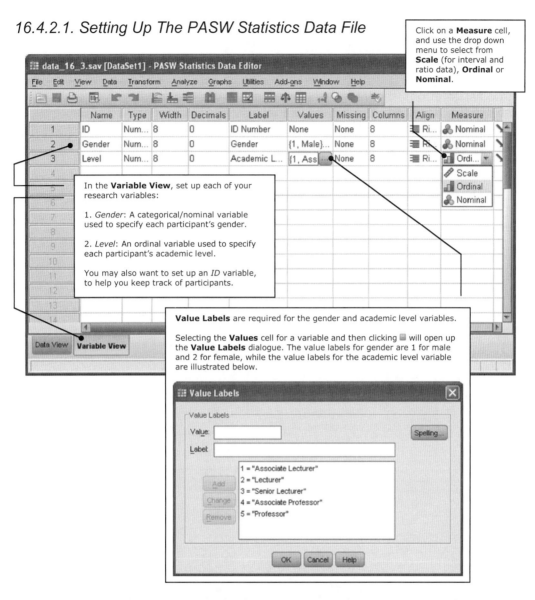

Click on a **Measure** cell, and use the drop down menu to select from **Scale** (for interval and ratio data), **Ordinal** or **Nominal**.

In the **Variable View**, set up each of your research variables:

1. *Gender*: A categorical/nominal variable used to specify each participant's gender.

2. *Level*: An ordinal variable used to specify each participant's academic level.

You may also want to set up an *ID* variable, to help you keep track of participants.

Value Labels are required for the gender and academic level variables.

Selecting the **Values** cell for a variable and then clicking ▦ will open up the **Value Labels** dialogue. The value labels for gender are 1 for male and 2 for female, while the value labels for the academic level variable are illustrated below.

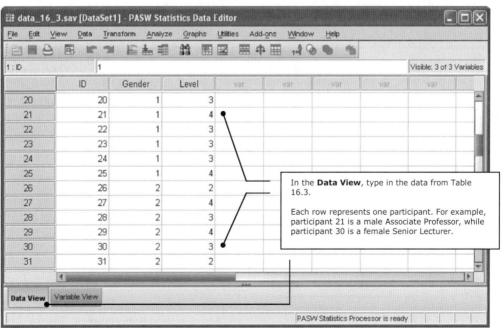

In the **Data View**, type in the data from Table 16.3.

Each row represents one participant. For example, participant 21 is a male Associate Professor, while participant 30 is a female Senior Lecturer.

🖥 *Syntax:*
Run these
analyses with
syntax_16_3.sps
on the companion
website.

16.4.2.2. Analysing the Data

16.4.2.2.1. Assumptions

Three assumptions should be met when using the Mann-Whitney U test. The first two are dealt with during the planning stages of research.

1. **Independence.** Each participant should participate only once in the research, and should not influence the participation of others.

2. **Scale of Measurement.** The dependent variable (DV) should be at least ordinal.

3. **Shape of the Distributions.** The two distributions of data should look the same (i.e., same shape and spread).

16.4.2.2.2. PASW Statistics Procedure

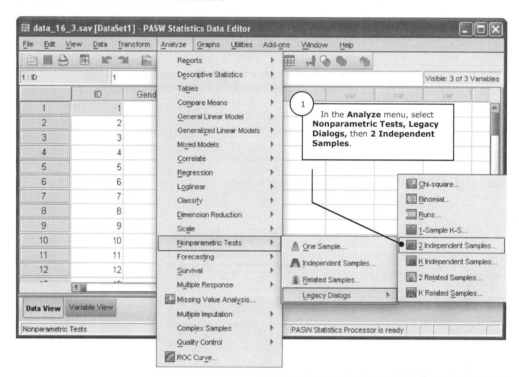

In the **Analyze** menu, select **Nonparametric Tests, Legacy Dialogs**, then **2 Independent Samples**.

🗗 *Included With:*
You will not have an
Exact button like the
one illustrated to the
right unless you have
PASW Exact Tests
add-on installed on
your system.

Exact tests are
preferred by
researchers working
with small samples
and rare events.

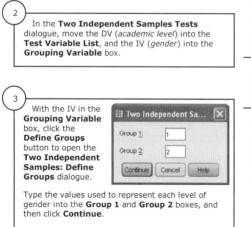

2 In the **Two Independent Samples Tests** dialogue, move the DV (*academic level*) into the **Test Variable List**, and the IV (*gender*) into the **Grouping Variable** box.

3 With the IV in the **Grouping Variable** box, click the **Define Groups** button to open the **Two Independent Samples: Define Groups** dialogue.

Type the values used to represent each level of gender into the **Group 1** and **Group 2** boxes, and then click **Continue**.

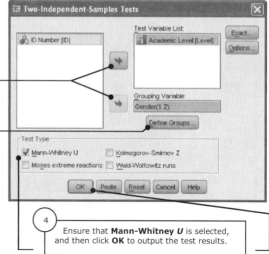

4 Ensure that **Mann-Whitney U** is selected, and then click **OK** to output the test results.

16.4.2.2.3. PASW Statistics Output

NPar Tests

Mann-Whitney Test

Ranks

	Gender	N	Mean Rank	Sum of Ranks
Academic Level	Male	25	30.58	764.50
	Female	25	20.42	510.50
	Total	50		

There are three columns in the **Ranks** table:

N. The number of participants in each group, and the total number of participants. Here, there were 25 males and 25 females in our sample.

Mean Rank. When computing the Mann-Whitney test, PASW Statistics first merges, and then rank orders the entire data set. Once a rank (or tied rank) has been assigned to each case, they are returned to their respective groups (i.e., male and female), and then summed and averaged. The **Mean Rank** values are these averages. In our data set, the average rank for the male academics was 30.58, while the mean rank for the female academics was 20.42. This indicates that the males appear more frequently in the higher ranked positions, whereas the women tend to predominate in the lower ranked positions.

Sum of Ranks. The sum of all the ranks within each group of scores. This value can easily be calculated by multiplying the **Mean Rank** for a group by its **N**. For example, the **Sum of Ranks** for the male academics is 30.58 x 25 = 764.50.

The sums of ranks are necessary for the computation of the Mann-Whitney U, but it's unlikely that you would ever need to report them in a research report.

Test Statistics[a]

	Academic Level
Mann-Whitney U	185.500
Wilcoxon W	510.500
Z	-2.547
Asymp. Sig. (2-tailed)	.011

a. Grouping Variable: Gender

In **Test Statistics**, the following are reported:

Mann-Whitney U. The basic test statistic. A U of zero indicates the greatest possible difference between the two groups. In our example, U = 0 would mean that no females were ranked higher than (or even equal to) any males. As the two samples become more alike, and the scores begin to intermix (so some females are ranked higher than some males, and vice versa), U gets larger. U should be reported in your results.

Wilcoxon W. This is the Wilcoxon rank sum (W) statistic, and is basically equivalent to the Mann-Whitney U. It can be ignored for our purposes.

Z. For larger data sets (where each group is $n > 20$), the U distribution approximates a normal (z) distribution, and a z-approximation of U (corrected for ties) is calculated and used to assess statistical significance. It should be reported when used (i.e., when each group size is >20).

Asymp. Sig (2-tailed). The two-tailed asymptotic probability of **Z**. As this value is below .05, we can conclude that there is a statistically significant difference between the academic levels of the males and females (or, if you prefer, that academic level is related to gender). The direction of this effect can be determined by comparing the mean ranks in the **Ranks** table. In our case, the males' mean rank was higher than the females'.

For smaller data sets (where each group is $n \leq 20$), PASW Statistics will provide an exact probability (**Exact Sig**) in addition to the asymptotic probability (**Asymp. Sig**) shown here. It is considered more accurate when group sizes are small, and should be reported if available.

ⓘ Tip:
The critical z for α = .05 (two-tailed) is ± 1.96. A calculated z larger than this can be considered statistically significant.

Refer to the normal (z) distribution tables included in most statistics texts (e.g., Clark-Carter, 2009; Howell, 2010b) for the critical z values associated with alternate alpha levels (e.g., .01, .001 etc.).

16.4.2.3. Follow Up Analyses

16.4.2.3.1. Effect Size

Clark-Carter (2009) recommends converting z into r with:

$$r = \frac{z}{\sqrt{N}}$$

Where z can be read off the **Test Statistics** table, and N is the total sample size. So,

$$r = \frac{-2.547}{\sqrt{50}}$$

$$= -0.36$$

By Cohen's (1988) conventions, this is a medium sized effect.

ⓘ Tip:
Cohen (1988) suggested that r = .1 could be considered small, r = .3 could be considered medium, and r = .5 could be considered large.

16.4.2.3.2. Graphing the Effect

As PASW Statistics does not provide a visual depiction of the relationship between gender and academic level as part of the Mann-Whitney output, we shall create one using the PASW Statistics **Chart Builder**.

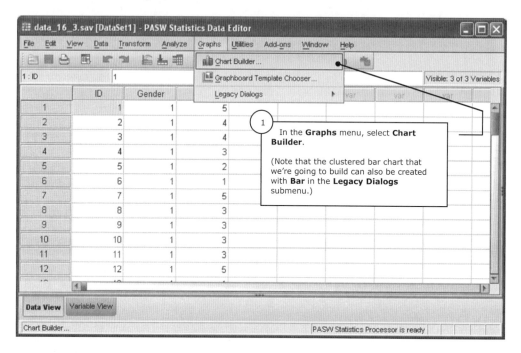

1. In the **Graphs** menu, select **Chart Builder**.

   (Note that the clustered bar chart that we're going to build can also be created with **Bar** in the **Legacy Dialogs** submenu.)

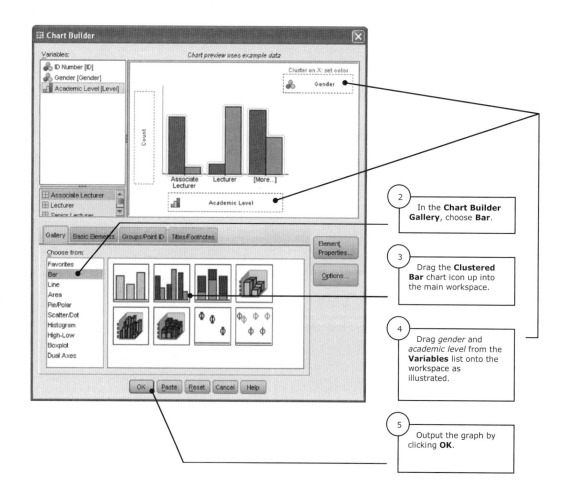

(i) Tip:
The **Chart Builder** is very flexible.

Customisation options can be accessed with the tabs and buttons below the **Chart preview**.

2. In the **Chart Builder Gallery**, choose **Bar**.

3. Drag the **Clustered Bar** chart icon up into the main workspace.

4. Drag *gender* and *academic level* from the **Variables** list onto the workspace as illustrated.

5. Output the graph by clicking **OK**.

GGraph

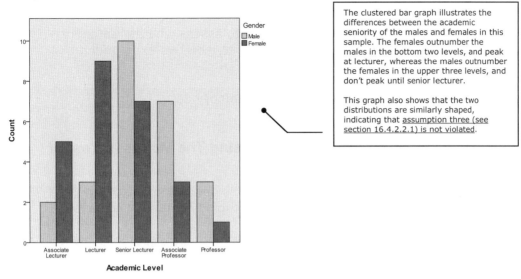

The clustered bar graph illustrates the differences between the academic seniority of the males and females in this sample. The females outnumber the males in the bottom two levels, and peak at lecturer, whereas the males outnumber the females in the upper three levels, and don't peak until senior lecturer.

This graph also shows that the two distributions are similarly shaped, indicating that assumption three (see section 16.4.2.2.1) is not violated.

16.4.2.4. APA Style Results Write-Up

Results

A Mann-Whitney U test indicated that the academic levels of the male participants (*Mean Rank* = 30.58, $n = 25$) were significantly higher than those of the female participants (*Mean Rank* = 20.42, $n = 25$), $U = 185.50$, $z = -2.55$ (corrected for ties), $p = .011$, two-tailed. This effect can be described as "medium" ($r = .36$), and is illustrated in Figure 1.

When there are 20 or fewer cases in each group, report U and the exact probability calculated by PASW Statistics.

For larger samples (i.e., when all groups have $n > 20$), report U, z and the asymptotic probability of z.

Also indicate whether or not your z test was corrected for ties. (PASW Statistics automatically makes this correction.)

Report the absolute value of r, rather than a negative value.

This clustered bar graph clearly illustrates how the males in the sample tend to hold higher academic positions than their female colleagues.

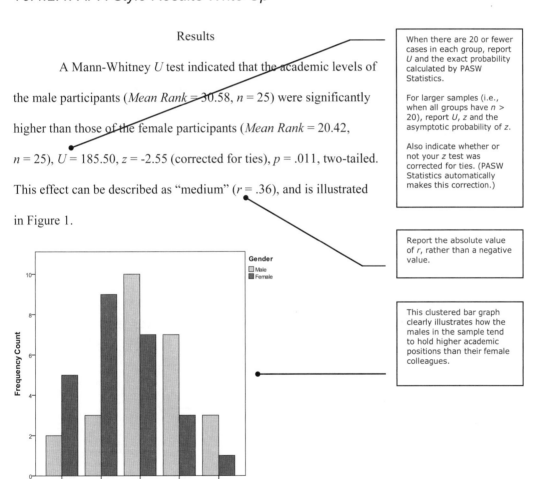

Figure 1. The distributions of male ($n = 25$) and female ($n = 25$) academic levels.

AKA:
McNemar's chi-square;
McNemar's test.
McNemar's test for
correlated proportions;
McNemar's test for
paired data.

16.5. McNemar Test of Change

The McNemar test is most commonly used to assess whether category membership on a binary variable changes between two experimental conditions or points in time.

It can also be used with matched or paired binary data.

16.5.1. Questions We Could Answer Using The McNemar Test of Change

1. Do students' responses to the question "is psychology a science (Y/N)?" change after taking an introductory psychology course at university?

This is a repeated measures question, which can be conceptualised within a 2 x 2 matrix:

		After taking the course.	
		Yes	No
Before taking	Yes	A	C
the course.	No	B	D

In this matrix, A and D represent students whose opinions did not change, and B and C represent students whose opinions did change. The McNemar test is only interested in cells B and C, and asks whether the proportion of students in B differs from proportion in C.

16.5.2. Illustrated Example

Before and after watching a persuasive television documentary about the scientific breakthroughs and future benefits that may be derived from human embryonic stem cell research, 200 participants were telephoned and asked "are you in favour of stem cell research (Y/N)?" Their responses are tabulated in Table 16.4.

Table 16.4

⬜ *Data:*
This is data file
data_16_4.sav
on the companion
website.

The Frequency of Participants For and Against Stem Cell Research Before and After Watching a Persuasive Documentary On This Topic

Before	After	Frequency
1	1	42
2	1	84
1	2	22
2	2	52

Note. Value labels for both the before and after conditions: 1 = Yes (in favour of stem cell research); 2 = No (not in favour of stem cell research).

16.5.2.1. Setting Up The PASW Statistics Data File

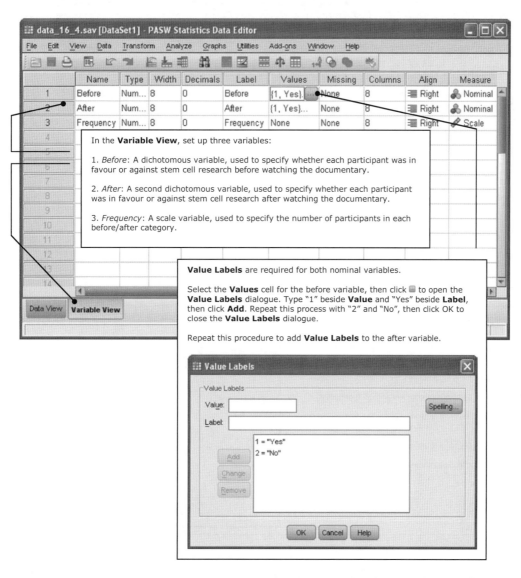

In the **Variable View**, set up three variables:

1. *Before*: A dichotomous variable, used to specify whether each participant was in favour or against stem cell research before watching the documentary.

2. *After*: A second dichotomous variable, used to specify whether each participant was in favour or against stem cell research after watching the documentary.

3. *Frequency*: A scale variable, used to specify the number of participants in each before/after category.

Value Labels are required for both nominal variables.

Select the **Values** cell for the before variable, then click ▦ to open the **Value Labels** dialogue. Type "1" beside **Value** and "Yes" beside **Label**, then click **Add**. Repeat this process with "2" and "No", then click OK to close the **Value Labels** dialogue.

Repeat this procedure to add **Value Labels** to the after variable.

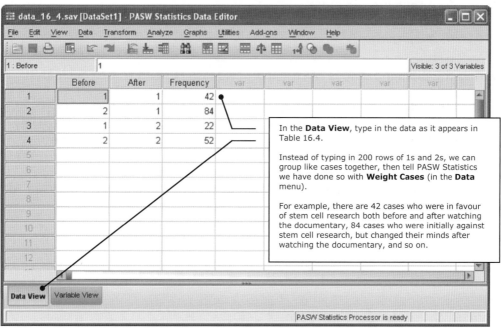

In the **Data View**, type in the data as it appears in Table 16.4.

Instead of typing in 200 rows of 1s and 2s, we can group like cases together, then tell PASW Statistics we have done so with **Weight Cases** (in the **Data** menu).

For example, there are 42 cases who were in favour of stem cell research both before and after watching the documentary, 84 cases who were initially against stem cell research, but changed their minds after watching the documentary, and so on.

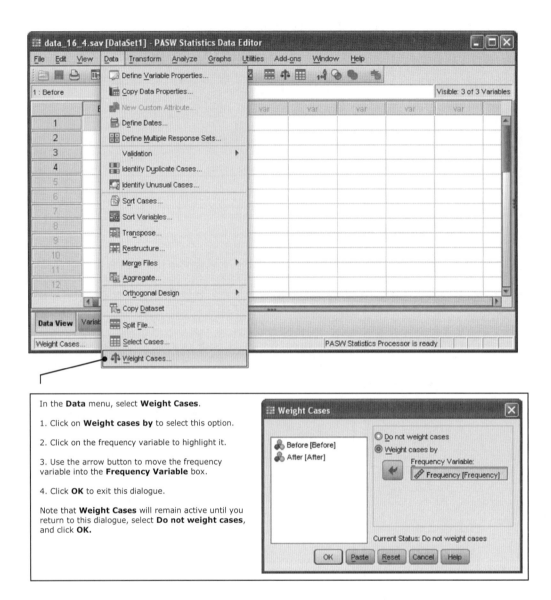

Syntax:
Run these
analyses with
syntax_16_4.sps
on the companion
website.

16.5.2.2. Analysing the Data

16.5.2.2.1. Assumptions

During the planning stages of research, two issues need to be considered.

1. **Independence.** Each participant (or pair of participants, in the case of a matched design) should participate only once in the research (i.e., they should provide only one set of data), and should not influence the participation of others.

2. **Scale of Measurement.** The McNemar test was designed to be used with binary data.

16.5.2.2.2. PASW Statistics Procedure

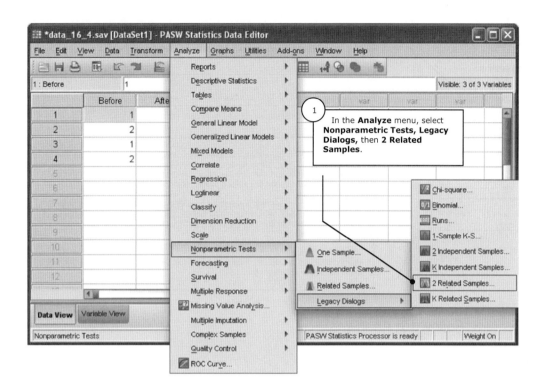

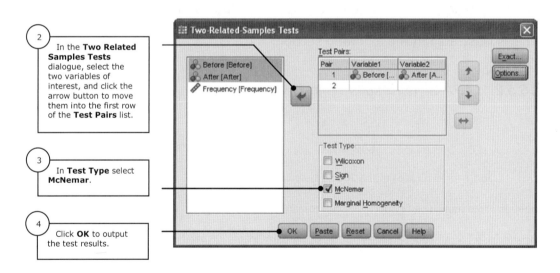

2 In the **Two Related Samples Tests** dialogue, select the two variables of interest, and click the arrow button to move them into the first row of the **Test Pairs** list.

3 In **Test Type** select **McNemar**.

4 Click **OK** to output the test results.

ⓘ *Tip:*
To select multiple variables, hold down the Shift key on your keyboard, then click each with your mouse.

One click will select a variable; a second click will deselect it.

16.5.2.2.3. PASW Statistics Output

NPar Tests

McNemar Test

This first table simply summarises the data set. Remember that 1 = Yes and 2 = No for both variables.

Crosstabs

Before & After

Before	After	
	Yes	No
Yes	42	22
No	84	52

Test Statistics[b]

	Before & After
N	200
Chi-square[a]	35.104
Asymp. Sig.	.000

a. Continuity Corrected

b. McNemar Test

In the **Test Statistics** table, the following are reported:

N. The total number of participants surveyed.

Chi-Square. When more than 25 cases swap categories between the first and second variables (e.g., were against stem cell research before the documentary, but in favour of it afterwards) PASW Statistics calculates a continuity corrected Chi-Square (χ^2) statistic, and associated asymptotic probability.

Asymp. Sig. The two-tailed asymptotic probability of χ^2. As this value is below .05 (our specified α level) we can conclude that significantly more participants altered their attitude towards stem cell research from negative to positive, than from positive to negative.

When 25 or fewer cases swap categories between the first and second variables, an exact probability is calculated using the binomial distribution. In these situations, there will be no χ^2 to report.

16.5.2.3. Follow Up Analyses

16.5.2.3.1. Effect Size

We are not aware of any commonly used measures of effect size for the McNemar test, but propose that a simple description/summary of your data will give readers an adequate sense of the size and direction of your effect. This is illustrated in the following section.

16.5.2.4. APA Style Results Write-Up

Results

Just over half (106) of the 200 participants changed their opinion about human embryonic stem cell research after viewing the persuasive television documentary. Of these, 84 participants changed in a positive direction (from "against" to "in favour"), whilst only 22 did the reverse. A McNemar test indicated that the observed tendency for those who changed their opinions to do so in favour of stem cell research was statistically significant, $\chi^2 (1, N = 200) = 35.10, p < .001$.

16.6. Wilcoxon Signed Rank Test

The Wilcoxon signed rank test is typically used to compare two related samples of ordinal (ranked) data. It is also useful to researchers who are unable to calculate and correctly interpret a paired samples t statistic (chapter 6) due to marked violations of the normality assumption.

> **ⓘ Tip:**
> Don't confuse the Wilcoxon signed rank test with the Wilcoxon rank sum test, which is used to compare two independent samples.

16.6.1. Questions We Could Answer Using the Wilcoxon Signed Rank Test

1. Before and after taking an introductory psychology course, university students were asked:

    Please tick the box that most closely represents the extent to which you agree with the following statement: "Psychology is a science".

    ☐   *Strongly disagree*
    ☐   *Somewhat disagree*
    ☐   *Neither agree nor disagree*
    ☐   *Somewhat agree*
    ☐   *Strongly agree*

    The researchers would like to know whether or not the students' attitudes changed during this time period.

This is obviously a repeated measures question, with time (before and after) as the IV, and agreement as the DV. As this DV is clearly ordinal, it would be inappropriate to perform a paired samples t test. Therefore, the researchers should analyse their data with a Wilcoxon signed rank test.

16.6.2. Illustrated Example

Before and after attending a public lecture on the health and environmental concerns associated with genetically modified (GM) food products, 40 people were asked the following question:

When purchasing breakfast cereals most people consider a number of different factors, including price, taste, brand, manufacturing location, and whether or not the product contains genetically modified ingredients. Please rank these factors from 1 to 5, where number 1 is the factor that <u>most</u> influences your purchasing decisions, and number 5 is the factor that <u>least</u> influences your purchasing decisions.

The lecture convenors would like to know whether the relative influence of GM ingredients on self-reported cereal purchasing decisions changed between the two testing sessions. At this stage, they are not interested in how participants ranked the other four factors.

Their data are summarised in Table 16.5.

Table 16.5

Pre- and Post-Lecture GM Ingredients Ranks (N = 40)

ID	Before	After	ID	Before	After
1	2	1	21	2	3
2	4	2	22	4	3
3	3	2	23	4	3
4	3	3	24	5	3
5	2	1	25	5	2
6	3	5	26	4	4
7	4	3	27	3	3
8	5	5	28	4	3
9	2	3	29	3	2
10	4	2	30	3	3
11	2	3	31	1	1
12	1	1	32	5	4
13	3	3	33	2	2
14	5	3	34	1	1
15	4	3	35	3	2
16	4	4	36	5	3
17	5	3	37	5	3
18	5	2	38	2	3
19	3	1	39	1	2
20	5	2	40	2	2

Note. Smaller values indicate higher ranks, which indicate a greater degree of influence on cereal purchasing decisions.

16.6.2.1. Setting Up the PASW Statistics Data File

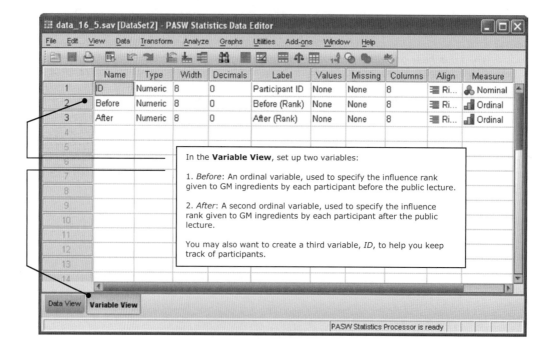

In the **Variable View**, set up two variables:

1. *Before*: An ordinal variable, used to specify the influence rank given to GM ingredients by each participant before the public lecture.

2. *After*: A second ordinal variable, used to specify the influence rank given to GM ingredients by each participant after the public lecture.

You may also want to create a third variable, *ID*, to help you keep track of participants.

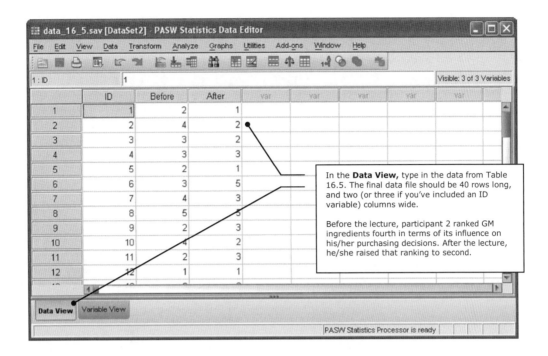

In the **Data View,** type in the data from Table 16.5. The final data file should be 40 rows long, and two (or three if you've included an ID variable) columns wide.

Before the lecture, participant 2 ranked GM ingredients fourth in terms of its influence on his/her purchasing decisions. After the lecture, he/she raised that ranking to second.

16.6.2.2. Analysing the Data

Syntax:
Run these analyses with **syntax_16_5.sps** on the companion website.

16.6.2.2.1. Assumptions

Three assumptions should be met when using the Wilcoxon signed rank test. The first two should be addressed during the planning stages of research.

1. **Independence.** Each participant (or pair of participants, in the case of a matched design) should participate only once in the research (i.e., they should provide only one set of data), and should not influence the participation of others.

2. **Scale of Measurement.** The DV should be at least ordinal.

3. **Symmetry of the Distribution of Difference Scores.** The distribution of the difference scores between the two levels of the IV should be roughly symmetrical.

   Calculating then graphing a distribution of difference scores using the procedures illustrated in chapter 6 (section 6.3.2.2) suggests that this assumption is not an unreasonable one to make regarding the current data.

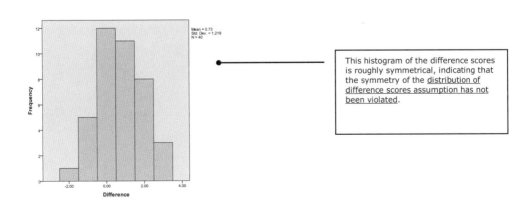

This histogram of the difference scores is roughly symmetrical, indicating that the symmetry of the distribution of difference scores assumption has not been violated.

16.6.2.2.2. PASW Statistics Procedure

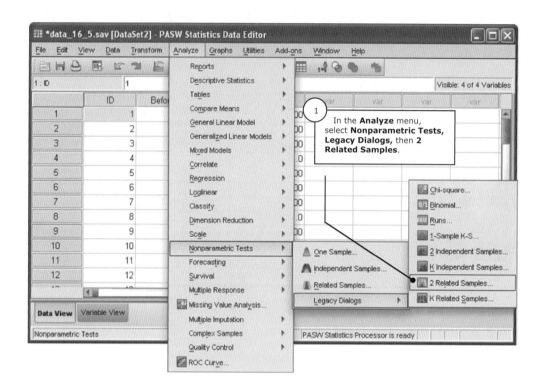

Tip:
To select multiple variables, hold down the Shift key on your keyboard, then click each with your mouse.

One click will select a variable; a second click will deselect it.

2   In the **Two Related Samples Tests** dialogue, select the two variables of interest, and click the arrow button to move them into the first row of the **Test Pairs** list.

3   In **Test Type** select **Wilcoxon**.

4   Click **OK** to output the test results.

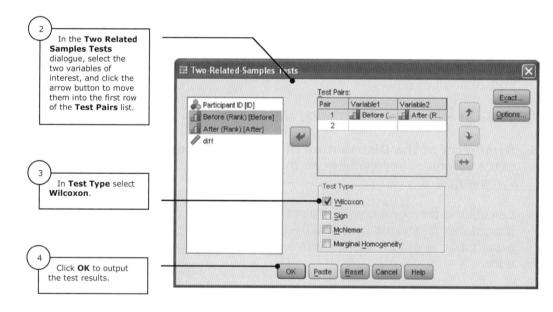

16.6.2.2.3. PASW Statistics Output

NPar Tests

Wilcoxon Signed Ranks Test

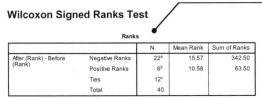

Ranks

		N	Mean Rank	Sum of Ranks
After (Rank) - Before (Rank)	Negative Ranks	22ᵃ	15.57	342.50
	Positive Ranks	6ᵇ	10.58	63.50
	Ties	12ᶜ		
	Total	40		

a. After (Rank) < Before (Rank)

b. After (Rank) > Before (Rank)

c. After (Rank) = Before (Rank)

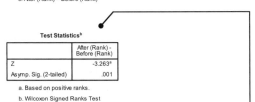

Test Statisticsᵇ

	After (Rank) - Before (Rank)
Z	-3.263ᵃ
Asymp. Sig. (2-tailed)	.001

a. Based on positive ranks.

b. Wilcoxon Signed Ranks Test

In **Test Statistics**, the following are reported:

Z. As the *T* distribution approximates a normal (*z*) distribution when *N* (excluding ties) > 20, PASW Statistics calculates a z-approximation of *T* (corrected for ties), which is then used to assess statistical significance.

Asymp. Sig (2-tailed). The two-tailed probability of **Z**. As this value is below .05, we can conclude that GM ingredients had significantly greater relative influence on purchasing decisions of the participants after the lecture.

There are three columns in the **Ranks** table:

N. This column contains four values:
- **Negative Ranks**: The number of participants reporting that GM ingredients had a greater relative influence on their purchasing decisions after the lecture (*n* = 22).
- **Positive Ranks**: The number of participants reporting that GM ingredients had a lesser relative influence on their purchasing decisions after the lecture (*n* = 6).
- **Ties**: The number of participants reporting that GM ingredients had the same relative influence on their purchasing decisions before and after the lecture (*n* = 12).
- **Total**: The total number of participants surveyed (*N* = 40).

Mean Rank. When performing the Wilcoxon signed rank test PASW Statistics first calculates a difference score for each case then, disregarding the positive and negative signs, ranks them according to size. Once this ranking process is complete, the positive and negative cases are separated out, and their respective ranks are averaged, providing the two figures shown in the **Mean Rank** column.

Here, the larger negative **Mean Rank** (15.57) indicates that the higher ranks (representing larger difference scores) were, relative to the lower ranks, more densely populated with participants who reported that GM ingredients had a greater influence on their purchasing decisions after the lecture.

Sum of Ranks. The smaller of these two values is the test statistic, *T*. Here, *T* = 63.5.

ⓘ *Tip:*
The critical *z* for α = .05 (two-tailed) is ± 1.96. A calculated z larger than this can be considered statistically significant.

Refer to the normal (*z*) distribution tables included in most statistics texts (e.g., Clark-Carter, 2009; Howell, 2010b) for the critical *z* values associated with alternate alpha levels (e.g., .01, .001 etc.).

16.6.2.3. Follow Up Analyses

16.6.2.3.1. Effect Size

Clark-Carter (2009) recommends converting *z* into *r* with:

$$r = \frac{z}{\sqrt{N}}$$

Where *z* can be read off the **Test Statistics** table, and *N* is the total sample size excluding ties. So,

$$r = \frac{-3.263}{\sqrt{28}}$$

$$= -0.62$$

By Cohen's (1988) conventions, this would be considered large.

ⓘ *Tip:*
Cohen (1988) suggested that *r* = .1 could be considered small, *r* = .3 could be considered medium, and *r* = .5 could be considered large.

16.6.2.4. APA Style Results Write-Up

Results

A Wilcoxon signed rank test indicated that GM ingredients had a significantly greater influence on the food purchasing decisions of participants after the lecture, $T = 63.5$, $z = -3.26$ (corrected for ties), $N -$ Ties $= 28$, $p = .001$, two-tailed.

Relative to their pre-lecture rankings, 22 participants ranked GM ingredients as more influential after attending the lecture (Sum of Ranks $= 342.50$), whilst only six ranked them as less influential (Sum of Ranks $= 63.50$). (Twelve participants reported that GM ingredients had the same degree of influence before and after the lecture.) This effect can be considered "large", $r = .62$.

For larger samples (i.e., where N excluding ties is 20+), the T distribution approximates a normal distribution, and z should be calculated and used to assess statistical significance.

PASW Statistics automatically applies a correction for ties (two or more cases with the same absolute difference scores) when calculating z.

A written description of your data should indicate the number of negative and positive ranks, as well as the number of tied cases excluded from the calculation of T and z.

Report the absolute value of r, rather than a negative value.

16.7. Kruskal-Wallis One-Way ANOVA

The Kruskal-Wallis one-way ANOVA is typically used to compare three or more independent samples of ordinal (ranked) data. It is also useful to researchers who are unable to calculate and interpret a regular one-way between groups ANOVA (see chapter 7) due to assumption violations.

16.7.1. Questions We Could Answer Using The Kruskal-Wallis One-Way ANOVA

1. Every year, the National Automotive Society publishes a "top 50 safest passenger vehicles" list. Vehicles on this list can be classified into one of four broad categories: small cars; family sized sedans; sports utility vehicles; and vans/people-movers. The society wants to know if vehicle safety is related to vehicle type.

In this example the IV is vehicle type, and has four levels. The DV is vehicle safety, which is indexed by the vehicles' positions/rankings in the top 50. As this DV is clearly ordinal, a Kruskal-Wallis ANOVA should be used to test whether some types of vehicles tend to achieve higher positions on the list than others.

2. Government statisticians with access to national census data would like to know whether religion is related to personal net wealth. Personal wealth is positively skewed in the population.

The DV (personal net wealth) in this example is ratio data, but its positive skew leaves it unsuitable for inclusion in a regular one-way between-groups ANOVA. As the Kruskal-Wallis ANOVA is not bound by the assumption of normality, it is a more appropriate statistical technique in circumstances like these.

(i) *Tip:*
When graphed using a histogram, a positively skewed distribution might look something like this:

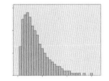

16.7.2. Illustrated Example

Researchers interested in the effects of teacher expectations on marking recruited 30 high-school English teachers, and randomly assigned them to three experimental conditions. Teachers in the first condition were asked to assign a letter grade (A, B, C, D, or F) to an essay written by "an extremely bright student"; teachers in the second condition were asked to grade an essay written by "a very poor student"; while teachers in the third condition (the control condition) were asked to grade an essay written by "a student". In reality, all 30 teachers were grading exactly the same essay.

The researchers' data are reproduced in Table 16.6.

Table 16.6

❑ *Data:*
This is data file
data_16_6.sav
on the companion
website.

Grades Assigned to an Essay Written by "an Extremely Bright Student", "a Very Poor Student" or "a Student" (N = 30)

Bright Student Condition		Poor Student Condition		Control Condition	
ID	Grade	ID	Grade	ID	Grade
1	2	11	4	21	5
2	3	12	2	22	1
3	4	13	3	23	3
4	3	14	5	24	3
5	2	15	1	25	1
6	4	16	2	26	2
7	4	17	1	27	4
8	1	18	3	28	4
9	1	19	4	29	2
10	2	20	4	30	3

Note. Value labels for grade variable: 1 = A; 2 = B; 3 = C; 4 = D; 5 = F.

16.7.2.1. Setting Up The PASW Statistics Data File

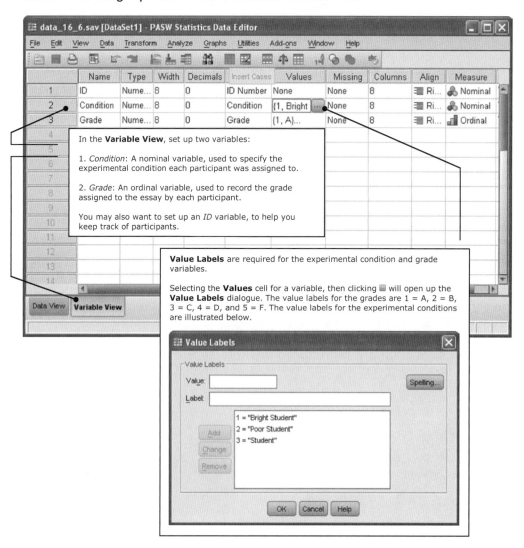

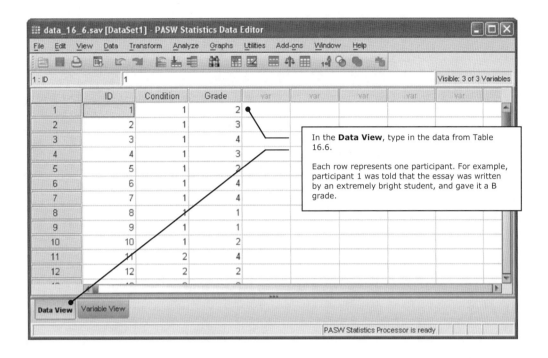

In the **Data View**, type in the data from Table 16.6.

Each row represents one participant. For example, participant 1 was told that the essay was written by an extremely bright student, and gave it a B grade.

16.7.2.2. Analysing the Data

16.7.2.2.1. Assumptions

Several issues must be considered when using the Kruskal-Wallis ANOVA. These include:

1. **Independence.** Each participant should participate only once in the research, and should not influence the participation of others.

2. **Scale of Measurement.** The DV should be at least ordinal.

3. **Shape of the Distributions.** The distributions of data should all look approximately the same (i.e., same shape and spread).

   Graphing each group of data (using, for example, the **Explore** procedure described in chapter 5, section 5.3.2.2) indicates that this is a reasonable assumption to make regarding this data set.

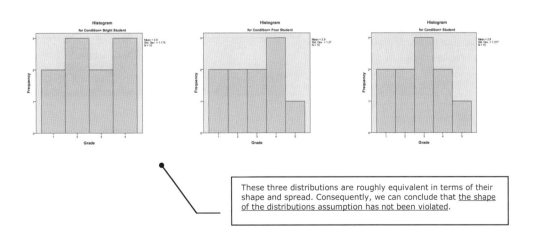

These three distributions are roughly equivalent in terms of their shape and spread. Consequently, we can conclude that the shape of the distributions assumption has not been violated.

Syntax:
Run these analyses with **syntax_16_6.sps** on the companion website.

16.7.2.2.2. PASW Statistics Procedure

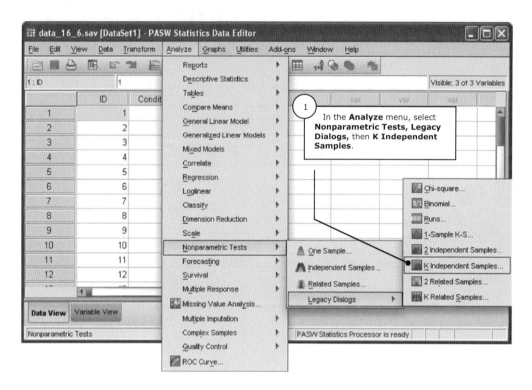

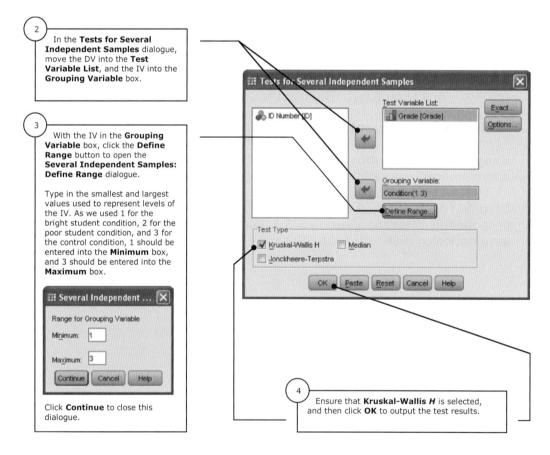

16.7.2.2.3. PASW Statistics Output

NPar Tests

Kruskal-Wallis Test

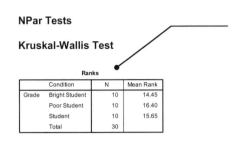

Ranks

	Condition	N	Mean Rank
Grade	Bright Student	10	14.45
	Poor Student	10	16.40
	Student	10	15.65
	Total	30	

There are two columns in the **Ranks** table:

N. The number of participants in each group, and the total number of participants. Here, there were 10 participants in each of the three experimental conditions.

Mean Rank. When computing the Kruskal-Wallis test, PASW Statistics first merges, and then rank orders the entire data set. Once a rank (or tied rank) has been assigned to each case, they are returned to their respective groups, summed, and then averaged. The **Mean Rank** values are these averages.

Here, the mean ranks are roughly equivalent, indicating that all three groups contain a relatively equal mix of high, low and middle ranking grades.

Test Statistics[a,b]

	Grade
Chi-square	.263
df	2
Asymp. Sig.	.877

a. Kruskal Wallis Test

b. Grouping Variable: Condition

In **Test Statistics**, the following are reported:

Chi-Square. The test statistic. This value is equivalent to H (corrected for ties), the formula for which appears in most textbook accounts of the Kruskal-Wallis ANOVA.

df. Degrees of freedom. The number of levels of the IV (k) minus one. Here, $k - 1 = 2$.

Asymp. Sig. The asymptotic probability of χ^2 at $k - 1$ degrees of freedom. As this value far exceeds .05, we must conclude that the Kruskal-Wallis ANOVA is non-significant. In other words, there are no differences between the mean ranks of the essay grades in the three experimental conditions.

16.7.2.3. Follow Up Analyses

16.7.2.3.1. Effect Size

Eta-squared (η^2), which is easily interpreted as the proportion of variance in the DV that can be attributed to the IV, can be calculated using the following formula:

$$\eta^2 = \frac{\chi^2}{N-1}$$

χ^2 is the same as H. They are used interchangeably.

Where χ^2 can be read from the **Test Statistics** table, and N is the total sample size. So,

$$\eta^2 = \frac{0.263}{30-1}$$

$$= .009$$

According to Cohen's (1988) conventions, this is a small effect size. Our study was woefully underpowered for detecting an effect of this magnitude.

If you want to convert η^2 to f, Cohen's (1988) measure of effect size for ANOVA, the following formula can be used:

$$f = \sqrt{\frac{\eta^2}{1-\eta^2}}$$

ⓘ *Tip:*
Cohen (1988) suggested that $\eta^2 = .01$ could be considered small, $\eta^2 = .059$ could be considered medium, and $\eta^2 = .138$ could be considered large.

Tip:
Cohen (1988)
suggested that f = .10
could be considered
small, f = .25 could be
considered medium,
and f = .40 could
be considered large.

In the current example,

$$f = \sqrt{\frac{.009}{1-.009}}$$

$$= .095$$

16.7.2.3.2. Pairwise Comparisons

Had the Kruskal-Wallis ANOVA been statistically significant, we'd be in the ambiguous situation of knowing that at least two experimental conditions differed, whilst not knowing specifically which two. (Or, indeed, if more than two differed.)

Tip:
The Bonferroni
adjustment controls
for the increased risk
of making a Type I
error when conducting
multiple comparisons
on a single set of data.

It involves dividing the
family-wise alpha level
by the number of
comparisons being
made.

For this example, the
significance of each U
would be evaluated
using an adjusted
alpha level of .017
(.05/3).

There are several methods that can be used to follow up on a significant omnibus H. The simplest of these involves conducting separate Mann-Whitney U tests on each pair of mean ranks (see section 16.4). For example, we could calculate a U for the difference between (a) the bright and poor conditions, (b) the bright and control conditions, and (c) the poor and control conditions. A Bonferroni adjusted alpha level should be used when determining the statistical significance of each test.

16.7.2.4. APA Style Results Write-Up

Results

A Kruskal-Wallis ANOVA indicated that there were no statistically significant differences between the grades assigned to the "bright" student (*Mean Rank* = 14.45), the "poor" student (*Mean Rank* = 16.40), and the control student (*Mean Rank* = 15.65), H (corrected for ties) = 0.263, $df = 2$, $N = 30$, $p = .877$, Cohen's $f = .095$.

At a minimum, provide a
brief description of your
data (e.g., mean ranks),
H (or χ^2), df, N, p, and a
measure of effect size
(e.g., f).

If the ANOVA is statistically
significant, also report the
results of any follow up
tests you ran, including
your adjusted alpha level.

16.8. Cochran's *Q* Test

Cochran's *Q* can be used to test whether proportions of category membership on a binary variable (e.g., yes or no; correct or incorrect; present or absent etc.) change over three or more experimental conditions or points in time.

It can also be used with matched or paired binary data.

16.8.1. Questions We Could Answer Using Cochran's *Q* Test

1. Does the proportion of students attending Statistics 101 lectures change over the semester?

To answer this question, we could record each enrolled student's attendance (as either present or absent) on multiple occasions throughout the semester, and then subject this data to a Cochran's *Q* test.

If the omnibus Cochran's *Q* test is significant, we can follow it up with a series of pairwise comparisons using the McNemar test (see section 16.5).

16.8.2. Illustrated Example

Market researchers interested in the reach of different types of news media telephoned randomly selected citizens and asked whether they had watched, read or listened to each of the following in the 24 hours prior: (a) a television news program; (b) a newspaper; (c) a news and current affairs website; or (d) a news radio program. A sample of their data is reproduced in Table 16.7.

Table 16.7

News Media Consumed by Each of 16 Participants Over 24 Hours

Data:
This is data file
data_16_7.sav
on the companion website.

ID	TV	Newspaper	Website	Radio
1	1	0	0	1
2	1	1	0	1
3	1	1	0	0
4	1	0	1	0
5	0	0	1	0
6	1	0	1	0
7	0	0	0	0
8	1	0	0	0
9	1	1	1	1
10	0	1	0	0
11	0	1	1	0
12	1	0	0	0
13	1	0	0	0
14	1	1	0	0
15	1	1	0	0
16	0	0	1	0

Note. 0 = No (not consumed) and 1 = Yes (consumed) for all for media types.

16.8.2.1. Setting Up The PASW Statistics Data File

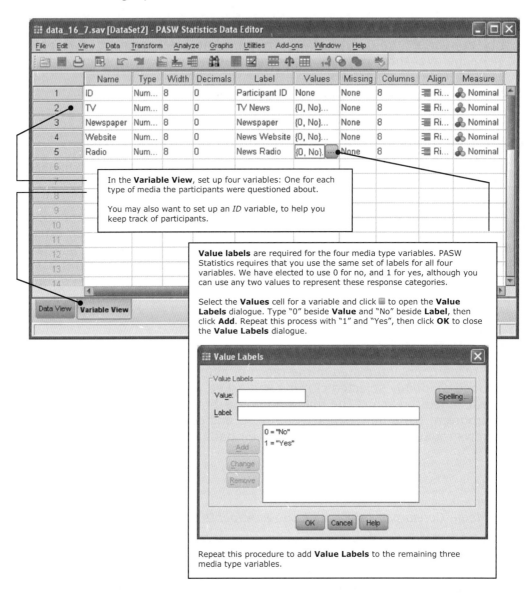

In the **Variable View**, set up four variables: One for each type of media the participants were questioned about.

You may also want to set up an *ID* variable, to help you keep track of participants.

Value labels are required for the four media type variables. PASW Statistics requires that you use the same set of labels for all four variables. We have elected to use 0 for no, and 1 for yes, although you can use any two values to represent these response categories.

Select the **Values** cell for a variable and click ▦ to open the **Value Labels** dialogue. Type "0" beside **Value** and "No" beside **Label**, then click **Add**. Repeat this process with "1" and "Yes", then click **OK** to close the **Value Labels** dialogue.

Repeat this procedure to add **Value Labels** to the remaining three media type variables.

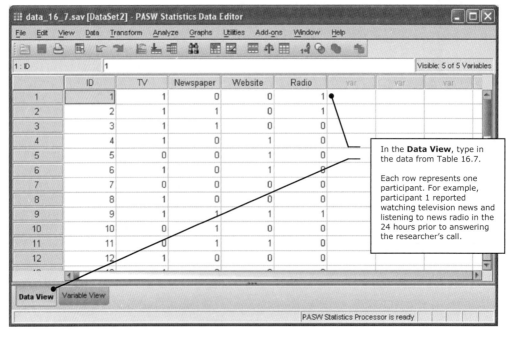

In the **Data View**, type in the data from Table 16.7.

Each row represents one participant. For example, participant 1 reported watching television news and listening to news radio in the 24 hours prior to answering the researcher's call.

16.8.2.2. Analysing the Data

16.8.2.2.1. Assumptions

Syntax:
Run these
analyses with
syntax_16_7.sps
on the companion
website.

During the planning stages of research, two issues need to be considered.

1. **Independence.** Each participant (or pair/trio/etc. of participants, in the case of a matched design) should participate only once in the research (i.e., they should provide only one set of data), and should not influence the participation of others.

2. **Scale of Measurement.** Cochran's Q test was designed to be used with binary data.

16.8.2.2.2. PASW Statistics Procedure

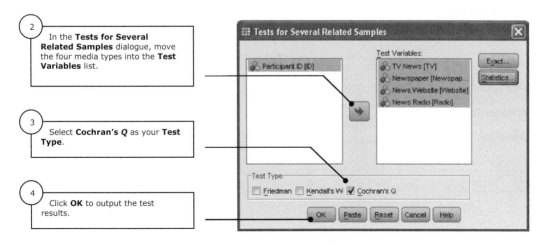

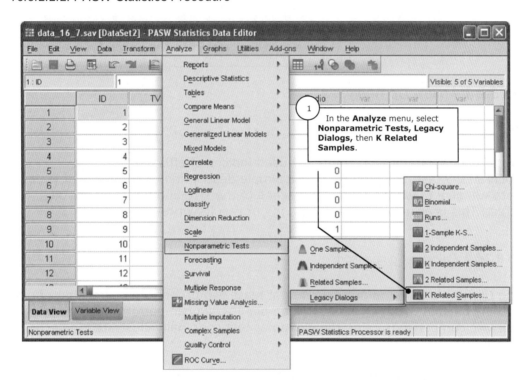

16.8.2.2.3. PASW Statistics Output

NPar Tests

Cochran Test

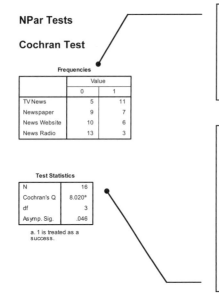

Frequencies

	Value	
	0	1
TV News	5	11
Newspaper	9	7
News Website	10	6
News Radio	13	3

Test Statistics

N	16
Cochran's Q	8.020ᵃ
df	3
Asymp. Sig.	.046

a. 1 is treated as a success.

> The **Frequencies** table simply summarises the data set, informing us that 11 participants reported watching TV news, 7 reported reading a newspaper, and so on.
>
> To convert these figures into percentages, simply divide them by the total sample size ($N = 16$) and then multiply by 100. For example: 11/16 * 100 = 68.75%. We can report that nearly 70% of the sample watched TV news in the 24 hours prior to answering the researcher's call.

> In **Test Statistics**, the following are reported:
>
> **N.** The total sample size. Here, $N = 16$.
>
> **Cochran's Q.** The test statistic. Here, $Q = 8.020$.
>
> **df.** Degrees of freedom. The number of levels of the IV (k) minus one. Here, $k - 1 = 3$.
>
> **Asymp. Sig.** The asymptotic probability of Q at $k - 1$ degrees of freedom. As this value is less than .05 (our specified α level) we can conclude that <u>some types of media were consumed by a significantly greater proportion of the sample than others</u>.
>
> All of these values should be reported in your results.

16.8.2.3. Follow Up Analyses

16.8.2.3.1. Effect Size

We are not aware of any commonly used indices of effect size for the Cochran's Q test, but propose that a simple description or summary of your data will give readers an adequate sense of the size and direction of your effect. This is achieved with a graph in our results write-up.

16.8.2.3.2. Pairwise Comparisons

The Cochran's Q test informs us that at least two proportions differ, but offers nothing beyond this. To reduce this ambiguity, we can perform a series of pairwise comparisons using the McNemar test (originally illustrated in section 16.5).

Using the procedure illustrated in section 16.5.2.2.2 we made the following six comparisons:

> These **Crosstabs** tables summarise the data analysed in the six comparisons.
>
> Looking at **TV News & News Radio** table we can see that eight participants watched television news during the survey period, but did not listen to news radio. Furthermore, no participants reported listening to news radio but not watching TV news, three participants reported doing both, and five participants did neither.
>
> Remember that the McNemar test ignores the cases with identical scores on both variables (i.e., the participants who consumed neither type of media, and the participants who consumed both types of media).

NPar Tests

McNemar Test

Crosstabs

TV News & Newspaper

TV News	Newspaper	
	No	Yes
No	3	2
Yes	6	5

TV News & News Radio

TV News	News Radio	
	No	Yes
No	5	0
Yes	8	3

Newspaper & News Radio

Newspaper	News Radio	
	No	Yes
No	8	1
Yes	5	2

TV News & News Website

TV News	News Website	
	No	Yes
No	2	3
Yes	8	3

Newspaper & News Website

Newspaper	News Website	
	No	Yes
No	5	4
Yes	5	2

News Website & News Radio

News Website	News Radio	
	No	Yes
No	8	2
Yes	5	1

Test Statistics[b]

	TV News & Newspaper	TV News & News Website	TV News & News Radio	Newspaper & News Website	Newspaper & News Radio	News Website & News Radio
N	16	16	16	16	16	16
Exact Sig. (2-tailed)	.289[a]	.227[a]	.008[a]	1.000[a]	.219[a]	.453[a]

a. Binomial distribution used.

b. McNemar Test

The only statistically significant comparison is between **TV News** and **News Radio**.

The other five comparisons are non-significant (at a Bonferroni corrected alpha level of .0083).

(i) *Tip:*
The Bonferroni adjustment is used to maintain a family-wise alpha rate of .05 over multiple comparisons.

It requires dividing the family-wise alpha level by the number of comparisons being made. Here,

$.05/6 = .0083$

A comparison can be considered statistically significant if p is less than .0083.

16.8.2.4. APA Style Results Write-Up

Results

The relative popularity of each type of news media during the survey period is illustrated in Figure 1.

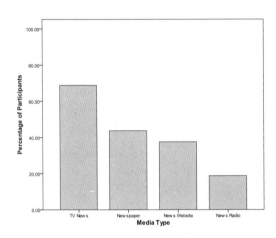

Figure 1. Bar chart illustrating the percentages of the total sample who consumed each of four types of news media ($N = 16$).

Cochran's Q (with $\alpha = .05$), which tests for differences between related proportions, was statistically significant, Q (3, $N = 16$) = 8.02, $p = .046$. To locate the source of this significance, a series of pairwise comparisons using the McNemar test were undertaken. The only comparison to achieve statistical significance at the Bonferroni corrected α of .0083 was between the proportion of the total sample who watched television news and the proportion who listened to news radio ($p = .008$).

16.9. Friedman Two-Way ANOVA

AKA:
Friedman rank test for *k* correlated samples.

The Friedman two-way ANOVA is typically used to compare three or more related samples of ordinal (ranked) data. It is also useful to researchers who are unable to calculate and correctly interpret a regular one-way repeated measures ANOVA (see chapter 9) due to assumption violations.

16.9.1. Questions We Could Answer Using The Friedman Two-Way ANOVA

1. In the lead-up to the Academy Awards, many magazines begin polling their readership for opinions on "best director", "best film" and so on. One particular magazine has asked readers to rank (from 1 to 5) the five "best film" nominees, and wants to know whether some of the nominated films tend to rank higher (i.e., are more popular) than others.

In this example, the IV is film, and has five levels. The DV is popularity ranking, a clearly ordinal variable. As each reader was asked to rank all five films, we are looking at a repeated measures design. Consequently, the Friedman two-way ANOVA should be the magazine editor's statistical procedure of choice.

2. A group of violin virtuosos have been asked to rate (on a scale from 1 to 10) each of three extremely rare 18th century violins that a wealthy music collector is considering adding to his private museum. After compiling the data, the ratings for each violin are clearly negatively skewed.

Tip:
When graphed using a histogram, a negatively skewed distribution might look something like this:

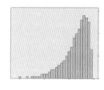

On reflection, the negative skew is not surprising. The three violins are all fine examples of 18th century artistry, and are all in beautiful condition; of course the violinists would tend to rate them highly!

As the data does not meet the one-way repeated measures ANOVA assumption of normality, the collector should instead use a Friedman test to assess whether some violins were rated consistently higher than others.

16.9.2. Illustrated Example

Researchers have long been interested in whether one side of the face is perceived as more emotionally expressive than the other.

In one recent experiment, researchers used a photograph of an "angry" face to create a "left-composite" (where the right side of the face was actually a mirror image of the left side) and a "right-composite" (where the left side of the face was actually a mirror image of the right side). These composite faces (along with the original) were then shown to a group of 21 participants, who were asked to order them from most to least angry.

The researchers' data are reproduced in Table 16.8.

□ *Data:*
This is data file
data_16_8.sav
on the companion
website.

Table 16.8

The Ranked Emotional Intensity of Left-Composite, Right-Composite and Control Faces

ID	Left	Right	Control
1	1	3	2
2	1	2	3
3	2	1	3
4	2	3	1
5	1	3	2
6	1	3	2
7	3	2	1
8	2	1	3
9	1	3	2
10	1	3	2
11	1	2	3
12	2	3	1
13	3	2	1
14	1	2	3
15	1	3	2
16	2	1	3
17	2	3	1
28	1	2	3
19	1	3	2
20	2	1	3
21	1	2	3

Note. Smaller values represent higher ranks (i.e., greater perceived emotional intensity).

16.9.2.1. Setting Up The PASW Statistics Data File

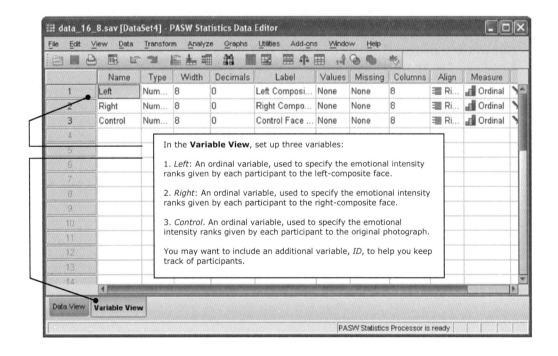

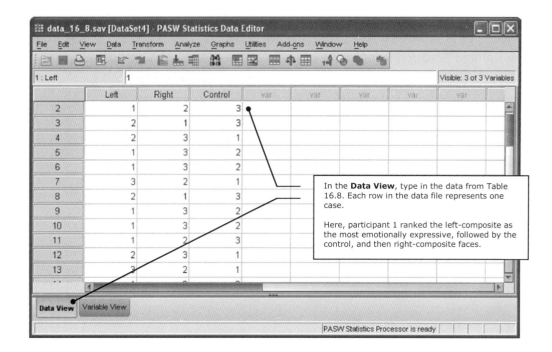

Syntax:
Run these
analyses with
syntax_16_8.sps
on the companion
website.

16.9.2.2. Analysing the Data

16.9.2.2.1. Assumptions

When using the Friedman two-way ANOVA, the following assumptions should be considered.

1. **Independence.** Each participant (or pair/trio/etc. of participants, in the case of a matched design) should participate only once in the research (i.e., they should provide only one set of data), and should not influence the participation of others.

2. **Scale of Measurement.** The DV should be at least ordinal.

3. **Symmetry of the Distributions of Difference Scores.** The Friedman test assumes that the distributions of difference scores between pairs of levels of the IV are roughly symmetrical. This assumption can be tested in the manner described in section 16.6.2.2.1.

16.9.2.2.2. PASW Statistics Procedure

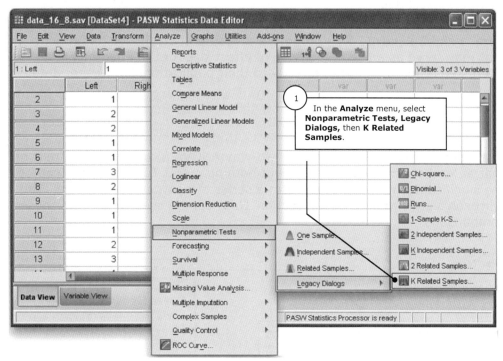

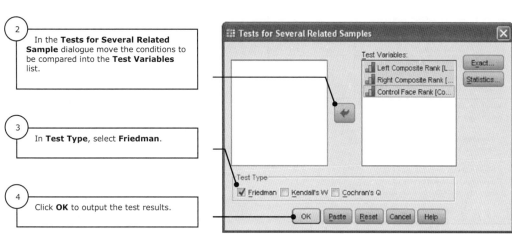

2 In the **Tests for Several Related Sample** dialogue move the conditions to be compared into the **Test Variables** list.

3 In **Test Type**, select **Friedman**.

4 Click **OK** to output the test results.

⊞ Included With:
You will not have an **Exact** button like the one illustrated to the left unless you have *PASW Exact Tests* add-on installed on your system.

Exact tests are preferred by researchers working with small samples and rare events.

16.9.2.2.3. PASW Statistics Output

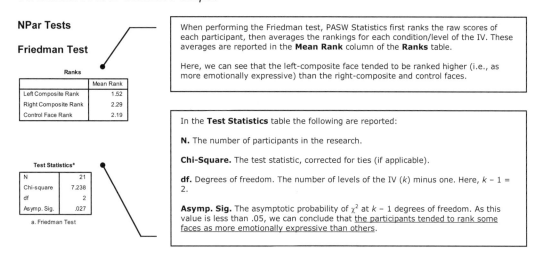

NPar Tests

Friedman Test

Ranks

	Mean Rank
Left Composite Rank	1.52
Right Composite Rank	2.29
Control Face Rank	2.19

When performing the Friedman test, PASW Statistics first ranks the raw scores of each participant, then averages the rankings for each condition/level of the IV. These averages are reported in the **Mean Rank** column of the **Ranks** table.

Here, we can see that the left-composite face tended to be ranked higher (i.e., as more emotionally expressive) than the right-composite and control faces.

Test Statistics[a]

N	21
Chi-square	7.238
df	2
Asymp. Sig.	.027

a. Friedman Test

In the **Test Statistics** table the following are reported:

N. The number of participants in the research.

Chi-Square. The test statistic, corrected for ties (if applicable).

df. Degrees of freedom. The number of levels of the IV (k) minus one. Here, $k - 1 = 2$.

Asymp. Sig. The asymptotic probability of χ^2 at $k - 1$ degrees of freedom. As this value is less than .05, we can conclude that <u>the participants tended to rank some faces as more emotionally expressive than others.</u>

16.9.2.3. Follow Up Analyses

16.9.2.3.1. Effect Size

Vargha and Delaney (2000) propose a measure of stochastic heterogeneity that may serve as a useful measure of effect size for the Friedman test. The interested reader is referred directly to their publications for further detail.

16.9.2.3.2. Pairwise Comparisons

A statistically significant Friedman two-way ANOVA is ambiguous. It tells us that at least two conditions differ, but nothing more! To identify the source of the significance, we need to perform a series of pairwise comparisons. The Wilcoxon signed rank test (see section 16.6) can be used to perform these comparisons.

NPar Tests

Wilcoxon Signed Ranks Test

In the **Two Related Samples Tests** dialogue (see section 16.6.2.2.2), multiple test pairs can be specified. We specified three pairs to arrive at this output:

1. Right-composite vs left-composite.
2. Control vs left-composite.
3. Control vs. right-composite.

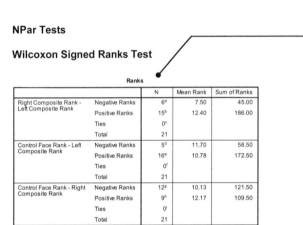

Ranks

		N	Mean Rank	Sum of Ranks
Right Composite Rank - Left Composite Rank	Negative Ranks	6[a]	7.50	45.00
	Positive Ranks	15[b]	12.40	186.00
	Ties	0[c]		
	Total	21		
Control Face Rank - Left Composite Rank	Negative Ranks	5[d]	11.70	58.50
	Positive Ranks	16[e]	10.78	172.50
	Ties	0[f]		
	Total	21		
Control Face Rank - Right Composite Rank	Negative Ranks	12[g]	10.13	121.50
	Positive Ranks	9[h]	12.17	109.50
	Ties	0[i]		
	Total	21		

a. Right Composite Rank < Left Composite Rank
b. Right Composite Rank > Left Composite Rank
c. Right Composite Rank = Left Composite Rank
d. Control Face Rank < Left Composite Rank
e. Control Face Rank > Left Composite Rank
f. Control Face Rank = Left Composite Rank
g. Control Face Rank < Right Composite Rank
h. Control Face Rank > Right Composite Rank
i. Control Face Rank = Right Composite Rank

> (i) **Tip:**
> The Bonferroni adjustment is used to maintain a family-wise alpha rate of .05 over multiple comparisons.
>
> It requires dividing the family-wise alpha level by the number of comparisons being made. Here,
>
> .05/3 = .017.
>
> A comparison can be considered statistically significant if *p* is less than .017.

Test Statistics[c]

	Right Composite Rank - Left Composite Rank	Control Face Rank - Left Composite Rank	Control Face Rank - Right Composite Rank
Z	-2.551[a]	-2.062[a]	-.217[b]
Asymp. Sig. (2-tailed)	.011	.039	.828

a. Based on negative ranks.
b. Based on positive ranks.
c. Wilcoxon Signed Ranks Test

At a Bonferroni corrected alpha level of .017, the difference between the rankings of the right- and left-composite faces is statistically significant (p = .011); the difference between the rankings of the control and left-composite faces is *approaching* significance (p = .039); and the difference between rankings of the control and right-composite faces is clearly non-significant (p = .828).

16.9.2.3.3. Effect Sizes for the Pairwise Comparisons

As explained in section 16.6.2.3.1, z can be converted into r using:

$$r = \frac{z}{\sqrt{N}}$$

Where z can be read off the **Test Statistics** table, and N is the total sample size excluding ties. Applying this formula to the three pairwise comparisons gives us:

Right-composite vs left-composite:	Control vs left-composite:	Control vs. right-composite:
$r = \dfrac{-2.551}{\sqrt{21}}$	$r = \dfrac{-2.062}{\sqrt{21}}$	$r = \dfrac{-0.217}{\sqrt{21}}$
$= -0.56$	$= -0.45$	$= -0.05$

> **ⓘ Tip:**
> Cohen (1988) suggested that $r = .1$ could be considered small, $r = .3$ could be considered medium, and $r = .5$ could be considered large.

These first two pairwise effects can be considered large, whilst the third is very small.

16.9.2.4. APA Style Results Write-Up

Results

A Friedman two way ANOVA indicated that rankings of emotional intensity varied significantly across the three face types,

> PASW Statistics automatically applies a correction for ties, if applicable.

$\chi^2_F = 7.24$ (corrected for ties), $df = 2$, $N - \text{Ties} = 21$, $p = .027$.

Follow-up pairwise comparisons with the Wilcoxon Signed Rank test and a Bonferroni adjusted α of .017 indicated that the left-

> Refer to section 16.6 for a full description of the Wilcoxon signed rank test.

composite face (*Mean Rank* = 1.52) was perceived as significantly more emotionally intense than the right-composite face (*Mean Rank* = 2.29), $T = 45$, $z = -2.55$ (corrected for ties), $N - \text{Ties} = 21$, $p = .011$. This effect can be described as "large", $r = .56$.

The difference between the ranked emotional intensity of the left-composite and control faces (*Mean Rank* = 2.19) approached significance ($p = .039$, $r = .045$), whereas the difference between the right-composite and control faces was clearly non-significant ($p = .828$), and trivial ($r = .05$).

AKA:
Cramer's phi (φ).

16.10. Cramer's *V*

Cramer's *V* is a measure of the degree of association (i.e., relationship strength) between two nominal variables. It was first introduced in section 16.3, where it was used as an index of effect size for the chi-square test of contingencies.

16.10.1. Questions We Could Answer Using Cramer's *V*

1. Is there a relationship between support for abortion rights and voting intentions (e.g., Labor, Liberal, Democrat, or other) in the next federal election?

2. Is there a relationship between whether or not participants smoke cigarettes and whether or not they catch at least one cold during the winter?

In each of these examples, we're simply asking about the strength of the association between two categorical variables.

16.10.2. Illustrated Example

In section 16.3.2, we asked whether or not developing aphasia following a stroke is related to handedness (i.e., whether the participant is left or right handed). The data we analysed are reproduced in Table 16.9.

Table 16.9

Data:
This is data file
data_16_9.sav
on the companion
website.

*The Frequency of Left and Right Handed Stroke Survivors
Who Were and Were Not Also Diagnosed with Aphasia*

Aphasia Diagnosis	Handedness	Frequency
1	1	36
1	2	74
2	1	44
2	2	276

Note. Value labels for the aphasia diagnosis variable: 1 = diagnosed with aphasia; 2 = not diagnosed with aphasia. Value labels for the handedness variable: 1= left handed; 2 = right handed.

After running the analyses described in section 16.3.2.2, we end up with a considerable amount of output, including this **Symmetric Measures** table:

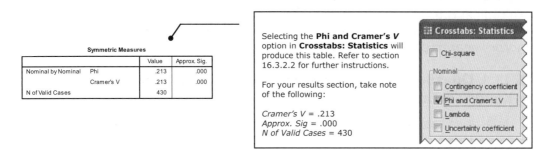

Syntax:
Run these analyses with **syntax_16_9.sps** on the companion website.

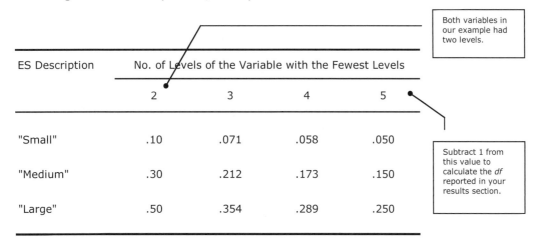

16.10.3. Follow Up Analyses

16.10.3.1. Effect Size

As an index of effect size, Cramer's *V* can be interpreted according to the following conventions (Cohen, 1988):

Both variables in our example had two levels.

ES Description	No. of Levels of the Variable with the Fewest Levels			
	2	3	4	5
"Small"	.10	.071	.058	.050
"Medium"	.30	.212	.173	.150
"Large"	.50	.354	.289	.250

Subtract 1 from this value to calculate the *df* reported in your results section.

In our example, a Cramer's *V* of .213 can be described as "small-to-medium".

16.10.4. APA Style Results Write-Up

Results

Cramer's *V* test indicated that the association between handedness and post-stroke aphasia was statistically significant, $V(1, N = 430) = .21, p < .001$. An association of this magnitude can be described as "small-to-medium".

You will sometimes see Cramer's *V* reported in the literature as ϕ' or ϕ_c. Regardless of which you use, remain consistent.

16.11. Spearman's Rho and Kendall's Tau-B

AKA:
Spearman's rho is
sometimes referred to
as Spearman's rank
order correlation, or
Spearman's correlation
coefficient for ranked
data.

Spearman's rho and Kendall's tau-b both measure the association between two ordinal (i.e., ranked) variables. They are suitable alternatives to the Pearson's correlation (see chapter 12) when the assumptions of normality and/or linearity cannot be met.

16.11.1. Questions We Could Answer Using Either Spearman's Rho or Kendall's Tau-B

1. Is a city's total electricity consumption related to daily temperature variations?

If the relationship between these two variables is not linear (perhaps it looks something like the relationship depicted below), a Pearson's correlation coefficient will not adequately capture it. Therefore, a researcher asking this question should consider either Spearman's rho or Kendall's tau-b instead.

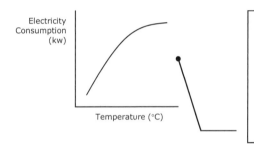

Electricity
Consumption
(kw)

Temperature (°C)

The hypothetical relationship between electricity consumption and temperature depicted here is obviously not linear.

At lower daily maximum temperatures (say between 20°C and 35°C) small temperature increases result in relatively large increases in electricity use. However, at higher temperatures (say between 35°C and 45°C) electricity consumption appears to level off. This is presumably because, by the time the mercury hits 35°C, most air-conditioners in the city will already be running.

2. Are students' letter grades (i.e., A, B, C, D or F) on a mid-semester Marketing 101 exam correlated with their letter grades on a mid-semester Epidemiology 100 exam?

16.11.2. Illustrated Example

Student researchers interested in the relationship between annual income and educational achievements surveyed 35 pedestrians on a busy city intersection. Their data are reproduced in Table 16.10.

Table 16.10

Data:
This is data file
data_16_10.sav
on the companion
website.

Highest Level of Education Achieved and Annual Income ($) Data (N = 35)

Education	Income	Education	Income	Education	Income	Education	Income
1	27,000	3	50,000	4	68,000	5	78,000
1	34,000	3	55,000	5	66,000	6	82,000
1	212,000	3	108,000	5	54,000	6	94,000
1	40,000	3	277,000	5	69,000	7	90,000
2	38,000	3	62,000	5	75,000	7	102,000
2	44,000	3	66,000	5	88,000	7	88,000
2	44,000	3	54,000	5	90,000	8	93,000
2	50,000	4	56,000	5	92,000	8	97,000
3	48,000	4	60,000	5	87,000		

Note. For the education variable, 1 = Primary school; 2 = Year 10; 3 = Year 12; 4 = Tertiary diploma; 5 = Bachelors degree; 6 = Postgraduate diploma; 7 = Masters degree; 8 = Doctorial degree.

16.11.2.1. Setting Up The PASW Statistics Data File

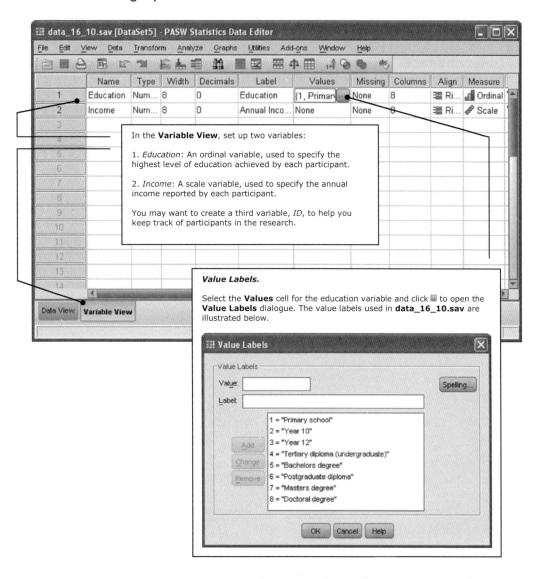

In the **Variable View**, set up two variables:

1. *Education*: An ordinal variable, used to specify the highest level of education achieved by each participant.

2. *Income*: A scale variable, used to specify the annual income reported by each participant.

You may want to create a third variable, *ID*, to help you keep track of participants in the research.

Value Labels.

Select the **Values** cell for the education variable and click ▦ to open the **Value Labels** dialogue. The value labels used in **data_16_10.sav** are illustrated below.

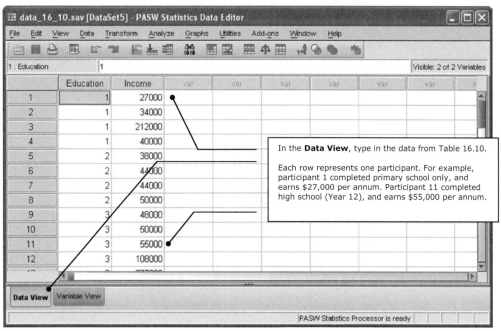

In the **Data View**, type in the data from Table 16.10.

Each row represents one participant. For example, participant 1 completed primary school only, and earns $27,000 per annum. Participant 11 completed high school (Year 12), and earns $55,000 per annum.

■ *Syntax:*
Run these
analyses with
syntax_16_10.sps
on the companion
website.

16.11.2.2. Analysing The Data

16.11.2.2.1. Assumptions

1. **Independence.** Each participant should provide data on one occasion only, and should not influence the participation of others.

2. **Scale of Measurement.** Both variables should be at least ordinal.

Both of these assumptions should be addressed at the planning and data collection stages of research.

16.11.2.2.2. PASW Statistics Procedure

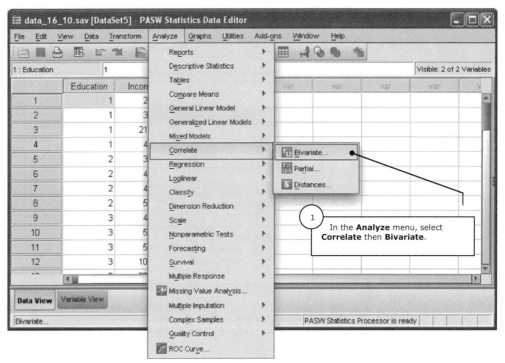

1 In the **Analyze** menu, select **Correlate** then **Bivariate**.

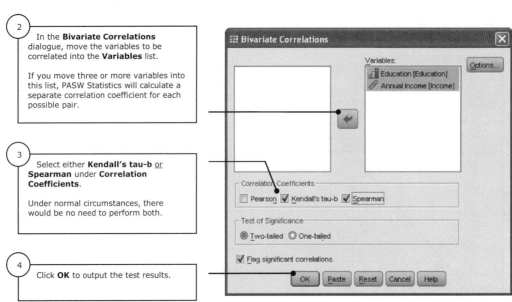

2 In the **Bivariate Correlations** dialogue, move the variables to be correlated into the **Variables** list.

If you move three or more variables into this list, PASW Statistics will calculate a separate correlation coefficient for each possible pair.

3 Select either **Kendall's tau-b** or **Spearman** under **Correlation Coefficients**.

Under normal circumstances, there would be no need to perform both.

4 Click **OK** to output the test results.

16.11.2.2.3. PASW Statistics Output

Nonparametric Correlations

Correlations

			Education	Annual Income
Kendall's tau_b	Education	Correlation Coefficient	1.000	.583**
		Sig. (2-tailed)	.	.000
		N	35	35
	Annual Income	Correlation Coefficient	.583**	1.000
		Sig. (2-tailed)	.000	.
		N	35	35
Spearman's rho	Education	Correlation Coefficient	1.000	.636**
		Sig. (2-tailed)	.	.000
		N	35	35
	Annual Income	Correlation Coefficient	.636**	1.000
		Sig. (2-tailed)	.000	.
		N	35	35

**. Correlation is significant at the 0.01 level (2-tailed).

The Nonparametric **Correlations** table presents the results of two separate tests:

1. **Kendall's tau-b**
2. **Spearman's rho**

In most circumstances, you would use one or the other, but not both.

Spearman's rho

Spearman's rho is essentially a Pearson's correlation (chapter 12) performed on ranked, rather than raw data.

Like tau-b, it can range from -1 to +1, with 0 signifying that the two variables are not correlated.

Here, **Spearman's rho** is .636 (*p* = .000), indicating a strong positive relationship between education and income levels.

For the write-up, take note of the following:

Correlation Coefficient = .636
Sig (2-tailed) = .000
N = 35

Kendall's tau-b

Kendall's tau-b is generally preferred over **Spearman's rho**, as it tends to provide a better estimate of the true population correlation, and is not artificially inflated by multiple tied ranks.

Kendall's tau-b can range from -1 (a perfect inverse correlation) to +1 (a perfect positive correlation), with 0 indicating that the two variables are uncorrelated.

Here, **Kendall's tau-b** is .583 (*p* = .000), which indicates that the relationship between education and annual income is positive, and quite strong. That is, people with higher levels of education also tend to have higher incomes; whereas people with lower levels of education tend to have lower incomes.

For the write-up, take note of the following:

Correlation Coefficient = .583
Sig (2-tailed) = .000
N = 35

ⓘ **Tip:**
The *Pearson's correlation* between these variables is non-significant (*r* = .150, *p* = .389), highlighting the importance of using tests that are appropriate for your data.

16.11.2.3. APA Style Results Write-Up

16.11.2.3.1. Spearman's Rho

Results

Spearman's rho indicated the presence of a strong positive correlation between ranked education and income levels, $r_s = .64$, $p < .001$, two-tailed, $N = 35$.

Indicate both the direction ("positive") and strength ("strong") of the relationship.

r_s can be used to represent Spearman's rho.

16.11.2.3.2. Kendall's Tau-B

Results

Kendall's tau-b indicated that the correlation between ranked education and income levels was strong and positive, $\tau = .58$, $p < .001$, two-tailed, $N = 35$.

The τ symbol can be used to represent tau.

16.12. Non-Parametric Checklist

Have you:

- ✔ Confirmed that your data does not meet the assumptions of an appropriate parametric test? Remember that parametric tests are typically more powerful than non-parametric tests when their assumptions are not violated.
- ✔ Considered the assumptions underpinning the non-parametric test selected?
- ✔ Performed the selected test and taken note of (if applicable) the test statistic and associated significance level, degrees of freedom and sample sizes (including or excluding ties as necessary)?
- ✔ Performed follow up tests (e.g., pairwise comparisons) if necessary?
- ✔ Calculated a measure of effect size, if necessary?
- ✔ Written up your results in the APA style?

Chapter 17: Working with Syntax

Chapter Overview

Available In:

PASW Statistics Student Version is unable to read the command syntax that is discussed in this chapter.

17.1. Purpose of Working with Syntax

In previous chapters, we've used drop-down menus and dialogue boxes to illustrate a range of common statistical procedures. This approach is very user friendly, and will meet your analytic needs in most instances. However, as you become increasingly familiar with PASW Statistics, and use it to run increasingly sophisticated analyses, you'll find yourself beginning to rely on an alternative method for getting the job done. This alternative involves utilising the PASW Statistics command syntax.

Command syntax is the language that PASW Statistics translates your instructions into each time you click an **OK** button in a dialogue box. With the **PASW Statistics Syntax Editor** you can capture, manipulate and save these instructions (which are written in the command syntax language) for future reference. The appeal of saving syntax is twofold. Firstly, it provides you with an exact record of the analyses you performed; a record you would otherwise have to handwrite (or dictate etc.) as you navigate through various menus and dialogues. Secondly, you can run (and re-run) analyses directly from saved syntax, with just a few mouse clicks.

In this chapter, we will illustrate how to produce and interact with basic PASW Statistics command syntax.

17.2. Using Syntax to Conduct an Independent Samples *t* Test

In the first illustrated example of chapter 5, thirty participants were shown video footage of a car accident, and were then asked one of the following questions: (1) "about how fast were the cars going when they hit each other?" or (2) "about how fast were the cars going when they smashed into each other?"

The speed estimates provided by the participants are reproduced in Table 17.1.

Table 17.1

⬜ **Data:**
This is data file
data_17_1.sav
on the companion
website.

Speed Estimates (in km/h) Given by Participants (N = 30) in Response to Either the "Hit" Question, or the "Smashed" Question

Hit Condition (Group 1)		Smashed Condition (Group 2)	
Participant ID	Est. Speed	Participant ID	Est. Speed
1	39	16	41
2	33	17	36
3	32	18	49
4	37	19	50
5	35	20	39
6	35	21	38
7	34	22	39
8	33	23	42
9	34	24	41
10	31	25	40
11	38	26	40
12	36	27	45
13	30	28	36
14	34	29	42
15	30	30	47

After the data file was set up, an independent samples *t* test was used to determine whether there was a statistically significant difference between the speed estimates provided by the 15 participants who were asked the "hit" question, and the 15 participants who were asked the "smashed" question.

In the sections that follow, we will re-run this *t* test (and assumption testing, of course) with PASW Statistics command syntax.

17.2.1. Generating Syntax to Test the Normality Assumption

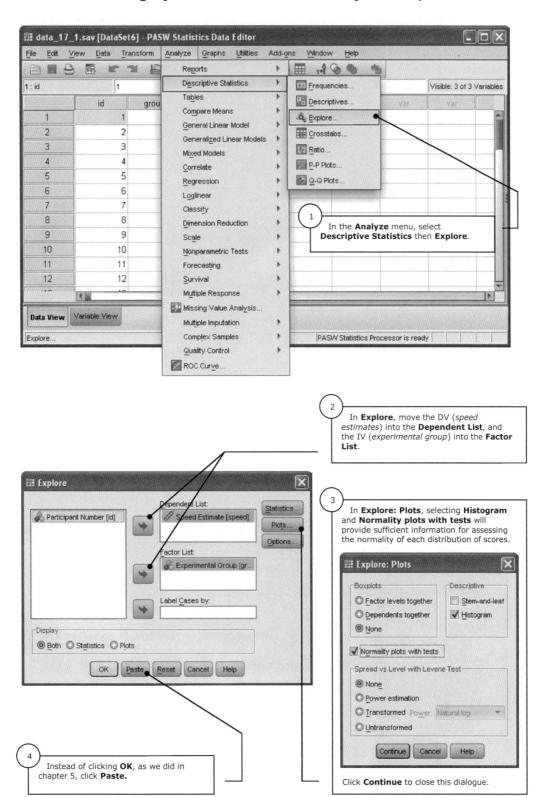

17.2.2. Command Syntax for Testing the Normality Assumption

 Syntax:
This is the first half of
syntax_17_1.sps
on the companion
website.

(i) **Tip:**
General information
about syntax:

1. Each command (e.g.
EXAMINE) begins on a
new line and ends with
a period (**.**).

2. Sub-commands are
always preceded by a
forward slash (**/**).

3. PASW Statistics
syntax is not case
sensitive.

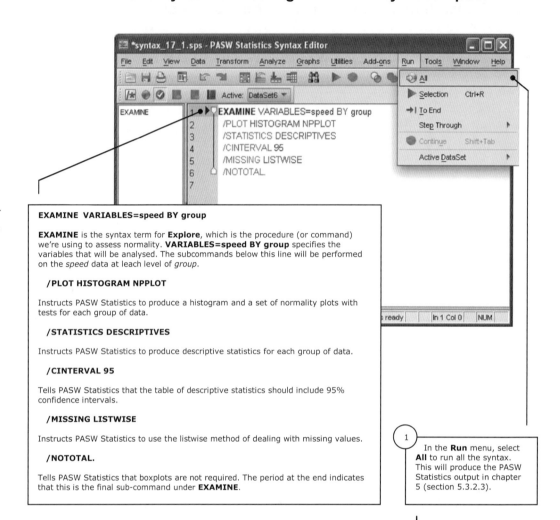

EXAMINE VARIABLES=speed BY group

EXAMINE is the syntax term for **Explore**, which is the procedure (or command)
we're using to assess normality. **VARIABLES=speed BY group** specifies the
variables that will be analysed. The subcommands below this line will be performed
on the *speed* data at leach level of *group*.

   /PLOT HISTOGRAM NPPLOT

Instructs PASW Statistics to produce a histogram and a set of normality plots with
tests for each group of data.

   /STATISTICS DESCRIPTIVES

Instructs PASW Statistics to produce descriptive statistics for each group of data.

   /CINTERVAL 95

Tells PASW Statistics that the table of descriptive statistics should include 95%
confidence intervals.

   /MISSING LISTWISE

Instructs PASW Statistics to use the listwise method of dealing with missing values.

   /NOTOTAL.

Tells PASW Statistics that boxplots are not required. The period at the end indicates
that this is the final sub-command under **EXAMINE**.

1 In the **Run** menu, select
All to run all the syntax.
This will produce the PASW
Statistics output in chapter
5 (section 5.3.2.3).

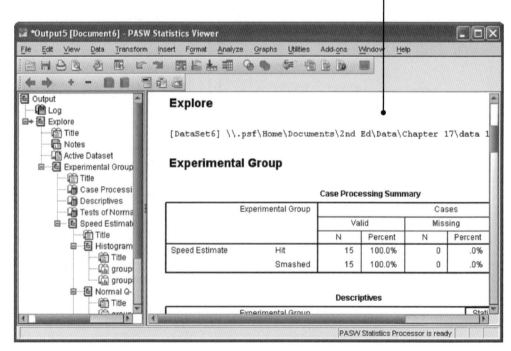

17.2.3. Generating Syntax for Assessing Homogeneity of Variance and Running the *t* Test

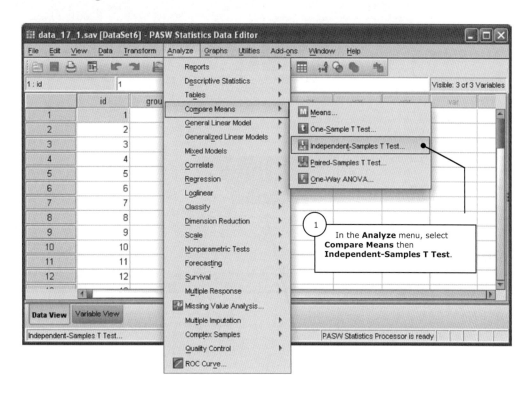

1 In the **Analyze** menu, select **Compare Means** then **Independent-Samples T Test**.

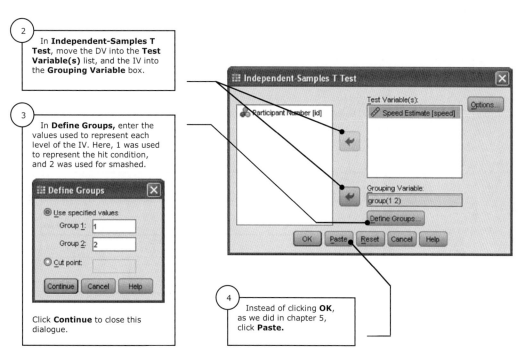

2 In **Independent-Samples T Test**, move the DV into the **Test Variable(s)** list, and the IV into the **Grouping Variable** box.

3 In **Define Groups**, enter the values used to represent each level of the IV. Here, 1 was used to represent the hit condition, and 2 was used for smashed.

Click **Continue** to close this dialogue.

4 Instead of clicking **OK**, as we did in chapter 5, click **Paste.**

17.2.4. Command Syntax for Assessing Homogeneity of Variance and Running the *t* Test

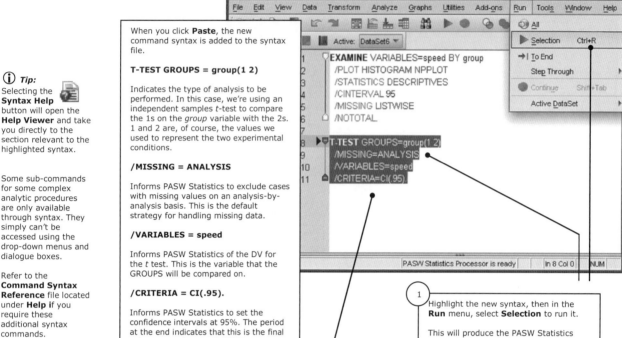

(i) Tip:
Selecting the **Syntax Help** button will open the **Help Viewer** and take you directly to the section relevant to the highlighted syntax.

Some sub-commands for some complex analytic procedures are only available through syntax. They simply can't be accessed using the drop-down menus and dialogue boxes.

Refer to the **Command Syntax Reference** file located under **Help** if you require these additional syntax commands.

When you click **Paste**, the new command syntax is added to the syntax file.

T-TEST GROUPS = group(1 2)

Indicates the type of analysis to be performed. In this case, we're using an independent samples *t*-test to compare the 1s on the *group* variable with the 2s. 1 and 2 are, of course, the values we used to represent the two experimental conditions.

/MISSING = ANALYSIS

Informs PASW Statistics to exclude cases with missing values on an analysis-by-analysis basis. This is the default strategy for handling missing data.

/VARIABLES = speed

Informs PASW Statistics of the DV for the *t* test. This is the variable that the GROUPS will be compared on.

/CRITERIA = CI(.95).

Informs PASW Statistics to set the confidence intervals at 95%. The period at the end indicates that this is the final sub-command under **T-TEST**.

1 Highlight the new syntax, then in the **Run** menu, select **Selection** to run it.

This will produce the PASW Statistics output in chapter 5 (section 5.3.2.5).

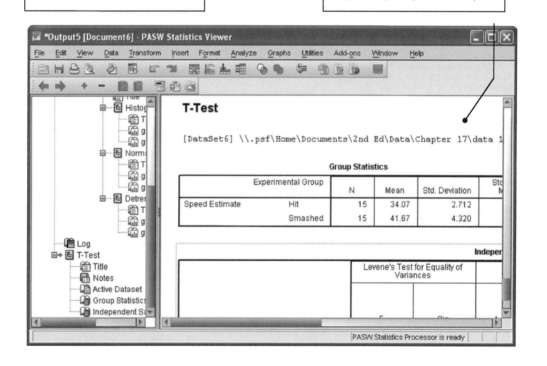

17.3. Summary

Syntax files can be saved alongside your data files, and are an important record of the analyses you've run. They are particularly important if it becomes necessary to re-run analysis at a later date if, for example, you collect additional data or need to provide copies of your output to support your research conclusions. In summary, generating syntax files as part of your analyses is a good habit to establish.

References

Ainsworth, M., Blehar, M., Waters, E., & Wall, S. (1978). *Patterns of attachment*. Hillsdale, NJ: Erlbaum.

American Psychological Association. (2010). *Publication manual of the American Psychological Association* (6th ed.). Washington, DC: Author.

Berryman, J. D., Neal, W. R., & Robinson, J. E. (1980). The validation of a scale to measure attitudes toward the classroom integration of disabled students. *Journal of Educational Research, 73*, 199–203.

Burger, J.M. (2008). *Personality* (7th ed.). Belmont, CA: Thomson-Wadsworth.

Cattell, R. B. (1966). The scree test for the number of factors. *Multivariate Behavioral Research, 1,* 245-276

Clark-Carter, D. (2004). *Quantitative psychological research: A student's handbook*. New York, NY: Psychology Press.

Clark-Carter, D. (2009). *Quantitative psychological research: The complete student's companion* (3rd ed.). New York, NY: Psychology Press.

Cohen, J. (1988). *Statistical power analysis for the behavioral sciences* (2nd ed.). Hillsdale, NJ: Erlbaum.

Cohen, J. (1992). A power primer. *Psychological Bulletin, 112*, 155–159.

Faul, F., Erdfelder, E., Lang, A. -G., & Buchner, A. (2007). G*Power 3: A flexible statistical power analysis program for the social, behavioral, and biomedical sciences. *Behavior Research Methods, 39*, 175-191.

Field, A. (2009). *Discovering statistics using SPSS* (3rd ed.). London, England: Sage.

Fleiss, J. L., Nee, J. C. M., & Landis, J R. (1979). Large sample variance of kappa in the case of different sets of raters. *Psychological Bulletin, 84*, 974-977.

Garson, D. (2009, Feburary 6). *Univariate GLM, ANOVA, and ANCOVA*. Retrieved from http://faculty.chass.ncsu.edu/garson/PA765/anova.htm

Goldberg, L. R., Johnson, J. A., Eber, H. W., Hogan, R., Ashton, M. C., Cloninger, C. R., & Gough, H. G. (2006). The International Personality Item Pool and the future of public-domain personality measures. *Journal of Research in Personality, 40*, 84–96. doi:10.1016/j.jrp.2005.08.007

Gravetter, F. J., & Wallnau, L. B. (2008). *Essentials of Statistics for the Behavioral Sciences* (6th ed.). Belmont, CA: Wadsworth/Cengage Learning.

Howell, D. C. (2002). *Statistical methods for psychology* (5th ed.). London, England: Duxbury.

Howell, D. C. (2010a). *Fundamental statistics for the behavioral sciences* (7th ed.). Belmont, CA: Wadsworth/Cengage Learning.

Howell, D. C. (2010b). *Statistical methods for psychology* (7th ed.). Belmont, CA: Wadsworth/Cengage Learning.

King, J. E. (2004, February). *Software solutions for obtaining a kappa-type statistic for use with multiple raters.* Paper presented at the Annual Meeting of the Southwest Educational Research Association, Dallas, TX. Retrieved from http://www.ccit.bcm.tmc.edu/jking/homepage/genkappa.doc

Krawitz, R. (2004). Borderline personality disorder: Attitudinal change following training. *Australian and New Zealand Journal of Psychiatry, 38*, 554–559. doi: 10.1111/j.1440-1614.2004.01409.x

Langdridge, D., & Hagger-Johnson, G. (2009). *Introduction to research methods and data analysis in psychology* (2nd ed.). Harlow, England: Pearson Education.

Lindman, H. R. (1974). *Analysis of variance in complex experimental designs.* San Francisco, CA: W. H. Freeman.

Loftus, E. F., & Palmer, J. C. (1974). Reconstruction of automobile destruction: An example of the interaction between language and memory. *Journal of Verbal Learning and Verbal Behavior, 13,* 585-589. doi:10.1016/S0022-5371(74)80011-3

Menard, S. (2002). *Applied logistic regression analysis* (2nd ed.). Thousand Oaks: CA: Sage.

Robson, C. (2002). *Real world research: A resource for social scientists and practitioner-researchers* (2nd ed.). Oxford, England: Blackwell.

Rosnow, R. L., & Rosenthal, R. (2008). *Beginning behavioral research: A conceptual primer* (6th ed.). Upper Saddle River, NJ: Pearson/Prentice Hall.

Tabachnick, B. G., & Fidell, L. S. (2007a). *Experimental design using ANOVA.* Belmont, CA: Duxbury/Thomson/Brooks/Cole.

Tabachnick, B. G., & Fidell, L. S. (2007b). *Using multivariate statistics* (5th ed.). Boston, MA: Pearson/Allyn & Bacon.

Tyler, T. R., & Huo, Y. J. (2002). *Trust in the law: Encouraging public cooperation with the police and courts.* New York, NY: Russell-Sage Foundation.

Vargha, A., & Delaney, H. D. (2000). A critique and improvement of the *CL* common language effect size statistics of McGraw and Wong. *Journal of Educational and Behavioral Statistics, 25*, 101–132. doi:10.3102/10769986025002101

Weinberg, S. L., & Abramowitz, S. K. (2008). *Statistics using SPSS: An integrative approach* (2nd ed.), New York, NY: Cambridge University Press.

Index